MINNESOTA STUDIES IN THE PHILOSOPHY OF SCIENCE

Minnesota Studies in the
PHILOSOPHY OF SCIENCE

HERBERT FEIGL AND GROVER MAXWELL, GENERAL EDITORS

VOLUME IV

Analyses of Theories and Methods of Physics and Psychology

EDITED BY

MICHAEL RADNER AND STEPHEN WINOKUR

FOR THE MINNESOTA CENTER FOR PHILOSOPHY OF SCIENCE

UNIVERSITY OF MINNESOTA PRESS, MINNEAPOLIS

PRINTED IN THE UNITED STATES OF AMERICA

Library of Congress Catalog Card Number: 57-12861
ISBN 0-8166-0591-2

PUBLISHED IN GREAT BRITAIN, INDIA, AND PAKISTAN BY THE
OXFORD UNIVERSITY PRESS, LONDON, BOMBAY, AND KARACHI
AND IN CANADA BY THE COPP CLARK PUBLISHING CO.
LIMITED, TORONTO

Preface

In their preface to the first volume of *Minnesota Studies in the Philosophy of Science*, Professors Feigl and Scriven noted the extensive concern of staff and guests of the Minnesota Center for Philosophy of Science with "the meaning of theoretical concepts as defined by their locus in the 'nomological net' and the related rejection of the reductionist forms of operationism and positivism." In this fourth volume of the series, several of our contributors are again concerned with the philosophical, logical, and methodological problems of psychology. As before, some papers are concerned with broader philosophical issues, others with more specific problems of method or interpretation. However, the deep concern for logical and methodological problems of special relevance to the physical sciences manifested in volume III persists and is reflected in a number of contributions to the present volume.

The first group of papers presented here evolved from one of the many conferences held at the Minnesota Center for Philosophy of Science; reports of other conferences will appear in future volumes of this series. Although all the papers resulting from this conference have been considerably modified as a result of further discussions, we have placed them together in this volume. Thus Professor Feigl's paper, which is based upon his opening remarks, serves to set the theme of the conference: the problem of correspondence rules. This is then followed by papers by the main speakers, arranged in alphabetical order. Finally, we conclude this section with a report of some of the discussions that followed the original presentation of the papers. Our readers should bear in mind that these remarks were made during the course of several days in May 1966 and that this report is an edited transcript of the recorded proceedings. Since that time, a number of those philosophers presented here have changed their views or styles of presenting them. However, we believe that publication of these

Preface

proceedings will be fruitful in leading to a wider and deeper appreciation of the problems which have been the concern of the Center's staff and guests.

The second group of papers in this volume consists of essays by various members of the staff of the Center and its visitors. Some of these are new and reflect work going on in 1969–70. Again, we have arranged them in alphabetical order. We believe these contributions are of particular timeliness since they reflect current issues and controversies of great interest. In this connection, our readers are referred to the December 1970 issue of *Philosophy of Science* in which Professor Adolf Grünbaum published his comments on Mr. Demopoulos's paper.

It is with profound gratitude that we acknowledge our indebtedness to the Hill Family Foundation, the Carnegie Corporation of New York, and the National Science Foundation for grants which made continued operation of the Center possible. We also thank Professor N. R. Hanson's literary executors for permission to publish his paper, and Herbert Feigl for his advice and encouragement.

Michael Radner
Stephen Winokur

January 1970

Contents

Contents

viii

Contents

PART ONE. PAPERS AND DISCUSSION FROM THE CONFERENCE ON THE PROBLEMS OF CORRESPONDENCE RULES

The "Orthodox" View of Theories:
Remarks in Defense as well as Critique

The purpose of the following remarks is to present in outline some of the more important features of scientific theories. I shall discuss the "standard" or "orthodox" view, mainly in order to set up a target for criticisms, some of which I shall briefly sketch by way of anticipation. The standard account of the structure of scientific theories was given quite explicitly by Norman R. Campbell [7], as well as independently in a little-known article by R. Carnap [12]. A large part of the voluminous literature in the philosophy of science of the logical empiricists and related thinkers contains, though with a great many variations, developments, modifications, and terminological diversities, essentially similar analyses of the logical structure and the empirical foundations of the theories of physics, biology, psychology, and some of the social sciences. Anticipating to some extent Campbell and Carnap, Moritz Schlick, in his epoch-making *Allgemeine Erkenntnislehre* [38], championed the doctrine of "implicit definition." In this he was influenced by David Hilbert's axiomatization of geometry, as well as by Henri Poincaré's and Albert Einstein's conceptions of theoretical physics and the role of geometry in physics. These matters were then developed more fully and precisely in the work of H. Reichenbach, R. Carnap, C. G. Hempel, R. B. Braithwaite, E. Nagel, and many other logicians and methodologists of science.

In order to understand the aim of this important approach in the philosophy of science it is essential to distinguish it from historical, sociological, or psychological studies of scientific theories. Since a good deal of regrettable misunderstanding has occurred about this, I shall try to defend the legitimacy and the fruitfulness of the distinction before I discuss what, even in my own opinion, are the more problematic points in the "orthodox" logico-analytic account.

3

Herbert Feigl

It was Hans Reichenbach [36] who coined the labels for the important distinction between "analyses in the context of discovery" and "analyses in the context of justification." Even if this widely used terminology is perhaps not the most felicitous, its intent is quite clear: It is one thing to retrace the historical origins, the psychological genesis and development, the social-political-economic conditions for the acceptance or rejection of scientific theories; and it is quite another thing to provide a logical reconstruction of the conceptual structures and of the testing of scientific theories.

I confess I am dismayed by the amount of—it seems almost deliberate—misunderstanding and opposition to which this distinction has been subjected in recent years. The distinction and, along with it, the related idea of a rational reconstruction are quite simple, and are as old as Aristotle and Euclid. In Aristotle's account of deductive logic, mainly in his syllogistics, we have an early attempt to make explicit the rules of validity of necessary inference. For this purpose it was indispensable for Aristotle to disregard psychological factors such as plausibility and to formulate explicitly some of the forms of the propositions involved in deductive reasoning. This also required transforming the locutions of ordinary language into standard formal expressions. For an extremely simple example, remember that "Only adults are admitted" has to be rendered as "All those admitted are adults." Only after the standard forms have replaced the expressions of common discourse can the validity of deductive inferences be checked "automatically," e.g., nowadays by electronic computers.

Furthermore, Euclid already had a fairly clear notion of the difference between purely logical or "formal" truths and extralogical truths. This is explicit in his distinction between the axioms and the postulates of geometry. From our modern point of view it is still imperative to distinguish between the correctness (validity) of a derivation, be it in the proof of a theorem in pure mathematics or a corresponding proof in applied mathematics (such as in theoretical physics), and the empirical adequacy (confirmation or corroboration) of a scientific theory. In fairly close accordance with the paradigm of Euclid's geometry, theories in the factual sciences have for a long time been viewed as hypothetico-deductive systems. That is to say that theories are sets of assumptions, containing "primitive," i.e., undefined terms. The most important of these assumptions are lawlike, i.e., universal, propositions in their logi-

4

cal form. And, just as in geometry, definitions are needed in order to derive theorems of a more specific character. These definitions may be of a variety of kinds: explicit, contextual, coordinative, etc. They are indispensable for the derivation of empirical laws from the more general and usually more abstract assumptions (postulates). The "primitive" concepts serve as the definientia of the "derived" ones. The primitives themselves remain undefined (by explicit definition). They may be regarded as only "implicitly" defined by the total set of axioms (postulates). But it is important to realize that implicit definition thus understood is of a purely syntactical character. Concepts thus defined are devoid of empirical content. One may well hesitate to speak of "concepts" here, since strictly speaking even "logical" meaning as understood by Frege and Russell is absent. Any postulate system if taken as (erstwhile) *empirically uninterpreted* merely establishes a network of symbols. The symbols are to be manipulated according to preassigned formation and transformation rules and their "meanings" are, if one can speak of meanings here at all, purely formal. From the point of view of classical logic implicit definitions are circular. But as C. I. Lewis once so nicely put it, a circle is the less vicious the larger it is. I take this to mean that a "fruitful" or "fertile" postulate set is one from which a great (possibly unlimited) number of theorems can be (nontrivially) derived, and this desirable feature is clearly due to the manner in which the primitive terms are connected with one another in the network formed by the postulates, and also by aptness of the definitions of the derived (defined) terms.

In the picturesque but illuminating elucidations used, e.g., by Schlick, Carnap, Hempel, and Margenau, the "pure calculus," i.e., the uninterpreted postulate system, "floats" or "hovers" freely above the plane of empirical facts. It is only through the "connecting links," i.e., the "coordinative definitions" (Reichenbach's terms, roughly synonymous with the "correspondence rules" of Margenau and Carnap, or the "epistemic correlations" of Northrop, and only related to but not strictly identical with Bridgman's "operational definitions"), that the postulate system acquires empirical meaning. A simple diagram (actually greatly oversimplified!) will illustrate the logical situation. As the diagram indicates, the basic theoretical concepts (primitives) are implicitly defined by the postulates in which they occur. These primitives ($\bigcirc$), or more usually derived concepts ($\triangle$) explicitly defined in terms of them, are then

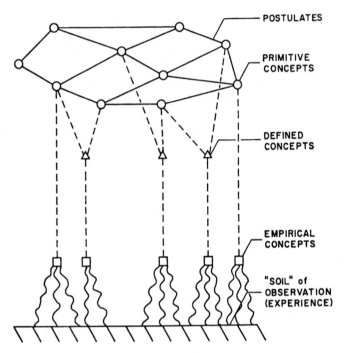

POSTULATES

PRIMITIVE
CONCEPTS

DEFINED
CONCEPTS

EMPIRICAL
CONCEPTS

"SOIL" of
OBSERVATION
(EXPERIENCE)

linked ("coordinated") by correspondence rules to concepts (□) referring to items of observation, e.g., in the physical sciences usually fairly
directly measurable quantities like mass, temperature, and light intensity. These empirical concepts are in turn "operationally defined," i.e.,
by a specification of the rules of observation, measurement, experimentation, or statistical design which determine and delimit their applicability and application.

Bridgman distinguished between "physical" and "mental" operations.
What he had in mind is perhaps more clearly but also more cumbersomely expressed by distinguishing observational (cum mensurational-
experimental) from logico-mathematical, e.g., computational, procedures.
Conceived broadly enough, these two types of "operations" cover the
entire variety of specifications of meaning of any sort of scientific concept. But Bridgman's examples indicate that he focused his attention
primarily upon concepts that are fairly close to the "plane of observation." One very elementary case is the concept of (average) velocity
of a moving body for a given distance in space, and a corresponding
interval of time: Determine, with yardstick or tape measure, etc., the
distance, and with the help of a stopwatch or other chronometric devices the duration in question; these are examples of Bridgman's "physi-

6

cal" operations. Then divide the numerical result of the first by the numerical result of the second ("mental" operation of arithmetic division), and you have arrived at your result: the (average) velocity.

Clearly, highly theoretical concepts such as, for example, that of the "spin" in quantum mechanics, involve much more complex operations—of both types. Hence, I think it is advisable to speak of operational definitions only for "empirical" concepts. The meaning of theoretical concepts can be specified only by their place in the entire theoretical system involving the postulates, definitions, correspondence rules, and finally the operational definitions. These last are indicated by the "rootlets" that "anchor" the empirical concepts in the "soil" of experience, i.e., mensurational-experimental observations.

In view of the "orthodox" logical analysis of scientific theories it is generally held that the concepts ("primitives") in the postulates, as well as the postulates themselves, can be given no more than a partial interpretation. This presupposes a sharp distinction between the language of observation (observational language; O.L.) and the language of theories (theoretical language; T.L.). It is asserted that the O.L. is fully understood. Indeed, in the view of Carnap, for example, the O.L. is not in any way theory-laden or "contaminated" with theoretical assumptions or presuppositions. In an earlier phase of positivism, for example in Carnap's [8], something like a language of sense data (actually a language of momentary total immediate experience) was proposed as the testing ground of all interpretive, inferential, or theoretical propositions. This was clearly the Humean doctrine of "impressions" brought up to date with the help of modern logic. Carnap, very likely influenced by Otto Neurath's and Karl R. Popper's criticisms, later proposed an intersubjective "physicalistic" O.L. as preferable to an essentially subjectivistic ("methodologically solipsistic") O.L. Hence, pointer readings and other similarly objective or intersubjectively concordant "data" would serve as an observation basis. Sharply in contrast to terms thus referring to intersubjectively observable qualities and relations are the theoretical concepts. Terms like "electromagnetic field," "neutron," "neutrino," and "spin" are understood only partially, i.e., with the help of postulates, explicit definitions, correspondence rules, and operational definitions. In the picturesque description of our diagram, it was said that there is an "upward seepage" of meaning from the observational terms to the theoretical concepts.

7

This, in brief outline, is the "orthodox" account of theories in the factual sciences. It has provided the guidelines for numerous axiomatizations of empirical theories. Various branches of theoretical physics [35], biology [40], especially genetics, psychology, especially learning theory [23], and the more recent voluminous output of P. Suppes and his collaborators at Stanford University in a wide range of subjects—all furnish examples of the many ways in which such reconstructions can be pursued. It is a matter of controversy just how fruitful or helpful strict axiomatizations are for the ongoing creative work of the theoretical scientists. If we disregard such relatively informal and "halfway" axiomatizations as can be found in the work of the great scientific innovators, such as Newton, Maxwell, and Einstein, it may well be said that the logicians of science work primarily by way of hindsight. That is, they analyze a given theory in regard to its logical structure and its empirical basis, but do not in any way add to the content of the theory in question. It seems to me that even this relatively modest endeavor can be useful in the following ways: (1) It enables us to understand a given theory more clearly; this is important at least in the teaching and learning procedures. (2) It provides a more precise tool for assessing the correctness of the logico-mathematical derivations on the one hand and the degree of evidential support (or else of disconfirmation) on the other. (3) Since no genuinely fruitful and important theory is "monolithic," but rather consists of a number of logically independent postulates, an exact reconstruction may well show which postulates rest on what empirical evidence.

It must be said immediately that all three of these contentions are being disputed nowadays. Regarding point (1), some criticisms concern the "partial interpretation" view. It is maintained, first of all, that the difference between observational and theoretical concepts is not all that sharp or fundamental; secondly, in this connection, it is urged that there are no observation statements that are free of theoretical presuppositions. Feyerabend even goes farther: He thinks that no neutral observation base exists and that none is needed for the testing of theories. He maintains that theories are tested against each other. If this were so, which I do not concede, then even the most liberal empiricism would have to be abandoned in favor of a, to me, highly questionable form of rationalism. But Feyerabend's construal of the history of scientific theories seems to me rather extravagant!

Furthermore, it is contended that we can understand scientific theories quite fully and that therefore the doctrine of "upward seepage" is all wrong. One reason why this criticism may seem justified is that the understanding of theoretical concepts and postulates rests on the use of analogies or analogical models. I would immediately admit the enormous importance of analogical conception and inference in heuristic and didactic matters. But it is a moot question whether analogical conception is part of the actual cognitive content of theories.

In regard to point (2), i.e., the separation of the assessment of the validity of derivations from the appraisal of the empirical adequacy of theories, I can hardly see any good grounds for criticism. To be sure, it is conceivable that the use of alternative, e.g., many-valued, logics might raise some questions here. But ever since the analyses of scientific explanation given on the basis of *statistical* postulates, especially by C. G. Hempel [20, 21], we have known how to explicate nondeductive derivations, which are actually the *rule* rather than the exception in recent science. Much more weighty are the questions regarding the precise analysis of the notion of evidential support or the "substantiation" of theories by observations (implemented wherever feasible by measurement, experimentation, or statistical design). I shall only mention here the radically different points of view of Carnap and Popper. Carnap has proposed a "logical" concept of probability, or of degree of confirmation of a hypothesis on the basis of a given body of evidence. Popper believes that the growth of scientific knowledge occurs through the severe testing of proposed hypotheses and that those hypotheses which survive such tests are "corroborated." Popper's "degree of corroboration," unlike Carnap's "degree of confirmation," is not a probability; it does not conform to the principles of the calculus of probabilities. The dispute between these two schools of thought still continues, but it is fairly clear that they are really reconstructing different concepts, each holding some promise of genuine illumination. There are also basic disagreements among the various schools of thought in statistical method. The controversies between the "Bayesians" or "subjectivists" and the "objectivists," e.g., those taking the Neyman-Pearson approach, might be mentioned here. It would lead us too far afield even to sketch in outline the various important points at issue.

Finally, in regard to point (3) we face the issues raised originally by Pierre Duhem, and more recently by W. V. O. Quine. Their contention is that theories can be tested only globally, in that it is (usually) the conjunction of all the postulates of a theory from which a conclusion is derived which is then either verified or refuted by observation. This contention is not to be confused with the (rather incredible) sort of assertion made on occasion by Sigmund Freud or his disciples that, as far as they are concerned, psychoanalytic theory is "monolithic," i.e., to be accepted or else rejected in its entirety. Duhem and Quine do not deny that the theories in the empirical sciences consist of *logically independent* postulates, or that they can at least be so reconstructed. What they deny is that the postulates can be independently *tested*. Prima facie this seems plausible, for in testing one postulate others are presupposed. The very use of instruments of observation and experimentation involves assumptions about the functioning of those instruments. In the formal reconstruction of theory testing there are then always assumptions, or auxiliary hypotheses, or parts of general background knowledge that are, in the given context, taken for granted. A closer look at the actual history and procedures of scientific research, however, indicates that the auxiliary hypotheses, etc., have usually been "secured" by previous confirmation (or corroboration). And while, of course, even the best established hypotheses are in principle kept open for revision, it would be foolish to call them into doubt when some other more "risky" hypotheses are under critical scrutiny. Thus, for example, the astronomer relies on the optics of his telescopes, spectroscopes, cameras, and so on in testing a given ("far-out") astrophysical hypothesis. Similarly, the functioning of the instruments of experimental atomic and subatomic physics (cloud or bubble chambers, Geiger counters, accelerators, etc.) is taken for granted in the scrutiny of a given hypothesis in quantum mechanics or nuclear theory. All this is simply the practical wisdom that enjoins us not to doubt everything equally strongly at the same time. Even more to the point: It seems that the "pinpointing of the culprit," i.e., the spotting of the false assumptions, is one of the primary aims as well as virtues of the experimental or statistical techniques. Thus the hypothesis of the stationary ether was refuted by the Michelson-Morley and analogous experiments. It was definitively refuted provided the theoretical physicists did not resort to special ad hoc hypotheses. Ritz's "ballistic" hypothesis regarding the propagation of light, and electromagnetic radiation generally, was refuted by the observations of de Sitter on double stars. *Both* of these pieces of evidence are needed for a justification of

Einstein's postulates in the special theory of relativity. Einstein's genius characteristically manifested itself when he guessed correctly in 1905 what de Sitter demonstrated only six years later. And there is some reason to believe that he did not explicitly utilize even the outcome of the Michelson-Morley experiment. Nevertheless, an objective confirmation of Einstein's theory does depend on these types of evidence.

Along with the "orthodox" view of the structure of scientific theories there goes an account of the levels of scientific explanation which, though often implicit, I formulated explicitly in one of my early articles [14]. This account has been, perhaps somewhat sarcastically, referred to by Feyerabend as the "layer-cake" view of theories. I still think that this account is illuminating even though it needs some emendations. As a first crude approximation the account in question maintains that the ground level consists of descriptions; whether they are based on observation or inference does not matter in this context. On this first level we place the explanandum, i.e., the individual fact or event to be explained, or rather its linguistic or mathematical formulation. Logically speaking only singular sentences or conjunctions thereof should appear on this level. Immediately above this level are the empirical laws (deterministic or statistical, as the case may be). We can utilize these empirical (or experimental) laws in the explanation of the facts or events described on the ground level. These explanations usually strike us as rather trivial because they amount to simply subsuming the individual fact or event under a class specified in the empirical law. For example, the fact that a lens functions as a magnifying glass can be explained by Snell's laws of the refraction of light rays. Snell's law specifies the relation of the angle of incidence to the angle of refraction in terms of a simple mathematical function. Snell's law in turn can be derived from the wave theory of light. This theory already enables us to derive not only the laws of refraction but also those of propagation, reflection, diffraction, interference, and polarization. A still higher level of explanation is attained in Maxwell's principles of electrodynamics (electromagnetics). Here the phenomena of light are explained as a small subclass of electromagnetic waves, along with radio waves, infrared, ultraviolet, X rays, gamma radiations, etc. But in order to understand such optical phenomena as reflection and refraction, a theory of the interaction of the electromagnetic waves with various types of material substances is called for. In order to achieve that, the atomic and electron theories were in-

11

troduced toward the end of the last century. But for a fuller and more precise explanation we can ascend to the next, and thus far "highest," level, viz. the theories of quantum physics.

This level-structure analysis makes clear, I think, the progress from empirical laws to theories of greater and greater explanatory power. To speak very informally, it is the fact-postulate ratio that represents the explanatory power of theories. The aim of scientific explanation throughout the ages has been *unification*, i.e., the comprehending of a maximum of facts and regularities in terms of a minimum of theoretical concepts and assumptions. The remarkable success achieved, especially in the theories of physics, chemistry, and to some extent recent biology, has encouraged pursuit of a unitary system of explanatory premises. Whether this aim is attainable depends, of course, both on the nature of the world and on the ingenuity of the scientists. I think this is what Einstein had in mind in his famous sayings: "God is subtle but He is not malicious"; "The only thing that is incomprehensible, is that the world is comprehensible." (There is serious doubt about the contention of a third well-known bon mot of Einstein's, "God does not play with dice.") Einstein's deep conviction of the basic determinism—at "rock bottom"—of nature is shared by very few theoretical physicists today. There may be no rock bottom; moreover there is no criterion that would tell us that we have reached rock bottom (if indeed we had!).

The plausibility of the level-structure model has been, however, drastically affected by Feyerabend's criticisms. He pointed out quite some years ago that there is hardly an example which illustrates strict deducibility of the lower from the higher levels, even in theories with 100 percent deterministic lawlike postulates. The simple reason is that in straightforward deductive inference there can be no concepts in the conclusion that are not present in the premises and definitions. Most of us thought that definitions, or else bridge laws, would accomplish the job. In fact, however, the lower levels which (historically) usually precede, in their formulation, the construction of the higher levels are, as a rule, incisively revised in the light of the higher level theory. This certainly was the case in the relations of Newtonian to Einsteinian physics, of Maxwellian to quantum electrodynamics, etc. When presenting the level scheme in my philosophy of science courses I have, for more than thirty years, spoken of "corrections from above" accruing to the lower level lawlike assertions. It is also to be admitted that while some of those corrections, within a certain range of the relevant variables, are so minute as to be practically negligible, they become quite significant and even indefinitely large outside that range. Moreover, and this is important, the conceptual frameworks of the theories of different levels are so radically different as to exclude any deductive relationships. Only if bridge laws help in defining the lower level concepts can the derivations be rendered deductive.

In disagreement with Feyerabend, I remain convinced that in the testing of a new theory, the relevant observation language must not be contaminated by that theory; nor need there be a competing alternative theory. If he contends that in most concerns of empirical testing there are presuppositions of a pervasive theoretical character, I would argue that those pervasive presuppositions, for example, regarding the relative permanence of the laboratory instruments, of the experimental records, are "theoretical" only from a deep epistemological point of view and are not called into question when, for example, we try to decide experimentally between rival theories in the physical, biological, or social sciences.

In conclusion I wish to say that the "orthodox" view of scientific theories can help in clarifying their logico-mathematical structure, as well as their empirical confirmation (or disconfirmation). It should be stressed, and not merely bashfully admitted, that the rational reconstruction of theories is a highly artificial hindsight operation which has little to do with the work of the creative scientist. No philosopher of science in his right mind considers this sort of analysis as a recipe for the construction of theories. Yet even the creative scientist employs, at least informally and implicitly, some of the criteria of logico-empirical analysis and appraisal which the logician of science endeavors to make fully explicit. Perhaps there is here an analogy with the difference between a creative composer of music and a specialist in musical theory (counterpoint, harmony, etc.). *Psychologically* the creation of a work of art and the creation of a scientific theory may have much in common. But logically, the standards and criteria of appraisal are radically different, if for no other reason than that the *aims* of art and science are so different.

According to the standard view correspondence rules are semantic designation rules. They merely provide an empirical interpretation of an

13

erstwhile completely uninterpreted postulate system (pure calculus). Let me emphasize once more that this manner of regarding theories is a matter of highly artificial reconstruction. It does not in the least reflect the way in which theories originate. Correspondence rules thus understood differ from bridge laws in that the latter make empirical assertions. For example, if a bridge law states the relation between the mean kinetic energy of gas molecules and the thermometrically determined gas temperature, then this is, logically speaking, a matter of contingent empirical regularity. Nevertheless, in a complete theory of heat, i.e., statistical and quantum mechanical, the behavior of thermometric substances, e.g., alcohol, mercury, and gases, should in principle be derivable. Hence the bridge laws are to be regarded as *theorems* of the respective theories. This can also be formulated by saying that a logically contingent identification of empirical with theoretical concepts is thus achieved. This is surely part of what occurs in the *reduction* of empirical laws to theories, or of theories of lower level to a theory of higher level. Thus the theory of light rays (optics) is reduced to the theory of electromagnetic waves. Or light rays are identified with electromagnetic waves of certain wavelengths and frequencies. Similarly, ordinary (crystalline) table salt is identified with a three-dimensional lattice of sodium and chlorine atoms, etc., etc. The reduction of (parts of) psychology to neurophysiology is still scientifically and philosophically problematic and controversial, but if it were to succeed, it would involve the identification of the qualities of immediate experience with certain patterns of neural processes. In a unitary theory of perception the data of observation could then be characterizable as the direct-acquaintance aspect of brain states.

REFERENCES

1. Achinstein, Peter. *Concepts of Science.* Baltimore: Johns Hopkins Press, 1968.
2. Achinstein, Peter, and Stephen F. Barker, eds. *The Legacy of Logical Positivism.* Baltimore: Johns Hopkins Press, 1969.
3. Braithwaite, R. B. *Scientific Explanation.* Cambridge: Cambridge University Press, 1968.
4. Bridgman, P. W. *The Logic of Modern Physics.* New York: Macmillan, 1927.
5. Bridgman, P. W. *The Nature of Physical Theory.* Princeton, N.J.: Princeton University Press, 1936.
6. Bunge, Mario. *The Foundations of Physics.* Berlin, Heidelberg, and New York: Springer, 1967.
7. Campbell, Norman Robert. *Physics: The Elements.* Cambridge: Cambridge University Press, 1920.
8. Carnap, Rudolf. *Der Logische Aufbau der Welt.* Berlin-Schlachtensee: Weltkreis-Verlag, 1928.

9. Carnap, Rudolf. *Foundations of Logic and Mathematics*, vol. I, no. 3 of the *International Encyclopedia of Unified Science*. Chicago: University of Chicago Press, 1939.
10. Carnap, Rudolf. "The Methodological Character of Theoretical Concepts," in H. Feigl and M. Scriven, eds., *Minnesota Studies in the Philosophy of Science*, vol. I. Minneapolis: University of Minnesota Press, 1956.
11. Carnap, Rudolf. *Philosophical Foundations of Physics*. New York: Basic Books, 1966.
12. Carnap, Rudolf. "Ueber die Aufgabe der Physik und die Anwendung des Grundsatzes der Einfachstheit," *Kant-Studien*, 28 (1923), 90–107.
13. Colodny, Robert G., ed. *Beyond the Edge of Certainty*. Englewood Cliffs, N.J.: Prentice-Hall, 1965.
14. Feigl, Herbert. "Some Remarks on the Meaning of Scientific Explanation," in H. Feigl and W. Sellars, eds., *Readings in Philosophical Analysis*. New York: Appleton-Century-Crofts, 1949.
15. Feigl, Herbert. "Confirmability and Confirmation," in P. P. Wiener, ed., *Readings in Philosophy of Science*. New York: Scribner's, 1953.
16. Feigl, Herbert. *The "Mental" and the "Physical": The Essay and a Postscript*. Minneapolis: University of Minnesota Press, 1967.
17. Feyerabend, Paul K. "Problems of Empiricism," in R. G. Colodny, ed., *Beyond the Edge of Certainty*. Englewood Cliffs, N.J.: Prentice-Hall, 1965.
18. Feyerabend, Paul K. "How to Be a Good Empiricist—A Plea for Tolerance in Matters Epistemological," in B. Baumrin, ed., *Philosophy of Science: The Delaware Seminar*, vol. 2. New York: Wiley, 1963.
19. Grünbaum, Adolf. *Philosophical Problems of Space and Time*. New York: Knopf, 1963.
20. Hempel, Carl G. *Aspects of Scientific Explanation*. New York: Free Press, 1965.
21. Hempel, Carl G. "Deductive-Nomological vs. Statistical Explanation," in H. Feigl and G. Maxwell, eds., *Minnesota Studies in the Philosophy of Science*, vol. III. Minneapolis: University of Minnesota Press, 1962.
22. Hempel, Carl G. *Fundamentals of Concept Formation in Empirical Science*. Chicago: University of Chicago Press, 1952.
23. Hull, Clark L., et al. *Mathematico-Deductive Theory of Rote Learning*. New Haven, Conn.: Yale University Press, 1940.
24. Körner, Stephan. *Experience and Theory*. New York: Humanities, 1966.
25. Lenzen, V. F. *The Nature of Physical Theory*. New York: Wiley, 1931.
26. Margenau, Henry. *The Nature of Physical Reality*. New York: McGraw-Hill, 1950.
27. Mehlberg, Henryk. *The Reach of Science*. Toronto: University of Toronto Press, 1958.
28. Nagel, Ernest. *The Structure of Science*. New York: Harcourt, Brace and World, 1961.
29. Northrop, F. S. C. *The Logic of the Sciences and the Humanities*. New York: Macmillan, 1947.
30. Pap, Arthur. *An Introduction to the Philosophy of Science*. New York: Free Press, 1962.
31. Poincaré, Henri. *Science and Hypothesis*. New York: Dover, 1952.
32. Poincaré, Henri. *Science and Method*. New York: Dover, 1952.
33. Popper, Karl R. *Conjectures and Refutations*. New York: Basic Books, 1962.
34. Popper, Karl R. *The Logic of Scientific Discovery*. New York: Harper, 1959.
35. Reichenbach, Hans. *Axiomatik der relativistischen Raum-Zeit-Lehre*. Braunschweig: Vieweg, 1924.

Herbert Feigl

36. Reichenbach, Hans. *Experience and Prediction.* Chicago: University of Chicago Press, 1938.
37. Scheffler, Israel. *The Anatomy of Inquiry.* New York: Knopf, 1963.
38. Schlick, Moritz. *Allgemeine Erkenntnislehre.* 2nd ed. Berlin: Springer, 1925 (1st ed. 1918).
39. Smart, J. J. C. *Between Science and Philosophy.* New York: Random House, 1968.
40. Woodger, Joseph Henry. *The Techniques of Theory Construction.* Chicago: University of Chicago Press, 1939.

Against Method: Outline of
an Anarchistic Theory of Knowledge

What is all this commotion good for? The most it can achieve is to ruin one's peace of mind. There one has one's little rooms. Everything in them is known, has been added, one item after another, has become loved, and well esteemed. Need I fear that the clock will breathe fire into my face or that the bird will emerge from its cage and greedily attack the dog? No. The clock strikes six when it is six like it has been six for three thousand years. This is what I call *order*. This is what one loves, this is what one can identify with. CARL STERNHEIM, *Die Hose*

Preface

The following essay has been written in the conviction that anarchism, while perhaps not the most attractive *political* philosophy, is certainly an excellent foundation for *epistemology*, and for the *philosophy* of science.

The reason is not difficult to find.

"History generally, and the history of revolutions in particular, is always richer in content, more varied, more manysided, more lively and 'subtle' than even" the best historian and the best methodologist can imagine.[1] * "Accidents and conjunctures, and curious juxtapositions of events"[2] are the very substance of history, and the "complexity of human change and the unpredictable character of the ultimate consequences of any given act or decision of men"[3] its most conspicuous feature. Are we really to believe that a bunch of rather naive and simpleminded rules will be capable of explaining such a "maze of interactions"?[4] And is it not clear that a person who *participates* in a complex process of this kind will succeed only if he is a ruthless *opportunist*, and capable of quickly changing from one method to another?

This is indeed the lesson that has been drawn by intelligent and

AUTHOR'S NOTE: For support of research I am indebted to the National Science Foundation.

* The notes for this essay begin on p. 94.

17

thoughtful observers. "From this [character of the historical process]," writes Lenin, continuing the passage just quoted, "follow two very important practical conclusions: first, that in order to fulfill its task, the revolutionary class [i.e., the class of those who want to change either a part of society, such as science, or society as a whole] must be able to master *all* forms and sides of social activity [it must be able to understand, and to apply not only one particular methodology, but any methodology, and any variation thereof it can imagine], without exception; second, [it] must be ready to pass from one to another in the quickest and most unexpected manner." [5] "The external conditions," writes Einstein, "which are set for [the scientist] by the facts of experience do not permit him to let himself be too much restricted in the construction of his conceptual world by the adherence to an epistemological system. He therefore must appear to the systematic epistemologist as a type of unscrupulous opportunist . . ." [6]

The difference between epistemological (political, theological) *theory* and scientific (political, religious) *practice* that emerges from these quotations is usually formulated as a difference between "certain and infallible" (or, at any rate, clear, systematic, and objective) *rules*, or *standards*, and "our fallible and uncertain faculties [which] depart from them and fall into error." [7] Science as it should be, third-world science,[8] agrees with the proscribed rules. Science as we actually find it in history is a combination of such rules and of error. It follows that the scientist who works in a particular historical situation must learn how to recognize error and how to live with it, always keeping in mind that he himself is liable to add fresh error at any stage of the investigation. He needs a *theory of error* in addition to the "certain and infallible" rules which define the "approach to the truth."

Now error, being an expression of the idiosyncrasies of an individual thinker, observer, even of an individual measuring instrument, *depends on circumstances, on the particular phenomena or theories one wants to analyze, and it develops in highly unexpected ways. Error is itself a historical phenomenon.* A theory of error will therefore contain rules of thumb, useful hints, heuristic suggestions rather than general laws, and it will relate these hints and these suggestions to historical episodes so that one sees in detail how some of them have led some people to success in some situations. It will develop the imagination of the student without ever providing him with cut-and-dried prescriptions and procedures. It will be more a collection of stories than a theory in the proper sense and it will contain a sizable amount of aimless gossip from which everyone may choose what fits in with his intentions. Good books on the art of recognizing and avoiding error will have much in common with good books on the art of singing, or boxing, or making love. Such books consider the great variety of character, of vocal (muscular, glandular, emotional) equipment, of personal idiosyncrasies, and they pay attention to the fact that each element of this variety may develop in most unexpected directions (a woman's voice may bloom forth after her first abortion). They contain numerous rules of thumb, useful hints, and they leave it to the reader to choose what fits his case. Clearly the reader will not be able to make the correct choice unless he has already some knowledge of vocal (muscular, emotional) matters and this knowledge he can acquire only by throwing himself into the process of learning and hoping for the best. In the case of singing he must start using his organs, his throat, his brain, his diaphragm, his buttocks before he really knows how to use them, and he must learn from their reactions the way of learning most appropriate to him. And this is true of all learning: choosing a certain way the student, or the "mature scientist," creates a situation as yet unknown to him from which he must learn how best to approach situations of this kind. This is not as paradoxical as it sounds as long as we keep our options open and as long as we refuse to settle for a particular method, including a particular set of rules, without having examined alternatives. "Let people emancipate themselves," says Bakunin, "and they will instruct themselves of their own accord." [9] In the case of *science* the necessary tact can be developed only by *direct participation* (where "participation" means something different for different individuals) or, if such direct participation cannot be had, or seems undesirable, from a study of past episodes in the *history* of the subject. *Considering their great and difficult complexity these episodes must be approached with a novelist's love for character and for detail,* or with a gossip columnist's love for scandal and for surprising turns; they must be approached with insight into the positive function of strength as well as of weakness, of intelligence as well as of stupidity, of love for truth as well as of the will to deceive, of modesty as well as of conceit, rather than with the crude and laughably inadequate instruments of the logician. For nobody can say in abstract terms, without paying attention to idiosyncrasies of person and circumstance, what precisely it was that led to progress in the past, and nobody can say what moves will succeed in the future.

Now it is of course possible to simplify the historical medium in which a scientist works by simplifying its main actors. The history of science, after all, consists not only of facts and conclusions drawn therefrom. It consists also of ideas, interpretations of facts, problems created by a clash of interpretations, actions of scientists, and so on. On closer analysis we even find that there are no "bare facts" at all but that the facts that enter our knowledge are already viewed in a certain way and are therefore essentially ideational. This being the case the history of science will be as complex, as chaotic, as full of error, and as entertaining as the ideas it contains and these ideas in turn will be as complex, as chaotic, as full of error, and as entertaining as are the minds of those who invented them. Conversely, a little brainwashing will go a long way in making the history of science more simple, more uniform, more dull, more "objective," and more accessible to treatment by "certain and infallible" rules: a theory of errors is superfluous when we are dealing with well-trained scientists who are kept in place by an internal slave master called "professional conscience" and who have been convinced that it is good and rewarding to attain, and then to forever keep, one's "professional integrity." [10]

Scientific education as we know it today has precisely this purpose. It has the purpose of carrying out a rationalistic simplification of the process "science" by simplifying its participants. One proceeds as follows. First, a domain of research is defined. Next, the domain is separated from the remainder of history (physics, for example, is separated from metaphysics and from theology) and receives a "logic" of its own.[11] A thorough training in such a logic then conditions those working in the domain so that they may not unwittingly disturb the purity (read: the sterility) that has already been achieved. An essential part of the training is the inhibition of intuitions that might lead to a blurring of boundaries. A person's religion, for example, or his metaphysics, or his sense of humor must not have the slightest connection with his scientific activity. His imagination is restrained[12] and even his language will cease to be his own.[13]

It is obvious that such an education, such a cutting up of domains and of consciousness, cannot be easily reconciled with a humanitarian attitude. It is in conflict "with the cultivation of individuality which [alone] produces, or can produce well developed human beings";[14] it "maim[s] by compression, like a Chinese lady's foot, every part of human nature which stands out prominently, and tends to make a person markedly dissimilar in outline" [15] from the ideal of rationality that happens to be fashionable with the methodologists.

Now it is precisely such an ideal that finds expression either in "certain and infallible rules" or else in standards which separate what is correct, or rational, or reasonable, or "objective" from what is incorrect, or irrational, or unreasonable, or "subjective." Abandoning the ideal as being unworthy of a free man means abandoning standards and relying on theories of error entirely. Only these theories, these hints, these rules of thumb must now be renamed. Without universally enforced standards of truth and rationality we can no longer speak of universal error. We can only speak of what does, or does not, seem appropriate when viewed from a particular and restricted point of view, different views, temperaments, attitudes giving rise to different judgments and different methods of approach. Such an anarchistic epistemology—for this is what our theories of error now turn out to be—is not only a better means for improving knowledge, or of understanding history. It is also more appropriate for a free man to use than are its rigorous and "scientific" alternatives.

We need not fear that the diminished concern for law and order in science and society that is entailed by the use of anarchistic philosophies will lead to chaos. The human nervous system is too well organized for that.[16] Of course, there may arrive an epoch when it becomes necessary to give reason a temporary advantage and when it is wise to defend its rules to the exclusion of everything else. I do not think we are living in such an epoch today.

When we see that we have arrived at the utmost extent of human [understanding] we sit down contented. HUME[17]

The more solid, well defined, and splendid the edifice erected by the understanding, the more restless the urge of life . . . to escape from it into freedom. [Appearing as] reason it is negative and dialectical, for it dissolves into nothing the detailed determinations of the understanding. HEGEL[18]

Although science taken as whole is a nuisance, one can still learn from it. BENN[19]

1. Introduction; The Limits of Argument

The idea of a method that contains firm, unchanging, and absolutely binding principles for conducting the business of science gets into considerable difficulty when confronted with the results of historical research. We find, then, that there is not a single rule, however plausible, and however firmly grounded in epistemology, that is not violated at some time or other. It becomes evident that such violations are not accidental events, they are not the results of insufficient knowledge or of inattention which might have been avoided. On the contrary, we see that they are necessary for progress. Indeed, one of the most striking features of recent discussions in the history and philosophy of science is the realization that developments such as the Copernican Revolutions, or the rise of atomism in antiquity and recently (kinetic theory; dispersion theory; stereochemistry; quantum theory), or the gradual emergence of the wave theory of light occurred either because some thinkers *decided* not to be bound by certain "obvious" methodological rules or because they *unwittingly broke them.*[20]

This liberal practice, I repeat, is not just a *fact* of the history of science. It is not merely a manifestation of human inconstancy and ignorance. It is reasonable *and absolutely necessary* for the growth of knowledge. More specifically, the following can be shown: considering any rule, however "fundamental," there are always circumstances when it is advisable not only to ignore the rule, but to adopt its opposite. For example, there are circumstances when it is advisable to introduce, elaborate, and defend ad hoc hypotheses, or hypotheses which contradict well-established and generally accepted experimental results, or hypotheses whose content is smaller than the content of the existing and empirically adequate alternatives, or self-inconsistent hypotheses, and so on.[21]

There are even circumstances—and they occur rather frequently—when *argument* loses its forward-looking aspect and becomes a hindrance to progress. Nobody wants to assert[22] that the teaching of *small children* is exclusively a matter of argument (though argument may enter into it and should enter into it to a larger extent than is customary[23]), and almost everyone now agrees that what looks like a result of reason—the mastery of a language, the existence of a richly articulated perceptual world,[24] logical ability—is due partly to indoctrination, partly to a process of *growth* that proceeds with the force of natural law. And where arguments *do* seem to have an effect this must often be ascribed to their *physical repetition* rather than to their *semantic content.*[25] This much having been admitted, we must also concede the possibility of non-argumentative growth in the *adult* as well as in (the theoretical parts of) *institutions* such as science, religion, and prostitution. We certainly cannot take it for granted that what is possible for a small child—to acquire new modes of behavior on the slightest provocation, to slide into them without any noticeable effort—is beyond the reach of his elders. One should expect that catastrophic changes of the physical environment, wars, the breakdown of encompassing systems of morality, political revolutions, will transform adult reaction patterns, too, including important patterns of argumentation.[26] This may again be an entirely natural process and rational argument may but increase the mental tension that precedes and causes the behavioral outburst.

Now, if there are events, not necessarily arguments, which cause us to adopt new standards, including new and more complex forms of argumentation, will it then not be up to the defenders of the status quo to provide, not just arguments, but also contrary causes? (Virtue without terror is ineffective, says Robespierre.) And if the old forms of argumentation turn out to be too weak a cause, must not these defenders either give up or resort to stronger and more "irrational" means? (It is very difficult, and perhaps entirely impossible, to combat the effects of brainwashing by argument.) Even the most puritanical rationalist will then be forced to stop reasoning and to use, say, *propaganda* and *coercion*, not because some of his *reasons* have ceased to be valid, but because the *psychological conditions* which make them effective, and capable of influencing others, have disappeared. And what is the use of an argument that leaves people unmoved?[27]

Of course, the problem never arises quite in this form. The teaching of standards never consists in merely putting them before the mind of the student and making them as *clear* as possible. The standards are supposed to have maximal *causal efficacy* as well. This makes it very difficult to distinguish between the *logical force* and the *material effect* of an argument. Just as a well-trained pet will obey his master no matter how great the confusion he finds himself in and no matter how urgent the need to adopt new patterns of behavior, in the very same way a well-trained rationalist will obey the mental image of *his* master, he will conform to the standards of argumentation he has learned, he will adhere to these standards no matter how great the difficulty he finds himself in, and he will be quite unable to discover that what he regards as the "voice of reason" is but a *causal aftereffect* of the training he has received. We see here very clearly how the appeal to "reason" works. At first sight this appeal seems to be to some *ideas* which convince a man instead of *pushing* him. But conviction cannot remain an ethereal state; it is supposed to lead to *action*. It is supposed to lead to the *appropriate* action, and it is supposed to *sustain* this action as long as necessary. What is the force that upholds such a development? It is the causal efficacy of the standards to which appeal was made and this causal efficacy in turn is but an effect of training, as we have seen. It follows that appeal to argument either has no content at all, and can be made to agree with any procedure,[28] or else will often have a conservative function: it will set limits to what is about to become a natural way of behavior.[29] In the latter case, however, the appeal is nothing but a concealed *political* maneuver. This becomes very clear when a rationalist wants to restore an earlier point of view. Basing his argument on natural habits of reasoning which either have become extinct or have no point of attack in the new situation, such a champion of "rationality" must first restore the earlier material and psychological conditions. This, however, involves him in "a struggle of interests and forces, not of argument." [30]

That interests, forces, propaganda, brainwashing techniques play a much greater role in the growth of our knowledge and, a fortiori, of science than is commonly believed can also be seen from an analysis of the *relation between idea and action*. One often takes it for granted that a clear and distinct understanding of new ideas precedes and should precede any formulation and any institutional expression of them. (An investigation starts with a problem, says Popper.) *First*, we have an idea, or a problem; *then* we act, i.e., either speak, or build, or destroy.[31] This is certainly not the way in which small children develop. They use words, they combine them, they play with them until they grasp a meaning that so far has been beyond their reach. And the initial playful activity is an essential presupposition of the final act of understanding.[32] There is no reason why this mechanism should cease to function in the adult. On the contrary, we must expect, for example, that the idea of liberty could be made clear only by means of the very same actions which were supposed to *create* liberty. Creation of a *thing*, and creation plus full understanding of a *correct idea* of the thing, very often are parts of one and the same indivisible process and they cannot be separated without bringing the process to a standstill. The process itself is not guided by a well-defined program; it cannot be guided by such a program for it contains the conditions of the realization of programs. It is rather guided by a vague urge, by a "passion" (Kierkegaard). The passion gives rise to specific behavior which in turn creates the circumstances and the ideas necessary for analyzing and explaining the whole development, for making it "rational." [33]

The development of the Copernican point of view from Galileo up to the twentieth century is a perfect example of the situation we want to describe. We start with a strong belief that runs counter to contemporary reason. The belief spreads and finds support from other beliefs which are equally unreasonable, if not more so (law of inertia; telescope). Research now gets deflected in new directions, new kinds of instruments are built, "evidence" is related to theories in new ways until there arises a new ideology that is rich enough to provide independent arguments for any particular part of it and mobile enough to find such arguments whenever they seem to be required. Today we can say that Galileo was on the right track, for his persistent pursuit of what once seemed to be a silly cosmology created the material needed for the defense of this cosmology against those of us who accept a view only if it is told in a certain way and who trust it only if it contains certain magical phrases, called "observational reports." [34] And this is not an exception—it is the normal case: theories become clear and "reasonable" only after incoherent parts of them have been used for a long time. Such unreasonable, nonsensical, unmethodical foreplay thus turns out to be an unavoidable precondition of clarity and of empirical success.[35]

Trying to describe developments of this kind in a general way, we are of course obliged to appeal to the existing forms of speech which do not take them into account and which must be distorted, misused, and beaten into new patterns in order to fit unforeseen situations (without a constant misuse of language there cannot be any discovery and any progress). "Moreover, since the traditional categories are the gospel of everyday thinking (including ordinary scientific thinking) and of everyday practice, [such an attempt at understanding] in effect presents rules and forms of false thinking and action—false, that is, from the standpoint of [scientific] commonsense." [36] This is how *dialectical thinking* arises as a form of thought that "dissolves into nothing the detailed determinations of the understanding." [37]

It is clear, then, that the idea of a fixed method, or of a fixed (theory of) rationality, arises from too naive a view of man and of his social surroundings. To those who look at the rich material provided by history, and who are not intent on impoverishing it in order to please their lower instincts, their craving for intellectual security as it is provided, for example, by clarity and precision, to such people it will seem that there is only one principle that can be defended under all circumstances, and in all stages of human development. It is the principle: *anything goes.*[38]

This abstract principle (which is the one and only principle of our anarchistic methodology) must now be elucidated, and explained in concrete detail.

2. Counterinduction I: Theories

It was said that when considering any rule, however fundamental or "necessary for science," one can imagine circumstances when it is advisable not only to ignore the rule, but to adopt its opposite. Let us apply this claim to the rule that "experience," or "the facts," or "experimental results," or whatever words are being used to describe the "hard" elements of our testing procedures, measure the success of a theory, so that agreement between the theory and "the data" is regarded as favoring the theory (or as leaving the situation unchanged), while disagreement endangers or perhaps even eliminates it. This rule is an essential part of all theories of induction, including even some theories of corroboration. Taking the opposite view, I suggest introducing, elaborating, and propagating hypotheses which are inconsistent either with well-established *theories* or with well-established *facts.* Or, as I shall express myself: *I suggest proceeding counterinductively in addition to proceeding inductively.*

There is no need to discuss the first part of the suggestion which favors hypotheses inconsistent with well-established *theories.* The main argument has already been published elsewhere.[39] It may be summarized by saying that evidence that is relevant for the test of a theory T can often be unearthed only with the help of an incompatible alternative theory T'. Thus, the advice to postpone alternatives until the first refutation has occurred means putting the cart before the horse. In this connection, I also advised increasing empirical contents with the help of a *principle of proliferation:* invent and elaborate theories which are inconsistent with the accepted point of view, even if the latter should happen to be highly confirmed and generally accepted. Considering the arguments just summarized, such a principle would seem to be an essential part of any critical empiricism.[40]

The principle of proliferation is also an essential part of a humanitarian outlook. Progressive educators have always tried to develop the individuality of their pupils, and to bring to fruition the particular and sometimes quite unique talents and beliefs that each child possesses. But such an education very often seemed to be a futile exercise in daydreaming. For is it not necessary to prepare the young for life? Does this not mean that they must learn one particular set of views to the exclusion of everything else? And, if there should still remain a trace of their youthful gift of imagination, will it not find its proper application in the arts, that is, in a thin domain of dreams that has but little to do with the world we live in? Will this procedure not finally lead to a split between a hated reality and welcome fantasies, science and the arts, careful description and unrestrained self-expression?[41] The argument for proliferation shows that this need not be the case. It is possible to retain what one might call the freedom of artistic creation and to use it to the full, not just as a road of escape, but as a necessary means for discovering and perhaps even changing the properties of the world we live in. For me this coincidence of the part (individual man) with the whole (the world we live in), of the purely subjective and arbitrary with the objective and lawful, is one of the most important arguments in favor of a pluralistic methodology.[42]

3. Philosophical Background: Mill, Hegel

The idea that a pluralistic methodology is necessary both for the advancement of knowledge and for the development of our individuality has been discussed by J. S. Mill in his admirable essay On Liberty. This essay, according to Mill, is "a kind of philosophical text book of a single truth, which the changes progressively taking place in modern society tend to bring out into ever stronger relief: the importance, to man and society, of a large variety in types of character, and of giving full freedom to human nature to expand itself in innumerable and conflicting directions." [43] Such variety is necessary both for the production of "well-developed human beings" (page 258) and for the improvement of civilization. "What has made the European family of nations an improving, instead of a stationary, portion of mankind? Not any superior excellence in them, which, when it exists, exists as the effect, not as the cause, but their remarkable diversity of character and culture. Individuals, classes, nations have been extremely unlike one another: they have struck out a great variety of paths, each leading to something valuable; and although at every period those who traveled in different paths have been intolerant of one another, and each would have thought it an excellent thing if all the rest would have been compelled to travel his road, their attempts to thwart each other's development have rarely had any permanent success, and each has in time endured to receive the good which the others have offered. Europe is, in my judgment, wholly indebted to this plurality of paths for its progressive and many-sided development" (pages 268–269).[44] The benefit to the individual derives from the fact that "[t]he human faculties of perception, judgment, discriminative feeling, mental activity, and even moral preference are exercised only in making a choice . . . [t]he mental and moral, like the muscular, powers are improved only by being used. The faculties are called into no exercise by doing a thing merely because others do it, no more than by believing a thing only because others believe it" (page 252). Choice presupposes alternatives between which to choose; it presupposes a society which contains and encourages "different opinions" (page 249), "antagonistic modes of thought," [45] as well as "different experiments of living" (page 249), so that the "worth of different modes of life is proved not just in the imagination, but practically" (page 250).[46] "[U]nity of opinion," however, "unless resulting from the fullest and freest comparison of opposite opinions, is not desirable, and diversity not an evil, but a good . . ." (page 249).

This is how proliferation is introduced by Mill. It is not the result of a detailed epistemological analysis, or, what would be worse, of a linguistic examination of the usage of such words as "to know" and "to have evidence for." Nor is proliferation proposed as a solution to *epistemological* problems such as Hume's problem, or the problem of the testability of general statements. (The idea that *experience* might be a basis for our knowledge is at once removed by the remark that "[t]here must be discussion to show how experience is to be interpreted," page 208.) Proliferation is introduced as the solution to a problem of *life*: how can we achieve full consciousness; how can we learn what we are capable of doing; how can we increase our freedom so that we are able to decide, rather than adopt by habit, the manner in which we want to use our talents? Considerations like these were common at a time when the connection between truth and self-expression was still regarded as a problem, and when even the arts were supposed not just to please, but to elevate and to instruct.[47] Today the only question is how science can improve *its* own resources, no matter what the human effect of its methods and of its results. For Mill the connection still exists. Scientific method is part of a general theory of man. It receives its rules from this theory and is built up in accordance with our ideas of a worthwhile human existence.

In addition, pluralism is supposed to lead to the truth: ". . . the peculiar evil of silencing the expression of an opinion is that it is robbing the human race, posterity as well as the existing generation—those who dissent from the opinion, still more than those who hold it. If the opinion is right, they are deprived of the opportunity of exchanging error for truth; if wrong, they lose, what is almost as great a benefit, the clearer perception and livelier impression of truth produced by its collision with error" (page 205).[48] "The beliefs which we have most warrant for have no safeguard to rest on but a standing invitation to the whole to prove them unfounded" (page 209). If "with every opportunity for contesting it [a certain opinion, or a hypothesis] has not been refuted" (page 207), then we can regard it as better than another opinion that has "not gone through a similar process" (page 208).[49] "If even the Newtonian philosophy were not permitted to be questioned, mankind could not feel as complete assurance of its truth as they now do" (page 209). "So essential is this discipline to a real understanding of moral and human subjects [as well as of natural philosophy—page 208] that, if opponents of all-important truths do not exist, it is indispensable to imagine them and to supply them with the strongest arguments which the most skillful devil's advocate can conjure up" (page 228). There is no harm if such opponents produce positions which sound absurd and eccentric: "Precisely because the tyranny of opinion is such as to make eccentricity a reproach, it is desirable, in order to break through that tyranny, that people should be eccentric" (page 267).[50] Nor should those who "admit the validity of the arguments for free discussion[s] . . . object to their being 'pushed to an extreme' . . . unless the reasons are good for an extreme case, they are not good for any case" (page 210).[51] Thus methodological and humanitarian arguments are intermixed in every part of Mill's essay,[52] and it is on both grounds that a pluralistic epistemology is defended, for the natural as well as for the social sciences.[53]

One of the consequences of pluralism and proliferation is that stability of knowledge can no longer be guaranteed. The support a theory receives from observation may be very convincing; its categories and basic principles may appear well founded; the impact of experience itself may be extremely forceful. Yet there is always the possibility that new forms of thought will arrange matters in a different way and will lead to a transformation even of the most immediate impressions we receive from the world. Considering this possibility, we may say that the long-lasting success of our categories and the omnipresence of a certain point of view is not a sign of excellence or an indication that the truth or part of the truth has at last been found. *It is, rather, the indication of a failure of reason to find suitable alternatives which might be used to transcend an accidental intermediate stage of our knowledge.* This remark leads to an entirely new attitude toward success and stability.

As far as one can see, the aim of all methodologies is to find principles and facts which, if possible, are not subjected to change. Principles which give the *impression* of stability are, of course, tested. One tries to refute them, at least in some schools. If all attempts at refutation fail, we have a *positive result*, nevertheless: we have succeeded in discovering a new stable feature of the world that surrounds us; we have come a step closer to the truth.

Moreover, the process of refutation itself rests on assumptions which are not further investigated. An instrumentalist will assume that there are stable facts, sensations, everyday situations, classical states of affairs, which do not change, not even as the result of the most revolutionary discovery. A "realist" may admit changes of the observational matter, but he will insist on the separation between subject and object and he will try to restore it wherever research seems to have found fault with it.[54] Believing in an "approach to the truth," he will also have to set limits to the development of concepts. For example, he will have to exclude incommensurable concepts from a series of succeeding theories.[55] This is the traditional attitude, up to, and *including*, Popper's critical rationalism.

As opposed to it, the attitude about to be discussed regards any prolonged stability, either of ideas and impressions which are capable of test or of background knowledge which one is not willing to give up (realism; separation of subject and object; commensurability of concepts), as an indication of *failure*, pure and simple. Any such stability indicates that *we have failed to transcend an accidental stage of knowledge, and that we have failed to rise to a higher stage of consciousness and of understanding.* It is even questionable whether we can still claim to possess knowledge in such a state. As we become familiar with the existing categories and with the alternatives that are being used in the examination of the received view, our thinking loses its spontaneity until we are reduced to the "bestial and goggle-eyed contemplation of the world around us." [56] "The more solid, well defined, and splendid the edifice erected by the understanding, the more restless the urge of life to escape from it into freedom." [57] Each successful refutation, by opening the way to a new and as yet untried system of categories, temporarily returns to the mind the freedom and spontaneity that is its essential property.[58] But complete freedom is never achieved. For any change, however dramatic, always leads to a new system of fixed categories. Things, processes, states are still separated from each other. The existence of different elements, of a manifold, is still "exaggerated into an opposition by the understanding." [59]

This "evil manner of reflection,[60] to always work with fixed categories," [61] is extended by the customary modes of research to the most widely presupposed and unanalyzed opposition between a subject and an entirely different world of objects.[62] The following assumptions which are important for a methodological realism have been made in this connection: "the object . . . is something finished and perfect that does not need the slightest amount of thought in order to achieve reality while thought itself is . . . something deficient that needs . . . material for its completion[63] and must be soft enough to adapt itself to the material in question." [64] "If thought and appearance do not completely correspond to each other, one has, to start with, a choice: the one or the other may be at fault. [Scientific empiricism] blames thought for not adequately mirroring experience . . ." [65] "These are the ideas which form the core of our customary views concerning the relation between subject and object," [66] and they are responsible for whatever immobility remains in science, even at a time of crisis.

How can this immobility be overcome? How can we obtain insight into the most fundamental assumptions, not only of science and common sense, but of our existence as thinking beings as well? Insight cannot be obtained as long as the assumptions form an unreflected and unchanging part of our life. But, if they are allowed to change, can we then finish the task of criticizing as identically the same persons who started it? Problems like these are raised not only by the abstract question of criticism, but also by more recent discoveries in anthropology, history of science, and methodology. I shall return to them when I discuss incommensurable theories. For the moment, I would like to indicate, very briefly,

31

how certain ideas of Hegel can be used to get a tentative first answer, and thus to make a first step in our attempt to reform the sciences.

Science, common sense, and even the refined common sense of critical rationalism use certain fixed categories ('subject'; 'object'; 'reality'), in addition to the many changing views they contain. They are therefore not fully rational. Full rationality can be obtained by extending criticism to the stable parts also. This presupposes the invention of alternative categories and their application to the whole rich material at our disposal. The categories, and all other stable elements of our knowledge, must be set in motion. "Our task is to make fluid the petrified material which we find, and to relight [wieder entzuenden] the concepts contained in this dead stuff." [67] We must "dissolve the opposition of a frozen subjectivity and objectivity and comprehend the origin of the intellectual and real world as a becoming, we must understand their being as a product, as a form of producing." [68] Such dissolving is carried out by reason, which is "the force of the negative absolute, that is, an absolute negation," [69] that "annihilates" [70] science and common sense, and the state of consciousness associated with both. This annihilation is not a conscious act of a scientist who has decided to eliminate some basic distinctions in his field. For although he may consciously try to overcome the limitations of a particular stage of knowledge, he may not succeed for want of objective conditions (in his brain, in his social surroundings, in the physical world[71]) favoring his wish.[72] Hegel's general theory of development, his cosmology, as one might call it, gives an account of such conditions.

According to this cosmology, every object, every determinate being, is related to everything else: "a well determined being, a finite entity is one that is related to others; it is a content that stands in the relation of necessity to another content and, in the last resort, to the world. Considering this mutual connectedness of the whole, metaphysics could assert . . . the tautology that the removal of a single grain of dust must cause the collapse of the whole universe." [73] The relation is not external. Every process, object, state, etc., actually contains part of the nature of every other process, object, state, etc.[74] Conceptually this means that the complete description of an object is self-contradictory. It contains elements which say what the object is; these are the elements used in the customary accounts provided by science and by common sense which consider part of its properties only and ascribe the rest to the outside.

And the description also contains other elements which say *what the object is not*. These are the elements which science and common sense put outside the object, attributing them to things which are supposed to be completely separated but which are actually contained in the object under consideration. The result is that "all things are beset by an internal contradiction." [75] This contradiction cannot be eliminated by using different words, for example, by using the terminology of a *process* and its *modifications*. For the process will again have to be separated, at least in thought, from something other than itself; otherwise it is pure being which is in no way different from pure *nothingness*.[76] It will contain part of what it is separated from, and this part will have to be described by ideas inconsistent with the ideas used for describing the original process, which therefore is bound to contain contradictions also.[77] Hegel has a marvelous talent for making visible the contradictions which arise when we examine a concept in detail, wishing to give a complete account of the state of affairs it describes. "Concepts which usually appear stable, unmoved, dead are analyzed by him and it becomes evident that they move." [78]

Now we come to a second principle of Hegel's cosmology. The motion of concepts is not merely a motion of the *intellect*, which, starting the analysis with certain determinations, moves away from them and posits their negation. It is an *objective* development as well, and it is caused by the fact that every finite (well-determined, limited) object, process, state, etc., has the tendency to emphasize the elements of the other objects present in it, and to become what it is not. The object, "being restless within its own limit," [79] "strives *not* to be what it is." [80] "Calling things finite, we want to say that they are not merely limited . . . but rather that the negative is essential to their nature and to their being . . . Finite things are, but the truth of their being is their end.[81] What is finite does not merely change . . . it passes away; nor is this passing away merely possible, so that the finite thing could continue to be, without passing away; quite the contrary, the being of a finite thing consists in its having in itself the seeds of passing away . . . the hour of its birth is the hour of its death." [82] "What is finite, therefore, can be set in motion." [83]

Moving beyond the limit, the object ceases to be what it is and becomes what it is not; it is *negated*. A third principle of Hegel's cosmology is that the result of the negation is "not a mere nothing; it is a *spe-*

cial content, for . . . it is the negation of a determined and well defined thing." [84] Conceptually speaking, we arrive at a "new concept which is higher, richer than the concept that preceded it, for it has been enriched by its negation or opposition, contains it *as well as* its negation, being the unity of the original concept and of its opposition." [85] This is an excellent description, for example, of the transition from the Newtonian conception of space to that of Einstein, *provided* we continue using the *unchanged* Newtonian concept. [86] "It is clear that no presentation can be regarded as scientific that does not follow the path and simple rhythm of this method, for this is the path pursued by the things themselves." [87]

Considering that the motion beyond the limit is not arbitrary, but is directed "towards its [i.e., the object's] end" [88] it follows that not *all* the aspects of other things which are present in a certain object are realized in the next stage. Negation, accordingly, "does not mean simply saying No, or declaring a thing to be non-existent, or destroying it in any way one may choose . . . Each kind of thing . . . has its own peculiar manner of becoming negated, and in such a way that a development results from it, and the same holds good for each type of ideas and conceptions . . . *This must be learned like everything else*." [89] What has to be learned, too, is that the "negation of the negation" does not lead further away from the original starting point but that it returns to it. [90] This is an "extremely universal and just on that account extremely far-reaching and important law of development in nature, history and thought; a law which . . . asserts itself in the plant and animal world, in geology, in mathematics, [91] in history, in philosophy." [92] Thus for example "a grain of barley falling under suitable conditions on suitable soil disappears, is negated, and in its place there arises out of it the plant, the negation of the grain . . . This plant grows, blossoms, bears fruit and finally produces other grains of barley, and as soon as these ripen, the stalk dies, is in turn negated. As a result of this negation of the negation, we again have the grain of barley we started with, not singly, but rather in ten, twenty or thirtyfold number . . . and perhaps even qualitatively improved . . ." [93] "It is evident that I say nothing whatever about the particular process of development which, for example, the grain of barley undergoes from its germination up to the dying off of the fruit-bearing plant, when I state that it is the negation of the negation . . . I rather comprise these processes altogether under this one *law of motion* and just for that reason disregard the peculiarities of each special process. *Dialectics*, however, is nothing else than the science of the general laws of motion and development in nature, human society and thought." [94]

In the foregoing account, concepts and real things have been treated as separate. Similarities and correspondences were noted: each thing *con-tains* elements of everything else, it *develops* by turning into these alien elements, it *changes*, and it finally tries to *return* to itself. The *notion* of each thing, accordingly, contains contradictory elements. It is negated, and it moves in a way corresponding to the movement of the thing. This presentation has one serious disadvantage: "Thought is here described as a mere subjective and formal activity while the world of objects, being situated vis-à-vis thought, is regarded as something fixed and as having independent existence. This dualism . . . is not a true account of things and it is pretty thoughtless to simply take over the said properties of subjectivity and objectivity without asking for their origin . . . Taking a more realistic view we may say that the subject is only a stage in the development of being and essence." [95] The concept, too, is then part of the general development of nature, in a materialistic interpretation of Hegel. "Life," for example, "or organic nature is that phase of nature when the concept appears on the stage; it enters the stage as a blind concept that does not comprehend itself, i.e., does not think." [96] Being part of the *natural* behavior, first of an organism, then of a thinking being, it not only mirrors a nature that "lies entirely outside of it," [97] it is not merely "something subjective and accidental," [98] it is not "merely a concept";[99] it participates in the general nature of all things, i.e., it contains an element of everything else, it has the tendency to be the end result of the development of a specific thing, so that, finally, the concept and this thing become one.[100] "That real things do not agree with the idea ["read: with the total knowledge of man"[101]] constitutes their *finitude*, their *untruth* because of which they are *objects*, each determined in its special sphere by the laws of mechanics, chemistry, or by some external purpose." [102] In this stage "there can be nothing more detrimental and more unworthy of a philosopher than to point, in an entirely vulgar fashion, to some experience that contradicts the idea . . . When something does not correspond to its concept, it must be led up to it" [103] (counterinduction!) until "concept and thing have become one." [104]

To sum up: Knowledge is part of nature and is subjected to its gen-

35

eral laws. The laws of dialectics apply to the motion of objects and concepts, as well as to the motion of higher units comprising objects and concepts. According to these general laws, every object participates in every other object and tries to change into its negation. This process cannot be understood by attending to those elements in our subjectivity which are still in relative isolation and whose internal contradictions are not yet revealed. (Most of the customary concepts of science, mathematics, and especially the rigid categories used by our modern axiomaniacs are of this kind.) To understand the process of negation we must attend to those other elements which are fluid, about to turn into their opposites, and which may, therefore, bring about knowledge and truth, "the identity of thing and concept." [105] The identity itself cannot be achieved mechanically, i.e., by arresting some aspect of reality and fiddling about with the remaining aspects, or theories, until agreement is achieved (the aspects one wants to arrest, being in motion, will soon be replaced by dogmatic opinions of them, rigid perceptions included). We must rather proceed dialectically, i.e., by an *interaction* of concept and fact (observation, experiment, basic statement, etc.) that affects *both* elements. The lesson for methodology is, however, this: Do not work with stable concepts. Do not eliminate counterinduction. Do not be seduced into thinking that you have at last found the correct description of "the facts" when all that has happened is that some new categories have been adapted to some older forms of thought, which are so familiar that we take their outlines to be the outlines of the world itself.

4. Counterinduction II: Experiments, Observations, "Facts"

Considering now the invention, the use, and the elaboration of theories which are inconsistent, not only with other theories, but even with *experiments, facts, observations*, we may start by pointing out that *not a single theory ever agrees with all the known facts in its domain.* And the trouble is not created by rumors, or by the results of sloppy procedure. It is created by experiments and measurements of the highest precision and reliability.

It will be convenient, at this place, to distinguish two different kinds of disagreement between theory and fact: numerical disagreements and qualitative failures.

The first case is quite familiar: a theory makes a certain numerical prediction and the value that is actually obtained differs from the prediction made outside the margin of error. Precision instruments are usually involved here. Numerical disagreements abound in science.

Thus the Copernican view at the time of Galileo was inconsistent with facts so plain and obvious that Galileo had to call it "surely false." [106] "There is no limit to my astonishment," he writes in a later work,[107] "when I reflect that Aristarchus and Copernicus were able to make reason so conquer sense that, in defiance of the latter, the former became mistress of their belief." Newton's theory of gravitation was beset, from the very beginning, by a considerable number of difficulties which were serious enough to provide material for refutations. Even today, and in the nonrelativistic domain, there exist "numerous discrepancies between observation and theory." [108] Bohr's atomic model was introduced and retained in the face of very precise and unshakable contrary evidence.[109] The special theory of relativity was retained, despite D. C. Miller's decisive refutation. (I call the refutation "decisive" because the experiment was, from the point of view of contemporary evidence, at least as well performed as the earlier experiment of Michelson and Morley.[110]) The general theory of relativity, though surprisingly successful in some domains, failed to explain about 10″ in the movement of the nodes of Venus and more than 5″ in the movement of the perihelion of Mars. All these are quantitative difficulties which can be resolved by discovering a better set of numbers but which do not force us to make qualitative adjustments.

The second case, the case of qualitative failures, is less familiar, but of much greater interest. In this case a theory is inconsistent not with a recondite fact that must be unearthed with the help of complex equipment and is known to experts only, but with circumstances which can be noticed with the unaided senses and which are familiar to everyone.

The first and to my mind the most important example of an inconsistency of this kind is Parmenides' theory of the unchanging One. The theory has much in its favor[111] and it plays its role even today, for example in the general theory of relativity.[112] Used in an undeveloped form by Anaximander it led to the insight, repeated by Heisenberg in his theory of elementary particles,[113] that the basic substance, or the basic elements of the universe, cannot obey the same laws as do the visible elements. Zeno's arguments, on the other hand, show the difficulties inherent in the idea of a continuum consisting of isolated elements. Aristotle took these arguments seriously and developed his own theory of the continuum.[114] Yet the idea of a collection of elements remained and continued to be used, despite the quite obvious difficulties, until these difficulties were removed early in the twentieth century.[115]

Another example of a theory with qualitative defects is Newton's theory of colors. According to this theory light consists of rays of different refrangibility which can be separated, reunited, refracted, but which are never changed in their internal constitution, and which have a very small lateral extension in space. Considering that the surface of mirrors is much rougher than is the lateral extension of the rays, the ray theory is found to be inconsistent with the existence of mirror images (as is admitted by Newton himself: *Opticks*, book II, part III, proposition viii) : if light con-

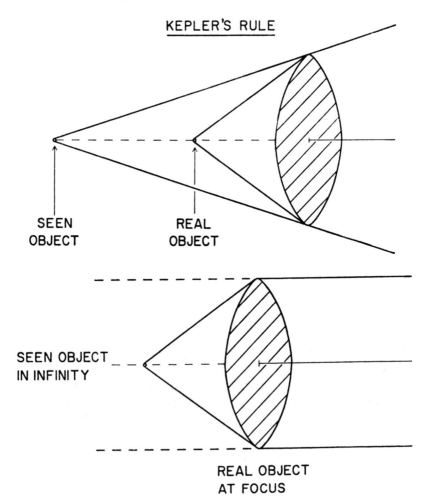

KEPLER'S RULE

SEEN
OBJECT

REAL
OBJECT

SEEN OBJECT
IN INFINITY

REAL OBJECT
AT FOCUS

sists of rays, then a mirror should behave like a rough surface, i.e., it should look to us like a wall. Newton retains his theory, eliminating the difficulty with the help of an ad hoc hypothesis: "the reflection of a ray is effected, not by a single point of the reflecting body, but by some power of the body which is evenly diffused all over its surface . . ." [116]

In Newton's case the qualitative discrepancy between theory and fact is removed by an ad hoc hypothesis. In other cases not even this very flimsy maneuver is used. One retains the theory and tries to forget its shortcomings. An example is the attitude toward Kepler's rule according to which an object seen through a lens is perceived at the distance at which the rays traveling from the lens toward the eye intersect (see the first diagram). [117] The rule implies that an object situated at the focus will be seen infinitely far away (see the second diagram). "But on the contrary," writes Barrow, Newton's teacher and predecessor in Cambridge, commenting on this predication, "we are assured by experience that [a point situated close to the focus] appears variously distant, according to the different situations of the eye . . . And it does almost never seem farther off than it would be if it were beheld with the naked eye; but, on the contrary, it does sometime appear much nearer . . . All which does seem repugnant to our principles. But for me," Barrow continues, "neither this nor any other difficulty shall have so great an influence on me, as to make me renounce that which I know to be manifestly agreeable to reason." [118]

Barrow mentions the qualitative difficulties, and he says that he will retain the theory nevertheless. This is not the usual procedure. The usual procedure is to forget about the difficulties, never to talk about them, and to proceed as if the theory were without fault. This attitude is very common today.

Thus classical electrodynamics contains the absurd consequence that the motion of a free particle is self-accelerated. [119] This consequence is little known though it makes it impossible to calculate even the simplest case of a motion in a homogeneous electric field. What one does is to make "an approximation" which neglects effects too small to be noticed but which also eliminates the quite noticeable absurd consequence. Theory plus "approximation" produces a reasonable prediction though the theory itself suffers from qualitative difficulties. The quantum theory of fields which one might want to consult in order to remove the troubles of classical electromagnetic theory has absurdities of its own such as the infinite self energies. The situation is not improved by the remark that these self energies can be corrected by renormalization. They can of course be corrected by this method, and in a consistent manner, but only *after* redefining certain terms in the calculations with an eye to the results to be achieved. This procedure, which is ad hoc, certainly does not establish the excellence of the theory; it shows that as it stands the theory is either refuted[120] or else woefully incomplete.

Another example of modern physics is quite instructive, for it might have led to an entirely different development of our knowledge concerning the microcosm. Ehrenfest has proved a theorem[121] according to which the classical electron theory of Lorentz taken together with the equipartition principle excludes induced magnetism. The reasoning is exceedingly simple: according to the equipartition principle the probability of a motion is proportional to $\exp[-U/kT]$, where U is the energy of the motion. Now the energy in a constant magnetic field is, according to Lorentz, $= q(E + [vB]) \cdot v$, where q is the charge of the moving parts, E the electric field, B the magnetic field, v the velocity of the moving parts. This magnitude reduces to qE in all cases *unless* one is prepared to admit the existence of single magnetic poles (given the proper context, this result strongly supports the ideas and the experimental findings of the late Felix Ehrenhaft[122]).

Occasionally it is impossible to survey all the interesting consequences and to discover the absurd results of a theory. This may be due to a deficiency in the existing mathematical methods; it may also be due to the ignorance of those who defend the theory.[123] Under such circumstances the most common procedure is to use an older theory up to a certain point (which is often quite arbitrary) and to add the new theory for calculating refinements. Seen from a methodological point of view the procedure is a veritable nightmare. Let us explain it, using the relativistic calculation of the path of Mercury as an example.

The perihelion of Mercury moves along at a rate of about 5600″ per century. Of this value, 5026″ are geometric, having to do with the movement of the reference system; 575″ are dynamical, due to perturbations in the solar system. Of these perturbations all but the famous 43″ are accounted for by classical celestial mechanics. And the remaining 43″ are accounted for by general relativity. This is how the situation is usually explained.

Now this explanation shows that the premise from which we derive the 43″ is not the general theory of relativity plus suitable initial conditions. The premise contains classical physics *in addition* to whatever relativistic assumptions are made. Furthermore, the relativistic calculation, the so-called "Schwarzschild solution," does not deal with the planetary system as it exists in the real world (i.e., our own asymmetric galaxy); it deals with the entirely fictional case of a central symmetrical universe containing a singularity in the middle and nothing else. What are the reasons for employing such an insane conjunction of premises?[124]

One reason, so the customary reply continues, is that we are dealing with approximations. The formulas of classical physics do not appear because relativity is incomplete. Nor is the central symmetrical case used because relativity does not offer anything better. Both schemata flow from the general theory under the special circumstances realized in our planetary system *provided* we omit magnitudes too small to be considered. Hence, we are using the *theory* of relativity throughout, and we are using it in an adequate manner.

Note, now, how this idea of an approximation differs from the legitimate idea: usually one has a theory, one is able to calculate the particular case one is interested in, one notes that this calculation leads to magnitudes below experimental precision, one omits such magnitudes, and one obtains a vastly simplified formalism. In the present case making the required approximations would mean calculating the n-body problem relativistically, omitting magnitudes smaller than the precision of observation reached, and showing that the theory thus curtailed coincides with classical celestial mechanics as corrected by Schwarzschild. This procedure has not been used by anyone simply because the relativistic n-body problem has as yet withstood solution.[125] There are not even approximate solutions for important problems such as, for example, the problem of stability (a first great stumbling stone for Newton's theory). This being the case, the classical part of the explanans is not only used for convenience, *it is absolutely necessary.* And the approximations made are not a *result* of relativistic calculation, they are introduced in order to make relativity fit the case. One may properly call them *ad hoc approximations.*

Ad hoc approximations abound in modern mathematical physics. They play a very important part in the quantum theory of fields and they are an essential ingredient of the correspondence principle. At the moment we are not concerned with the reasons for this fact, we are only concerned with its consequences: ad hoc approximations conceal, and even entirely

41

eliminate, qualitative difficulties. They create a false impression of the excellence of our science. It follows that a philosopher who wants to study the adequacy of science as a picture of the world, or who wants to build up a realistic scientific methodology, must look at modern science with special care. In most cases modern science is more opaque and much more deceptive than its sixteenth- and seventeenth-century ancestors have ever been.

As a final example of qualitative difficulties we mention again the heliocentric theory at the time of Galileo. We shall soon have occasion to show that this theory was inadequate both qualitatively and quantitatively, and that it was also philosophically absurd.

To sum up this brief and very incomplete list: Wherever we look, whenever we have a little patience and select our evidence in an unprejudiced manner, we find that theories fail to adequately reproduce certain *quantitative* results; and that they are *qualitatively incompetent* to a surprising degree.[126] Science gives us theories of high beauty and sophistication. Modern science has developed mathematical structures which exceed anything that has existed so far in coherence and generality. But in order to achieve this miracle all the existing troubles had to be pushed into the *relation* between theory and fact, and had to be concealed, by ad hoc approximations, and by other procedures.

This being the case—what shall we make of the methodological demand that a theory must be judged by experience and must be rejected if it contradicts accepted basic statements? What attitude shall we adopt toward the various theories of confirmation and corroboration which all rest upon the assumption that theories can be made to completely agree with the known facts and which use the amount of agreement reached as a principle of evaluation? This demand, these theories, are now all quite useless. They are as useless as a medicine that heals a patient only if he is bacteria free. In practice they are never obeyed by anyone. Methodologists may point to the importance of falsifications—but they blithely use falsified theories; they may sermonize how important it is to consider all the relevant evidence, and never mention those big and drastic facts which show that the theories which they admire and accept, the theory of relativity, the quantum theory, are at least as badly off as the older theories which they reject. In *practice* methodologists slavishly repeat the most recent pronouncements of the top dogs in physics, though in doing so they must violate some very basic rules of their trade. Is it possible to proceed in a more reasonable manner? Let us see!

According to Hume theories cannot be *derived from* facts. The demand to admit only those theories which follow from facts leaves us without any theory. Hence, a science *as we know it* can exist only if we drop the demand and revise our methodology.

According to our present results hardly any theory is *consistent with* the facts. The demand to admit only those theories which are consistent with the available and accepted facts again leaves us without any theory. (I repeat: *without any theory*, for there is not a single theory that is not in some trouble or other.) Hence, a science as we know it can exist only if we drop this demand also and again revise our methodology, *now admitting counterinduction in addition to admitting unsupported hypotheses.* The right method no longer consists of rules which permit us to choose between theories on the basis of falsifications. It must rather be modified so as to enable us to choose between theories which we have already tested and which are falsified.

To proceed further. Not only are facts and theories in constant disharmony, they are not even as neatly separated as everyone makes them out to be. Methodological rules speak of "theories" and "observations" and "experimental results" as if these were clear and well-defined objects whose properties are easy to evaluate and which are understood in the same sense by all scientists.

However, the material which a scientist *actually* has at his disposal, his laws, his experimental results, his mathematical techniques, his epistemological prejudices, his attitude toward the absurd consequences of the theories which he accepts, is indeterminate in many ways, it is ambiguous, *and never fully separated from the historical background.* This material is always contaminated by principles which he does not know and which, if known, would be extremely hard to test. Questionable views on cognition, such as the view that our senses, used in normal circumstances, give reliable information about the world, may invade the observation language itself, constituting the observational terms and the distinction between veridical and illusory appearances. As a result observation languages may become tied to older layers of speculation which affect, in this roundabout fashion, even the most progressive methodology. (Example: the absolute space-time frame of classical physics which was codified and consecrated by Kant.) The *sensory impression,* however simple, always contains a component that expresses the reaction of the perceiving subject and has no objective correlate. This subjective component often merges with the rest, and forms an unstructured whole which must then be subdivided from the outside with the help of counterinductive procedures. (An example of this is the appearance of a fixed star to the naked eye, which contains the subjective effects of irradiation, diffraction, diffusion, restricted by the lateral inhibition of adjacent elements of the retina.) Finally, there are the auxiliary premises which are needed for the derivation of testable conclusions, and which occasionally form entire *auxiliary sciences.*

Consider the case of the Copernican hypothesis, whose invention, defense, and partial vindication run counter to almost every methodological rule one might care to think of today. The auxiliary sciences here contained laws describing the properties and the influence of the terrestrial atmosphere (meteorology); optical laws dealing with the structure of the eye and telescopes, and with the behavior of light; and dynamical laws describing motion in moving systems. Most importantly, however, the auxiliary sciences contained a theory of cognition that postulated a certain simple relation between perceptions and physical objects. Not all these auxiliary disciplines were available in explicit form. Many of them merged with the observation language, and led to the situation described at the beginning of the preceding paragraph.

Consideration of all these circumstances, of observation terms, sensory core, auxiliary sciences, background speculation, suggests that a theory may be inconsistent with the evidence, not because it is not correct, *but because the evidence is contaminated.* The theory is threatened either because the evidence contains unanalyzed sensations which only partly correspond to external processes, or because it is presented in terms of antiquated views, or because it is evaluated with the help of backward auxiliary subjects. The Copernican theory was in trouble for *all* these reasons.

It is this *historico-physiologic character of the evidence,*[127] the fact that it does not merely describe some objective state of affairs, *but also expresses some subjective, mythical, and long-forgotten views* concerning this state of affairs, that forces us to take a fresh look at methodology. It shows that it would be extremely imprudent to let the evidence judge our theories directly, and without any further ado. A straightforward and unqualified judgment of theories by "facts" is bound to eliminate ideas *simply because they do not fit into the framework of some older cosmology.* Taking experimental results and observations for granted and putting the burden of proof on the theory means taking the observational ideology for granted without having ever examined it. (Note that the experimental results are supposed to have been obtained with the greatest possible care. Hence "taking observations, etc., for granted" means "taking them for granted *after* the most careful examination of their reliability"—for even the most careful examination of an observation statement does not interfere with the concepts in terms of which it is expressed, or with the structure of the sensory expression.)

Now—how can we possibly examine something we are using all the time? How can we criticize the terms in which we habitually express our observations? Let us see![128]

The first step in our criticism of commonly used concepts is to create a *measure of criticism*, something with which these concepts can be compared. Of course, we shall later want to know a little more about the measure stick itself, for example, we shall want to know whether it is better than, or perhaps not as good as, the material examined. But in order for this examination to start there must be a measure stick in the first place. Therefore the first step in our criticism of customary concepts and customary reactions is to step outside the circle and to invent a new conceptual system, a new theory, for example, that clashes with the most carefully established observational results and confounds the most plausible theoretical principles. This step is, again, counterinductive. Counterinduction, therefore, is both a *fact*—science could not exist without it—and a legitimate and much-needed move in the game of science.

5. The Tower Argument Stated: First Steps of Analysis

As a concrete illustration and as a basis for further discussion, I shall now briefly describe the manner in which Galileo defused an important counterargument against the idea of the motion of the earth. I say "defused," and not "refuted," because we are dealing with a changing conceptual system as well as with certain attempts at concealment.

According to the argument which convinced Tycho, and which is used against the motion of the earth in Galileo's own *Trattato della sfera*, observation shows that "heavy bodies . . . falling down from on high, go by a straight and vertical line to the surface of the earth. This is considered an irrefutable argument for the earth being motionless. For if it

45

made the diurnal rotation, a tower from whose top a rock was let fall, being carried by the whirling of the earth, would travel many hundreds of yards to the east in the time the rock would consume in its fall, and the rock ought to strike the earth that distance away from the base of the tower." [129]

In considering the argument, Galileo at once admits the correctness of the sensory content of the observation made, viz. that "heavy bodies . . . falling from a height, go perpendicularly to the surface of the earth." [130] Considering an author (Chiaramonti) who sets out to convert Copernicans by repeatedly mentioning this fact, he says: "I wish that this author would not put himself to such trouble trying to have us understand from our senses that this motion of falling bodies is simple straight motion and no other kind, nor get angry and complain because such a clear, obvious, and manifest thing should be called into question. For in this way he hints at believing that to those who say such motion is not straight at all, but rather circular, it seems they see the stone move visibly in an arc, since he calls upon their senses rather than their reason to clarify the effect. This is not the case, Simplicio; for just as I . . . have never seen nor ever expect to see the rock fall any way but perpendicularly, just so do I believe that it appears to the eyes of everyone else. It is therefore better to put aside the appearance, on which we all agree, and to use the power of reason either to confirm its reality or to reveal its fallacy." [131] The correctness of the observation is not in question. What *is* in question is its "reality" or "fallacy." What is meant by this expression?

The question is answered by an example that occurs in Galileo's next paragraph, and "from which . . . one may learn how easily anyone may be deceived by simple appearances, or let us say by the impressions of one's senses. This event is the appearance to those who travel along a street by night of being followed by the moon, with steps equal to theirs, when they see it go gliding along the eaves of the roofs. There it looks to them just as would a cat really running along the tiles and putting them behind it; an appearance which, if reason did not intervene, would only too obviously deceive the senses."

In this example we are asked to start with a sensory impression and consider a statement that is forcefully suggested by it. (The suggestion is so strong that it has led to entire systems of belief and rituals as becomes clear from a closer study of the lunar aspects of witchcraft and of other religions.) Now "reason intervenes": the statement suggested by the impression is examined, and one considers other statements in its place. The nature of the *impression* is not changed a bit by this activity. (This is only approximately true; but we can omit for our present purpose the complications arising from the interaction of impression and proposition.) But it enters new observation *statements* and plays new, better or worse, parts in our knowledge. What are the reasons and the methods which regulate such exchange?

To start with we must become clear about the nature of the total phenomenon: appearance plus statement. There are not two acts, one, noticing a phenomenon, the other, expressing it with the help of the appropriate statement, *but only one*, viz. saying, in a certain observational situation, "the moon is following me," or "the stone is falling straight down." We may of course abstractly subdivide this process into parts, and we may also try to create a situation where statement and phenomenon seem to be psychologically apart and waiting to be related. (This is rather difficult to achieve and is perhaps entirely impossible.[132]) But under normal circumstances such a division does not occur; describing a familiar situation is, for the speaker, an event in which statement and phenomenon are firmly glued together.

This unity is the result of a process of learning that starts in one's childhood. From our very early days we learn to react to situations with the appropriate responses, linguistic or otherwise. The teaching procedures both shape the 'appearance' or the 'phenomenon' and establish a firm connection with words, so that finally the phenomena seem to speak for themselves, without outside help or extraneous knowledge. They just *are* what the associated statements assert them to be. The language they 'speak' is of course influenced by the beliefs of earlier generations which have been held for such a long time that they no longer appear as separate principles, but enter the terms of everyday discourse, and, after the prescribed training, seem to emerge from the things themselves.

Now at this point we may want to compare, in our imagination and quite abstractly, the results of the teaching of different languages incorporating different ideologies. We may even want to consciously change some of these ideologies and adapt them to more 'modern' points of view. It is very difficult to say how this will change our situation, unless we make the further assumption that the quality and structure of sensations (perceptions), or at least the quality and structure of those sensations which enter the body of science, are independent of their linguistic expression.

I am very doubtful about even the approximate validity of this assumption which can be refuted by simple examples. And I am sure that we are depriving ourselves of new and surprising discoveries as long as we remain within the limits defined by it. Yet the present essay will remain quite consciously within these limits. (My first task, if I should ever resume writing, would be to explore these limits and to venture beyond them.)

Making the additional simplifying assumption, we can now distinguish between (a) sensations, and (b) those "mental operations which follow so closely upon the senses" [133] and are so firmly connected with their reactions that a separation is difficult to achieve. Considering the origin and the effect of such operations, I shall call them *natural interpretations.*

6. Natural Interpretations

In the history of thought, natural interpretations have been regarded either as *a priori presuppositions* of science or else as *prejudices* which must be removed before any serious examination can proceed. The first view is that of Kant, and, in a very different manner and on the basis of very different talents, that of some contemporary linguistic philosophers. The second view is due to Bacon (who had, however, predecessors, such as the Greek skeptics).

Galileo is one of those rare thinkers who neither wants to forever *retain* natural interpretations nor wants to altogether *eliminate* them. Wholesale judgments of this kind are quite alien to his way of thinking. He insists upon *critical discussion* to decide which natural interpretations can be kept and which must be replaced. This is not always clear from his writings. Quite the contrary, the methods of reminiscence, to which he appeals so freely, are designed to create the impression that nothing has changed and that we continue expressing our observations in old and familiar ways. Yet his attitude is relatively easy to ascertain: natural interpretations are necessary. The senses alone, without the help of reason, cannot give us a true account of nature. What is needed for arriving at such a true account are "the . . . senses, accompanied by reasoning." [134] Moreover, in the arguments dealing with the motion of the earth, it is this reasoning, it is the connotation of the observation terms, and not the message of the senses or the appearance, that causes trouble. "It is therefore better to put aside the appearance, on which we all agree, and to use the power of reason either to confirm [its] reality or to reveal [its]

48

fallacy." [135] "To confirm the reality or reveal the fallacy of appearances" means, however, to examine the validity of those natural interpretations which are so intimately connected with the appearances that we no longer regard them as separate assumptions. I now turn to the first natural interpretation implicit in the argument from falling stones.

According to Copernicus the motion of a falling stone should be "mixed straight-and-circular." [136] By the "motion of the stone" is meant, not just its motion relative to some visible mark in the visual field of the observer, or its *observed motion*, but rather its motion in the solar system, or in (absolute) space, or its *real motion*. The familiar facts appealed to in the argument assert a different kind of motion, a simple vertical motion. This result refutes the Copernican hypothesis only if the concept of motion that occurs in the observation statement is the same as the concept of motion that occurs in the Copernican prediction. The observation statement "the stone is falling straight down" must therefore, likewise refer to a movement in (absolute) space. It must refer to a real motion.

Now, the force of an "argument from observation" derives from the fact that the observation statements it involves are firmly connected with appearances. There is no use appealing to observation if one does not know how to describe what one sees, or if one can offer one's description with hesitation only, as if one had just learned the language in which it is formulated. An observation statement, then, consists of two very different psychological events: (1) a clear and unambiguous *sensation* and (2) a clear and unambiguous *connection* between this sensation and parts of a language. This is the way in which the sensation is made to speak. Do the sensations in the argument above speak the language of real motion?

They speak the language of real motion in the context of seventeenth-century everyday thought. At least this is what Galileo tells us. He tells us that the everyday thinking of the time assumes the "operative" character of *all* motion.[137] or, to use well-known philosophical terms, it assumes a *naive realism with respect to motion*: except for occasional and unavoidable illusions, apparent motion is identical with real (absolute) motion. Of course, this distinction is not explicitly drawn. One does not first distinguish the apparent motion from the real motion and then connect the two by a correspondence rule. Quite the contrary, one describes, perceives, acts toward the apparent motion as if it were already the real thing. Nor does one proceed in this manner under all circumstances. It is admitted that objects may move which are not seen to move; and it is also admitted that certain motions are illusory (see the example in section 7 above). Apparent motion and real motion are not always identified. However, there are *paradigmatic cases* in which it is psychologically very difficult, if not plainly impossible, to admit deception. It is from these paradigmatic cases, and not from exceptions, that naive realism with respect to motions derives its strength. These are also the situations in which we first learn our kinematic vocabulary. From our very childhood we learn to react to them with concepts which have naive realism built right into them, and which inextricably connect movement and the appearance of movement. The motion of the stone in the tower argument, or the alleged motion of the earth, is such a paradigmatic case. How could one possibly be unaware of the swift motion of a large bulk of matter such as the earth is supposed to be! How could one possibly be unaware of the fact that the falling stone traces a vastly extended trajectory through space! From the point of view of seventeenth-century thought and language, the argument is, therefore, impeccable and quite forceful. However, notice how *theories* ("operative character" of all motion: essential correctness of sense reports), which are not formulated explicitly, enter the debate in the guise of observational terms. We realize again that observational terms are Trojan horses which must be watched very carefully. How is one supposed to proceed in such a sticky situation?

The argument from falling stones seems to refute the Copernican view. This may be due to an inherent disadvantage of Copernicanism; but it may also be due to the presence of natural interpretations which are in need of improvement. The first task, then, is to discover and to isolate these unexamined obstacles to progress.

It was Bacon's belief that natural interpretations could be discovered by a method of analysis that peels them off, one after another, until the sensory core of every observation is laid bare. This method has serious drawbacks. First, natural interpretations of the kind considered by Bacon are not just *added to* a previously existing field of sensations. They are instrumental in *constituting* the field, as Bacon says himself. Eliminate all natural interpretations, and you also eliminate the ability to think and to perceive. Second, disregarding this fundamental function of natural interpretations, it should be clear that a person who faces a perceptual field without a single natural interpretation at his disposal would be *completely disoriented*; he could not even *start* the business of science. Third, the fact that we *do* start, even after some Baconian analysis, shows that the analysis has stopped prematurely. It has stopped at precisely those natural interpretations of which we are not aware and without which we cannot proceed. It follows that the intention to start from scratch, after a complete removal of all natural interpretations, is self-defeating.

Furthermore, it is not possible to even *partly* unravel the cluster of natural interpretations. At first sight the task would seem to be simple enough. One takes observation statements, one after the other, and analyzes their content. However, concepts that are hidden in observation statements are not likely to reveal themselves in the more abstract parts of language. If they do, it will still be difficult to nail them down; concepts, just as percepts, are ambiguous and dependent on background. Moreover, the content of a concept is determined also by the way in which it is related to perception. Yet how can this way be discovered without circularity? Perceptions must be identified, and the identifying mechanism will contain some of the very same elements which govern the use of the concept to be investigated. We never penetrate this concept completely, for we always use part of it in the attempt to find its constituents.[138] There is only one way to get out of this circle, and it consists in using an *external measure of comparison*, including new ways of relating concepts and percepts. Removed from the domain of natural discourse and from all those principles, habits, and attitudes which constitute its form of life, such an external measure will look strange indeed. This, however, is not an argument against its use. Quite the contrary, such an impression of strangeness reveals that natural interpretations are at work, and it is a first step toward their discovery. Let us explain this situation with the help of the tower example.

The example is intended to show that the Copernican view is not in accordance with 'the facts.' Seen from the point of view of these 'facts,' the idea of the motion of the earth appears to be outlandish, absurd, and obviously false, to mention only some of the expressions which were frequently used at the time, and which are still heard wherever professional squares confront a new and counterfactual theory. This makes us suspect that the Copernican view is an external measuring rod of precisely the kind described above.

We now can turn the argument around and use it as a *detecting device*

51

that helps us to discover the natural interpretations that exclude the motion of the earth. Turning the argument around, we *first assert* the motion of the earth and *then inquire* what changes will remove the contradiction. Such an inquiry may take considerable time, and there is a good sense in which one can say that it is not yet finished, not even today. The contradiction, therefore, may stay with us for decades or even centuries. Still, *it must be upheld* (Hegel!) until we have finished our examination or else the examination, the attempt to discover the antediluvian components of our knowledge, cannot even start. This, we have seen, is one of the reasons one can give for *retaining*, and, perhaps, even for *inventing*, theories which are inconsistent with the facts: Ideological ingredients of our knowledge and, more especially, of our observations, are discovered with the help of theories which are refuted by them. *They are discovered counterinductively.*

Let me repeat what has been asserted so far. Theories are tested and possibly refuted by facts. Facts contain ideological components, older views which have vanished from sight or were perhaps never formulated in an explicit manner. These components are highly suspicious, first, because of their age, because of their antediluvian origin; second, because their very nature protects them from a critical examination and always has protected them from such an examination. Considering a contradiction between a new and interesting theory and a collection of firmly established facts, the best procedure is, therefore, not to abandon the theory but to use it for the discovery of the hidden principles that are responsible for the contradiction. Counterinduction is an essential part of such a process of discovery. (Excellent historical example: the arguments against motion and atomicity of Parmenides and Zeno. Diogenes of Sinope, the Cynic, took the simple course that would be taken by many contemporary scientists and all contemporary philosophers: he refuted the arguments by rising and walking up and down. The opposite course, recommended here, led to much more interesting results, as is witnessed by the history of the case. One should not be too hard on Diogenes, however, for it is also reported that he beat a pupil who was content with his refutation, exclaiming that he had given reasons which the pupil should not accept without additional reasons of his own.[139])

Having *discovered* a particular natural interpretation, the next question is how it is to be *examined* and *tested*. Obviously, we cannot proceed in the usual way, i.e., derive predictions and compare them with "results of observation." These results are no longer available. The idea that the senses, employed under normal circumstances, produce correct reports of real events, for example reports of the real motion of physical bodies, has now been removed from all observational statements. (Remember that this notion was found to be an essential part of the anti-Copernican argument.) But without it our sensory reactions cease to be relevant for tests. This conclusion has been generalized by some rationalists, who decided to build their science on reason only and ascribed to observation a quite insignificant auxiliary function. Galileo does not adopt this procedure.

If one natural interpretation causes trouble for an attractive view, and if its elimination removes the view from the domain of observation, then the only acceptable procedure is to use other interpretations and to see what happens. The interpretation which Galileo uses restores the senses to their position as instruments of exploration, but only with respect to the reality of relative motion. Motion "among things which share it in common" is "nonoperative," that is, "it remains insensible, imperceptible, and without any effect whatever." [140] Galileo's first step in the joint examination of the Copernican doctrine, and of a familiar but hidden natural interpretation, consists therefore in replacing the latter by a different interpretation, or, considering the function of natural interpretations, he introduces a new observation language.

This is, of course, an entirely legitimate move. In general, the observation language which enters an argument has been in use for a long time and is quite familiar. Considering the structure of common idioms on the one hand, and of the Aristotelian philosophy on the other, neither this use nor the familiarity can be regarded as a test of the underlying principles. These principles, these natural interpretations, occur in every description. Extraordinary cases which might create difficulties are defused with the help of "adjuster words," [141] such as "like" or "analogous," which divert them so that the basic ontology remains unchallenged. A test is, however, urgently needed. It is needed especially in those cases where the principles seem to threaten a new theory. It is then quite reasonable to introduce alternative observation languages and to compare them both with the original idiom and with the theory under examination. Proceeding in this way, we must make sure that the comparison is fair. That is, we must not criticize an idiom that is supposed to function as an observation language because it is not yet well known and is therefore less strongly connected with our sensory reactions and less plausible than is another and more "common" idiom. Superficial criticisms of this kind, which have been elevated into an entire new "philosophy," abound in discussions of the mind-body problem. Philosophers who want to introduce and to test new views thus find themselves faced not with arguments, which they could most likely answer, but with an impenetrable stone wall of well-entrenched reactions. This is not at all different from the attitude of people ignorant of foreign languages, who feel that a certain color is much better described by "red" than by "rosso." As opposed to such attempts at conversion by appeal to familiarity ("I know what pains are, and I also know, from introspection, that they have nothing whatever to do with material processes!"), we must emphasize that a comparative judgment of observation languages, e.g., materialistic observation languages, phenomenalistic observation languages, objective-idealistic observation languages, theological observation languages, can start only when all of them are spoken equally fluently.

Let me assert at this point that while it is possible to consider and to actively apply various rules of thumb, and while we may in this way arrive at a satisfactory judgment, it is not at all wise to go further and to turn these rules of thumb into necessary conditions of science. For example, one might be inclined to say, following Neurath, that an observation language A is preferable to an observation language B, if it is at least as useful as B in our everyday life, and if more theories and more comprehensive theories are compatible with it than are compatible with B. Such a criterion takes into account that both our perceptions (natural interpretations included) and our theories are fallible, and it also pays attention to our desire for a harmonious and universal point of view. (One always seems to assume that observation languages should be employed not only in laboratories, but also at home, and in the "natural surroundings" of the scientist.) However, we must not forget that we find and improve the assumptions hidden in our observational reports by a method that makes use of inconsistencies. Hence, we might prefer B to A as a starting point of analysis, and we might in this way arrive at a language C which satisfies the criterion even better, but which cannot be reached from A. Conceptual progress like any other kind of progress depends on psychological circumstances, which may prohibit in one case what they encourage in another. Moreover the psychological factors which come into play are never clear in advance. Nor should the demand for practicality and sensory content be regarded as a *conditio sine qua non*. We possess detecting mechanisms whose performance outdistances our senses. Combining such detectors with a computer, we may test a theory directly, without intervention of a human observer. This would eliminate sensations and perceptions from the process of testing. Using hypnosis, one could eliminate them from the transfer of the results into the human brain also, and thus arrive at a science that is completely without experience.[142] Considerations like these, which indicate possible paths of development, should cure us once and for all of the belief that judgments of progress, improvement, etc., are based on rules which can be revealed *now* and will remain in action for all the years to come. My discussion of Galileo has not, therefore, the aim of arriving at the "correct method." It has rather the aim of showing that such a "correct method" does not and cannot exist. More especially, it has the limited aim of showing that counterinduction is very often a reasonable move. Let us now proceed a step further in our analysis of Galileo's reasoning!

7. The Tower Argument: Analysis Continued

Galileo replaces one natural interpretation by a very different and as yet (1630!) at least partly unnatural interpretation. How does he proceed? How does he manage to introduce absurd and counterinductive assertions such as the assertion that the earth moves, and how does he manage to get them a just and attentive hearing? One may anticipate that arguments will not suffice—an interesting, and highly important, limitation of rationalism—and Galileo's utterances are indeed arguments in appearance only. For Galileo uses *propaganda.* He uses *psychological tricks* in addition to whatever intellectual reasons he has to offer. These tricks are very successful; they lead him to victory. But they obscure the new attitude toward experience that is in the making, and postpone for centuries the possibility of a reasonable philosophy. They obscure the fact that the experience on which Galileo wants to base the Copernican view is nothing but the result of his own fertile imagination, that it has been *invented.* They obscure this fact by insinuating that the new results which emerge are known and conceded by all, and need only be called to our attention to appear as the most obvious expression of the truth.

Galileo "reminds" us that there are situations in which the nonoperative character of shared motion is just as evident and as firmly believed as the idea of the operative character of all motion is in other circumstances (this latter idea is therefore not the only natural interpretation of motion). The situations are events in a boat, in a smoothly moving carriage, and in any other system that contains an observer and permits him to carry out some simple operations.

Sagredo: There has just occurred to me a certain fantasy which passed through my imagination one day while I was sailing to Aleppo, where I was going as consul for our country . . . If the point of a pen had been on the ship during my whole voyage from Venice to Alexandretta and had had the property of leaving visible marks of its whole trip, what trace—what mark—what line would it have left?

Simplicio: It would have left a line extending from Venice to there; not perfectly straight—or rather, not lying in the perfect arc of a circle—but more or less fluctuating according as the vessel would now and again have rocked. But this bending in some places a yard or two to the right or left, up or down, in length of many hundreds of miles, would have made little alteration in the whole extent of the line. These would scarcely be sensible, and without an error of any moment it could be called part of a perfect arc.

Sagredo: So that if the fluctuation of the waves were taken away and the motion of the vessel were calm and tranquil, the true and precise motion of that pen point would have been an arc of a perfect circle. Now if I had had that same pen continually in my hand, and had moved it only a little sometimes this way or that, what alteration should I have brought into the main extent of this line?

Simplicio: Less than that which would be given to a straight line a thousand yards long which deviated from absolute straightness here and there by a flea's eye.

Sagredo: Then if an artist had begun drawing with that pen on a sheet of paper when he left the port and had continued doing so all the way to Alexandretta, he would have been able to derive from the pen's motion a whole narrative of many figures, completely traced and sketched in thousands of directions, with landscapes, buildings, animals, and other things. Yet the actual, real, essential movement marked by the pen point would have been only a line; long, indeed, but very simple. But as to the artist's own actions, these would have been conducted exactly the same as if the ship had been standing still. The reason that of the pen's long motion no trace would remain except the marks drawn upon the paper is that the gross motion from Venice to Alexandretta was common to the paper, the pen, and everything else in the ship. But the small motions back and forth, to right and left, communicated by the artist's fingers to the pen but not to the paper, and belonging to the former alone, could thereby leave a trace on the paper which remained stationary to those motions.[143]

Or:

Salviati: . . . imagine yourself in a boat with your eyes fixed on a point of the sail yard. Do you think that because the boat is moving along briskly, you will have to move your eyes in order to keep your vision always on that point of the sail yard and follow its motion?

Simplicio: I am sure that I should not need to make any change at all; not just as to my vision, but if I had aimed a musket I should never have to move it a hairsbreadth to keep it aimed, no matter how the boat moved.

Salviati: And this comes about because the motion which the ship confers upon the sail yard, it confers also upon you and upon your eyes, so that you need not move them a bit in order to gaze at the top of the sail yard, which consequently appears motionless to you. (And the rays of vision go from the eye to the sail yard just as if a cord were tied between the two ends of the boat. Now a hundred cords are tied at different fixed points, each of which keeps its place whether the ship moves or remains still).[144]

It is clear that these situations lead to a nonoperative concept of motion even within common sense.

On the other hand, common sense, and I mean seventeenth-century common sense, also contains the idea of the *operative* character of all motion. This latter idea arises when a limited object that does not contain too many parts moves in vast and stable surroundings, for example, when a camel trots through the desert, or when a stone descends from a tower.

Now, Galileo urges us to "remember" the conditions in which we assert the nonoperative character of shared motion in this case also, and to subsume the second case under the first.

Thus, the first of the two paradigms of nonoperative motion mentioned above is followed by the assertion that "it is likewise true that the earth being moved, the motion of the stone in descending is actually a long stretch of many hundred yards, or even many thousand; and had it been able to mark its course in motionless air or upon some other surface, it would have left a very long slanting line. But that part of all this motion which is common to the rock, the tower, and ourselves remains insensible and as if it did not exist. There remains observable only that part in which neither the tower nor we are participants; in a word, that with which the stone in falling measures the tower." [145]

And the second paradigm precedes the exhortation to "transfer this argument to the whirling of the earth and to the rock placed on top of the tower, whose motion you cannot discern because in common with the rock you possess from the earth that motion which is required for following the tower; you do not need to move your eyes. Next, if you add to the rock a downward motion which is peculiar to it and not shared by you, and which is mixed with this circular motion, the circular portion of the motion which is common to the stone and the eye continues to be imperceptible. The straight motion alone is sensible, for to follow that you must move your eyes downwards." [146]

This is strong persuasion indeed.

Yielding to this persuasion, we now *quite automatically* start confounding the conditions of the two cases and become relativists. This is the essence of Galileo's trickery! As a result the clash between Copernicus and "the conditions affecting ourselves and those in the air above us" [147] dissolves into thin air, and we finally realize "that all terrestrial events from which it is ordinarily held that the earth stands still and the sun and the fixed stars are moving would necessarily appear just the same to us if the earth moved and the others stood still." [148]

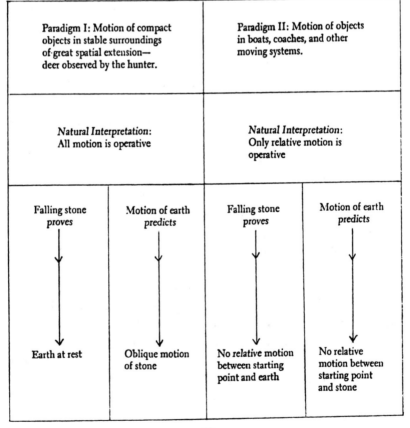

Paradigm I: Motion of compact objects in stable surroundings of great spatial extension— deer observed by the hunter.		Paradigm II: Motion of objects in boats, coaches, and other moving systems.	
Natural Interpretation: All motion is operative		*Natural Interpretation:* Only relative motion is operative	
Falling stone proves	Motion of earth predicts	Falling stone proves	Motion of earth predicts
↓	↓	↓	↓
Earth at rest	Oblique motion of stone	No relative motion between starting point and earth	No relative motion between starting point and stone

Let us now look at the situation from a more abstract point of view. We start with two conceptual subsystems of ordinary thought (see the preceding diagram). One of them regards motion as an absolute process which always has effects, effects on our senses included. The description of this conceptual system which appears in the present paper may be somewhat idealized, but the arguments of the opponents of Copernicus which are quoted by Galileo himself, and which according to him were "very plausible," [149] show that there was a widespread tendency to think in its terms, and that this tendency was a serious obstacle for the discussion of alternative ideas. Occasionally one finds even more primitive ways of thinking, where concepts such as "up" and "down" are used absolutely. Examples are the assertion "that the earth is too heavy to climb up over the sun and then fall headlong back down again," [150] or the assertion that "after a short time the mountains, sinking downward with the rotation of the terrestrial globe, would get into such a position that whereas a little earlier one would have had to climb steeply to their peaks, a few hours later one would have to stoop and descend in order to get there." [151] Galileo, in his marginal notes, calls these "utterly childish reasons [which] suffice[d] to keep imbeciles believing in the fixity of the earth" [152] and he thinks it unnecessary "to bother about such men as these, whose name is legion, or to take notice of their fooleries." [153] Yet it is clear that the absolute idea of motion was "well entrenched," and that the attempt to replace it was bound to encounter strong resistance.

The second conceptual system is built around the relativity of motion, and is also well entrenched in its own domain of application. Galileo aims at replacing the first system by the second in all cases, terrestrial as well as celestial. Naive realism with respect to motion is to be completely eliminated.

Now, we have seen that this naive realism is on occasions an essential part of our observational vocabulary. On these occasions (Paradigm I), the observation language contains the idea of the efficacy of all motion. Or, to express it in the material mode of speech, our experience in these situations is the experience of objects which move absolutely. Taking this into consideration, it is apparent that Galileo's proposal amounts to a partial revision of our observation language or of our experience. An experience which partly contradicts the idea of the motion of the earth is turned into an experience that confirms it, at least as far as "terrestrial things" are concerned.[154] This is what actually happens. But Galileo wants to persuade us that no change has taken place, that the second conceptual system is already universally *known*, even though it is not universally *used*. Both Salviati, his representative in the dialogue, and his opponent Simplicio, and also Sagredo, the intelligent layman, connect Galileo's method of argumentation with Plato's theory of anamnesis[155]—a clever tactical move, typically Galilean, one is inclined to say. Yet we must not allow ourselves to be deceived about the revolutionary development that is actually taking place.

The resistance against the assumption that shared motion is nonoperative was equated with the resistance which forgotten ideas exhibit toward the attempt to make them known. Let us accept this *interpretation* of the resistance! But let us not forget its *existence*. We must then admit that it restricts the use of the relativistic ideas, confining them to *part* of our everyday experience. Outside this part, and that means in interstellar space, they are "forgotten," and therefore not active. But outside this part there is not complete chaos. Other concepts are used, among them those very same absolutistic concepts which derive from the first paradigm. We not only use them, but must admit that they are entirely adequate. No difficulties arise as long as one remains within the limits of the first paradigm. "Experience," that is, the totality of all facts from all domains described with the concepts which are appropriate in these domains, cannot force us to carry out the change which Galileo wants to introduce. The motive for a change must come from a different source.

It comes, first, from the desire to see "the whole [correspond] to its parts with wonderful simplicity" [156] as Copernicus had already expressed himself. It comes from the "typically metaphysical urge" for unity of understanding and conceptual presentation. And the motive for a change is connected, secondly, with the intention to make room for the motion of the earth, which Galileo accepts and is not prepared to give up. The idea of the motion of the earth is closer to the first paradigm than to the second, or at least it was at the time of Galileo. This gave great strength to the Aristotelian arguments, and made them very plausible. To eliminate this plausibility, it was necessary to subsume the first paradigm under the second, and to extend the relative notions to all phenomena. The idea of *anamnesis* functions here as a psychological crutch, as a lever which smoothes the process of subsumption by concealing its existence. As a result we are now ready to apply the relative notions not only to boats, coaches, birds, but also to the "solid and well-established earth" as a whole.

And we have the impression that this readiness was in us all the time, although it took some effort to make it conscious. This impression is most certainly erroneous: it is the result of Galileo's propagandistic machinations. We would do better to describe the situation in a different way, as a change of our conceptual system. Or, because we are dealing with concepts which belong to natural interpretations, and which are therefore connected with sensations in a very direct way, we should describe it as a change of experience that allows us to accommodate the Copernican doctrine. The change corresponds perfectly to the pattern outlined in an earlier paper: an inadequate view, the Copernican theory, is supported by another inadequate view, the idea of the nonoperative character of shared motion, and both theories gain strength and give support to each other in the process. It is this change which constitutes the transition from the Aristotelian point of view to the epistemology of modern science.

For experience now ceases to be that unchangeable fundament which it is both in common sense and in the Aristotelian philosophy. The attempt to support Copernicus makes experience "fluid" in the very same manner in which it makes the heavens fluid, "so that each star roves around in it by itself." [157] An empiricist who starts from experience, and builds on it without ever looking back, now loses the very ground on which he stands. Neither the earth, "the solid, well-established earth," nor the facts on which he usually relies, can be trusted any longer. It is clear that a philosophy that uses such a fluid and changing experience needs new methodological principles which do not insist on an asymmetric judgment of theories by experience. Classical physics intuitively adopts such principles; at least the great and independent thinkers, such as Newton, Faraday, and Boltzmann, proceed in this way. But its official doctrine still clings to the idea of a stable and unchanging basis. The clash between this doctrine and the actual procedure is concealed by a tendentious presentation of the results of research that hides their revolutionary origin and suggests that they arose from a stable and unchanging source. These methods of concealment start with Galileo's attempt to introduce new ideas under the cover of anamnesis, and they culminate in Newton.[158] They must be exposed if we want to arrive at a better account of the progressive elements in science.

8. The Law of Inertia

Our discussion of the anti-Copernican argument is not yet complete. So far, we have tried to discover what assumption will make a stone that moves alongside a moving tower appear to fall "straight down," instead of being seen to move in an arc. The assumption, which I shall call the *relativity principle*, that our senses notice only relative motion, and are completely insensitive to a motion which objects have in common, was seen to do the trick. What remains to be explained is *why the stone stays with the tower*, and why it is not left behind. In order to save the Copernican view, one must explain not only why a motion that preserves the relation among visible objects remains unnoticed, but also why a common motion of various objects does not affect their relation. That is, one must explain why such a motion is not a causal agent. Turning the question around in the manner explained in section 6, it is now apparent that the anti-Copernican argument of section 5 rests on two natural interpretations,[159] viz. the *epistemological assumption* that absolute motion is always *noticed* and the *dynamical principle* that objects (such as the falling stone) which are not interfered with move toward their natural place. The present problem is to supplement the relativity principle with a new law of inertia in such a fashion that the motion of the earth can still be asserted. One sees at once that the following law, the *principle of circular inertia*, as I shall call it, provides the required solution: An object that moves with a given angular velocity on a frictionless sphere around the center of the earth will continue moving with the same angular velocity forever. Combining the appearance of the falling stone with the relativity principle, the principle of circular inertia, and some simple assumptions concerning the composition of velocities, yields an argument which no longer endangers Copernicus's view, but can be used to give it partial support.

The relativity principle was defended in two ways. The first was by showing how it helps Copernicus; this defense is truly ad hoc. The second was by pointing to its function in common sense, and by surreptitiously generalizing that function (see section 7). No independent argument was given for its validity.[160] Galileo's method of support for the principle of circular inertia is of exactly the same kind. He introduces it, again not by reference to experiment or to independent observation, but by reference to what everyone is already supposed to know.

Simplicio: So you have not made a hundred tests, or even one? And yet you so freely declare it to be certain? . . .

Salviati: Without experiment, I am sure that the effect will happen as I tell you, because it must happen that way; and I might add that you yourself also know that it cannot happen otherwise, no matter how you may pretend not to know it . . . But I am so handy at picking people's brains that I shall make you confess this in spite of yourself.[161]

Step by step Simplicio is forced to admit that a body that moves without friction on a sphere concentric with the center of the earth will carry out a "boundless," a "perpetual" motion.[162] We know, of course, especially after the analysis we have just completed of the nonoperative character of shared motion, that what Simplicio accepts is based neither on experiment nor on corroborated theory. It is a daring new suggestion involving a tremendous leap of the imagination. A little more analysis then shows that this suggestion is connected with experiments, such as the "experiments" of the Discorsi, by ad hoc hypotheses. (The amount of friction to be eliminated follows not from independent investigations—such investigations commence only much later, in the eighteenth century—but from the very result to be achieved, viz. the circular law of inertia.) Viewing natural phenomena in this way leads, as we have already said, to a complete reevaluation of all experience. We can now add that it leads to the invention of a new kind of experience that is not only more sophisticated but also far more speculative than is the experience of Aristotle or of common sense. Speaking paradoxically, but not incorrectly, one may say that Galileo invented an experience that has metaphysical ingredients.[163] It is by means of such an experience that the transition from a geostatic cosmology to the point of view of Copernicus and Kepler is achieved.

9. The Progressive Role of Ad Hoc Hypotheses

This is the place to briefly mention certain ideas which have been developed by Lakatos, and which throw new light on the problem of the growth of knowledge.

It is customary to assume that good scientists refuse to employ ad hoc hypotheses, and to assert that they are right in their refusal. New ideas, so it is thought, go far beyond the available evidence, and they must go beyond it in order to be of value. Ad hoc hypotheses are bound to creep in eventually, but they should be resisted and kept at bay. This is the customary attitude as it is expressed, for example, in the writings of K. R. Popper.

63

Paul K. Feyerabend

As opposed to this, Lakatos, in lectures, and now also in publications, has pointed out that "adhocness" is neither despicable nor absent from the body of science. New ideas, he emphasizes, are usually almost entirely ad hoc, they cannot be otherwise. And they are reformed only in a piecemeal fashion, by gradually stretching them, so that they apply to situations lying beyond their starting point. Schematically:

Popper: new theories have, and must have, excess content which is, but should not be, gradually infected by ad hoc adaptations.

Lakatos: new theories are, and cannot be anything but, ad hoc. Excess content is, and should be, created in a piecemeal fashion, by gradually extending them to new facts and domains.

The historical material I have just analyzed (and the more extensive material presented in "Problems of Empiricism, Part II") lends unambiguous support to the position of Lakatos. In what follows I shall try to show this in some detail.

First, *kinematic relativity* (cf. section 7, above):

Just like Newtonian physics, Aristotelian physics distinguishes between relative space and absolute space.[164] In addition, it allows one to "operationally" determine absolute places, directions, velocities. One may proceed in the following way: The *center* of the universe is found, for example, by backwardly elongating the direction of two flames, and it is tested by using a third flame. Flames function here as test bodies and not as reference bodies for relative motion. *Distance* from the center is determined by the strength of the upward motion of flames, or of suitable mixtures which may be enclosed in test capsules. Thus, space is traced out, in an entirely physical way, by using known physical laws. *Direction*, finally, is determined by determining the axis of rotation of the stellar sphere. This whole physical background is removed by Galileo. With it, we lose all means of testing for center, distance, and direction. The new relativistic principles (only relative motion is "operative") are therefore metaphysical, and, because adapted to the tower experiment, also ad hoc.

Considering now *dynamical relativity* (section 8), one should remember, first of all, that the natural character of circular motion was not first asserted by Galileo. It was an old assumption, concerning all supralunar entities. The new assumption introduced by Galileo (and by Copernicus, in chapter VIII of *De revolutionibus*) is that circular motion is a natural motion for terrestrial objects also. On the one hand, this is an immediate

64

consequence of having made the earth a star: Stars move in circles. Hence, if the earth is a star, its natural motion will be circular, both its motion around the sun and "its motion with respect to itself," as its rotation was described at the time. Now, does this particular assumption of the rotation of earth assert anything over and above what was known to happen at its surface at Galileo's own time? My attitude, which is in accordance with Lakatos's general theory, is that the answer must be no. The only consequence of the assertion is that it connects moving objects rigidly with the framework of the moving, i.e., rotating, earth. This leaves everything as it is, and it especially leaves the results of the tower experiment and the cannon experiment unchanged.[165] No further consequence was implied at the time. (It was different with the motion of the earth around the sun which led one to expect a sizable stellar parallax.) Even the later Newtonian argument that distant objects, moving with the same angular velocity, will hit the earth *ahead* of the tower cannot be used at this stage: it is not at all clear whether Galileo would want distant objects to move with the same angular velocity. (In the case of the planets he notices their *decreasing* angular velocity—the effect of Kepler's third law—and he might have been inclined to treat bodies circulating around the earth in the same way. On the other hand, he calculates the time a stone takes to drop from the moon to the earth by assuming a constant acceleration all the way.[166])

Furthermore, I do not think that bringing in the tradition of the impetus theory will improve matters. For this theory is again ad hoc, this time not with respect to the tower, but with respect to the behavior of objects thrown (which continue to move, contrary to Aristotle's law of inertia). When a circular law is asserted, as seems to be the case with Buridan, the problem is the same as for Galileo.[167] (Besides, the impetus theory is incompatible with Galileo's idea of the nonoperative character of all motion.[168])

Finally, one must not argue against "ad hocness" by pointing to the fact that *experiments* were made in boats, with cannon balls, on towers, and so on.[169] These experiments did not lead to any decisive result. And they did not test any *excess content* of the law of circular inertia, but tried to establish the *fact* which the law then explains in ad hoc fashion. Reference to the experiments with the inclined plane is also beside the point. These experiments test, if that is the right word, the law of free fall. But of course there still remains the task of subdividing *that* motion into an inertial motion and something else. However one looks at the matter, the best conjecture is that at the time in question the circular law of inertia, and to an even greater extent the idea of the relativity of motion, was an ad hoc hypothesis designed to get out of the trouble of the tower.

Now this is such an incredible situation that a little more argument seems to be required. We therefore take a brief look at Galileo's earlier work on mechanics and motion.

In *De motu* motions of spheres in the center of the universe, outside of it, homogeneous, nonhomogeneous, supported at the center of gravity, supported outside of it are discussed, and described as being either natural, or forced, or neither. But about the actual motion of such spheres we hear very little, and what we do hear is by implication only. Thus there appears the question[170] whether a homogeneous sphere made to move in the center of the universe would move forever. We read that "it seems that it should move perpetually," but an unambiguous answer is never given. A marble sphere supported on an axis through the center and set in motion is said to "rotate for a long time" [171] in *De motu* while a perpetual motion is said to be "quite out of keeping with the nature of the earth itself to which rest seems to be more congenial than motion" in the *Dialogue on Motion*.[172] Another argument against perpetual rotations is found in Benedetti's *Diverse Speculations*.[173] Rotations, says Benedetti, are "certainly not perpetual," for the parts of the sphere, wanting to move in a straight line, are constrained against their nature, "and so they come to rest naturally." Again, in *De motu*,[174] we find a criticism of the assertion that adding a star to the celestial sphere might slow it down by changing the relation between the force of the moving intelligences and the resistance of the sphere. This assertion, Galileo says, certainly applies to an excentric sphere. Adding weight to an excentric sphere means that a weight will occasionally be moved away from the center and be raised to a higher level. But "who would ever say that [a concentric sphere] was impeded by the weight, since the weight in its circular path would neither approach, nor recede from, the center." [175] Note that the original rotation is in this case said to be caused by an intelligence; it is not assumed to be taking place all by itself. This is in perfect agreement with Aristotle's general theory of motion[176] where a mover is postulated for every motion, and not just for violent motions. Galileo seems to accept this part of the theory both when letting rotating spheres slow down and when accepting the "force of the intelligences" in the present argument (he also accepts impetus—see below). But in objecting to the idea that a new star will increase resistance he adopts the entirely different view that resistance occurs only when a motion is forced, and is absent otherwise. This is neither Aristotelian nor compatible with the version of the impetus theory he holds at the time which attributes any prolonged motion to an internal moving force similar to the force of sound that resides in a bell long after it has been struck,[177] and which is again supposed to "gradually diminish." [178]

Looking at these few examples we see that Galileo ascribes a special position to motions which are neither violent nor forced. Such motions may last for a considerable time though they are not supported by the surrounding medium. But they do not last forever, and they need an internal driving force in order to even persist for a finite time.

Now if one wants to overcome the dynamical arguments against the motion of the earth (and we are here always thinking about its rotation rather than about its motion around the sun), then the two italicized principles must both be revised. It must be assumed that the "neutral" motions which Galileo discusses in his early dynamical writings may last forever, or at least for periods comparable to the age of historical records. And these motions must be regarded as "natural" in the entirely new and revolutionary sense that neither an outer nor an inner motor is needed to keep them going. The first assumption is necessary to allow the earth to rotate. The second assumption is necessary if we want to regard motion as a relative phenomenon, depending on the choice of a suitable coordinate system.[179] Copernicus, in his brief remarks on the problem,[180] makes both assumptions. Galileo never clearly resolves the problem. He formulates permanence along a horizontal line as a hypothesis in his Discorsi[181] and he seems to make both assumptions in the Dialogue.[182] Now my guess is that a clear statement of permanent motion with(out) impetus developed in Galileo only together with his gradual acceptance of the Copernican view. Galileo changed his view about the "neutral" motions—he made them permanent and "natural"—in order to make them compatible with the rotation of the earth and in order to evade the difficulties of the tower argument.[183] His new ideas concerning such motions are therefore at least partly ad hoc. Impetus in the old sense disappeared partly for methodological reasons (interest in the how, not in the why—this development itself deserves careful study), partly because of the vaguely perceived inconsistency with the idea of the relativity of all motion. The wish to save Copernicus plays a role in either case. This hypothesis must of course be tested by an examination of Galileo's published writings and his correspondence between 1590 and, say, 1630. Considering what we know already we must admit that it has much plausibility.

Now, if we are right in assuming that Galileo framed an ad hoc hypothesis at this point, then we can also praise him for his methodological acumen. It is obvious that the moving earth demands a new dynamics. One test of the old dynamics consists in the attempt to establish the motion of the earth. Trying to establish the motion of the earth is the same as trying to find a refuting instance for the old dynamics. The motion of the earth, however, is inconsistent with the tower experiment *interpreted in accordance with the old dynamics*. Interpreting the tower experiment in accordance with the old dynamics therefore means trying to save the old dynamics in an ad hoc fashion. If one does not want to do this one must find a different interpretation for the phenomena of free fall. What interpretation should be chosen? One wants an interpretation that turns the motion of the earth into a refuting instance of the old dynamics, without lending ad hoc support to the motion of the earth itself. The first step toward such an interpretation is to establish contact, however vague, with the "phenomena," i.e., with the falling stone, and to establish it in such a manner that the motion of the earth is not *obviously* contradicted. The most primitive element of this first step is to frame an ad hoc hypothesis with respect to the rotation of the earth. The next step would then be to elaborate the hypothesis, so that additional predictions become possible. Copernicus and Galileo take the first and most primitive step. Their procedure looks contemptible only if one forgets that the aim is *to test older views* rather than *to prove new ones*, and if one also forgets that developing a good theory is a complex process that has to start modestly and that takes time. But why, an impatient methodologist might ask, did it take so long before additional phenomena were added? It took so long *because the domain of possible phenomena had first to be circumscribed by the further development of the Copernican hypothesis.* It is much better to remain ad hoc for a while, and in the meantime to develop heliocentrism in all its astronomical ramifications which can then be used as guidelines for a further elaboration of dynamics.

Therefore: Galileo *did* use ad hoc hypotheses. It *was good* that he used them. Had he not been ad hoc, he would have been ad hoc anyway, but this time with respect to an older theory. Hence, as one cannot help being ad hoc, it is better to be ad hoc with respect to a new theory, for a new theory, like all new things, will give a feeling of freedom, excitement, and progress. Galileo is to be applauded because he preferred protecting an interesting hypothesis to protecting a dull one.

10. Summary of Analysis of Tower Argument

I repeat and summarize: An argument is proposed that refutes Copernicus by observation. The argument is inverted in order to discover those natural interpretations which are responsible for the contradiction. The offensive interpretations are replaced by others. Propaganda and appeal to distant and highly theoretical parts of common sense are used to defuse old habits and to enthrone new ones. The new natural interpretations which are also formulated explicitly as auxiliary hypotheses are established partly by the support they give to Copernicus and partly by plausibility considerations and ad hoc hypotheses. An entirely new "experience" arises in this way. Independent evidence is as yet entirely lacking, but this is no drawback as it is to be expected that independent support will take a long time appearing. For what is needed is a theory of solid objects, aerodynamics, hydrodynamics, and all these sciences are still hidden in the future. *But their task is now well defined,* for Galileo's assumptions, his ad hoc hypotheses included, are sufficiently clear and simple to prescribe the direction of future research. Let it be noted, incidentally, that Galileo's procedure drastically reduces the content of dynamics. Aristotelian dynamics was a general theory of change comprising locomotion, qualitative change, generation, and corruption, and it provided a theoretical basis for witchcraft also. Galileo's dynamics and its successors deal with *locomotion* only, and here again only with the locomotion of *matter.* The other kinds of motion are pushed aside with the promissory note, due to Democritos, that locomotion will eventually be capable of explaining *all* motion. Thus, a comprehensive empirical theory of motion is replaced by a much narrower theory[184] plus a metaphysics of motion, just as an "empirical" experience is replaced by an experience that contains strange and speculative elements. *Counterinduction,* however, is now justified both for theories and for facts. It clearly plays an important role in

69

the advancement of science. This concludes the considerations which started in section 2. For details and further examples the reader is again referred to my "Problems of Empiricism, Part II."

11. Discovery and Justification; Observation and Theory

Let us now use the material of the preceding sections to throw light on the following features of contemporary empiricism: first, the distinction between a context of discovery and a context of justification; second, the distinction between observational terms and theoretical terms; third, the problem of incommensurability.

One of the objections which may be raised against the preceding discussion is that it has confounded two contexts which are essentially separate, viz. a context of discovery and a context of justification. *Discovery may be irrational and need not follow any recognized method. Justification*, on the other hand, or, to use the Holy Word of a different school, *criticism*, starts only after the discoveries have been made and proceeds in an orderly way. Now, if the example given here and the examples I have used in earlier papers show anything, then they show that the distinction refers to a situation that does not arise in practice at all. And, if it does arise, it reflects a temporary stasis of the process of research. Therefore, it should be eliminated as quickly as possible.

Research at its best is an *interaction* between new theories which are stated in an explicit manner and older views which have crept into the observation language. It is not a one-sided *action* of the one upon the other. Reasoning within the context of justification, however, presupposes that one side of this pair, viz. observation, has frozen, and that the principles which constitute the observation concepts are preferred to the principles of a newly invented point of view. The former feature indicates that the discussion of principles is not carried out as vigorously as is desirable; the latter feature reveals that this lack of vigor may be due to some unreasonable and perhaps not even explicit preference. But is it wise to be dominated by an inarticulate preference of this kind? Is it wise to make it the raison d'être of a distinction that separates two entirely different modes of research? Or should we not rather demand that our methodology treat explicit and implicit assertions, doubtful and intuitively evident theories, known and unconsciously held principles, in exactly the same way, and that it provide means for the discovery and the criticism of the latter? Abandoning the distinction between a context of discovery and a context of justification is the first step toward satisfying this demand. Another distinction which is clearly related to the distinction between discovery and justification is the distinction between *observational terms* and *theoretical terms*. It is now generally admitted that the distinction is not as sharp as it was thought to be only a few decades ago. It is also admitted, in complete agreement with Neurath's original views, that *both* theories *and* observation statements are open to criticism. Yet the distinction is still held to be a useful one and is defended by almost all philosophers of science. But what is its point? Nobody will deny that the sentences of science can be classified into long sentences and short sentences, or that its statements can be classified into those which are intuitively obvious and others which are not. But nobody will put particular weight on these distinctions, or will even mention them, *for they do not now play any role in the business of science.* (This was not always so. Intuitive plausibility, for example, was once thought to be a most important guide to the truth; but it disappeared from methodology the very moment intuition was replaced by experience.) Does experience play such a role in the business of science? Is it as essential to refer to experience as it was once thought essential to refer to intuition? Considering what has been said in section 4, I think that these questions must be answered in the negative. True—much of our thinking arises from experience, but there are large portions which do not arise from experience at all but are firmly grounded on intuition, or on even deeper lying reactions. True— we often test our theories by experience, but we equally often *invert* the process; we *analyze* experience with the help of more recent views and we *change* it in accordance with these views (see the preceding discussion of Galileo's procedure). Again, it is true that we often rely on experience in a way that suggests that we have here a solid foundation of knowledge, but such reliance turns out to be just a psychological quirk, as is shown whenever the testimony of an eyewitness or of an expert crumbles under cross-examination. Moreover, we equaly firmly rely on general principles so that even our most solid *perceptions* (and not only our *assumptions*) become indistinct and ambiguous when they clash with these principles. The symmetry between observation and theory that emerges from such remarks is perfectly reasonable. Experience, just as our theories, contains natural interpretations which are abstract and even metaphysical ideas. For example, it contains the idea of an observer-independent existence. It is incontestable that these abstractions, these speculative ideas, are connected with sensations and perceptions. But, first of all, this does not give them a privileged position, unless we want to assert that perception is an infallible authority. And, secondly, it is quite possible to altogether *eliminate* perception from all the essential activities of science (see above, section 6 as well as the appendix). All that remains is that some of our ideas are accompanied by strong and vivid psychological processes, "sensations," while others are not. This, however, is just a peculiarity of human existence which is as much in need of examination as is anything else.

Now, if we want to be "truly scientific" (dreaded words!), should we then not regard the theses "experience is the foundation of our knowledge" and "experience helps us to discover the properties of the external world" as (very general) hypotheses? And must these hypotheses not be examined just like any other hypothesis, and perhaps even more vigorously, as so much depends on their truth? Furthermore, will not such an examination be rendered impossible by a method that either justifies or criticizes "on the basis of experience"? These are some of the questions which arise in connection with the customary distinctions between observation and theory, discovery and justification. None of them is really new. They are known to philosophers of science, and are discussed by them at length. But the inference that the distinction between theory and observation has now ceased to be relevant either is not drawn or is explicitly rejected.[185] Let us take a step forward, and let us abandon this last remainder of dogmatism in science!

12. Rationality Again

Incommensurability, which I shall discuss next, is closely connected with the question of the rationality of science. Indeed, one of the most general objections, either against the use of incommensurable theories or even against the idea that there are such theories to be found in the history of science, is the fear that they would severely restrict the efficacy of traditional, nondialectical argument. Let us, therefore, look a little more closely at the critical standards which, according to some people, constitute the content of a "rational' argument. More especially, let us look at the standards of the Popperian school with whose ratiomania we are here mainly concerned.

Critical rationalism is either a meaningful idea or a collection of slogans (such as "truth"; "professional integrity"; "intellectual honesty") designed to intimidate yellow-bellied opponents (who has the fortitude, or even the insight, to declare that Truth might be unimportant, and perhaps even undesirable?).

In the former case it must be possible to produce rules, standards, restrictions which permit us to separate critical behavior (thinking, singing, writing of plays) from other types of behavior so that we can *discover* irrational actions and correct them with the help of concrete suggestions. It is not difficult to produce the standards of rationality defended by the Popperian school.

These standards are standards of *criticism*: rational discussion consists in the attempt to criticize, and not in the attempt to prove, or to make probable. Every step that protects a view from criticism, that makes it safe, or "well founded," is a step away from rationality. Every step that makes it more vulnerable is welcome. In addition it is recommended that ideas which have been found wanting be abandoned, and it is forbidden to retain them in the face of strong and successful criticism unless one can present a suitable counterargument. Develop your ideas so that they can be criticized; attack them relentlessly; do not try to protect them, but exhibit their weak spots; and eliminate them as soon as such weak spots have become manifest—these are some of the rules put forth by our critical rationalists.

These rules become more definite and more detailed when we turn to the philosophy of science, and especially to the philosophy of the natural sciences.

Within the natural sciences criticism is connected with experiment and observation. The content of a theory consists in the sum total of those basic statements which contradict it; it is the class of its potential falsifiers. Increased content means increased vulnerability; hence theories of large content are to be preferred to theories of small content. Increase of content is welcome; decrease of content is to be avoided. A theory that contradicts an accepted basic statement must be given up. Ad hoc hypotheses are forbidden—and so on and so forth. A science, however, that accepts the rules of a critical empiricism of this kind will develop in the following manner.

We start with a *problem* such as the problem of the planets at the time of Plato. This problem is not merely the result of curiosity, it is a *theoretical result*, it is due to the fact that certain *expectations* have been disappointed: On the one hand it seemed to be clear that the stars must be divine; hence one expects them to behave in an orderly and lawful manner. On the other hand one cannot find any easily discernible regularity. The planets, to all intents and purposes, move in a quite chaotic fashion. How can this fact be reconciled with the expectation and with the principles that underlie the expectation? Does it show that the expectation is mistaken? Or have we failed in our analysis of the facts? This is the problem.

It is important to see that the elements of the problem are not simply given. The "fact" of irregularity, for example, is not accessible without further ado. It cannot be discovered by just anyone who has healthy eyes and a good mind. It is only through a certain expectation that it becomes an object of our attention. Or, to be more accurate: this fact of irregularity *exists* because there is an expectation of regularity. After all, the term "irregularity" makes sense only if we have a rule. In our case the rule (which is a more specific part of the expectation that has not yet been mentioned) asserts circular motion with constant angular velocity. The fixed stars agree with this rule and so does the sun if we trace its path relative to the fixed stars. The planets do not obey the rule, neither directly, with respect to the earth, nor indirectly, with respect to the fixed stars.

(In the case just discussed the rule is formulated explicitly, and it can be discussed. This need not be the case. Recognizing a color as red is made possible by deep-lying assumptions concerning the structure of our surroundings and recognition does not occur when these assumptions cease to be available.)

To sum up this part of the Popperian doctrine: Research starts with a problem. The problem is the result of a conflict between an expectation and an observation which in turn is constituted by the expectation. It is clear that this doctrine differs from the doctrine of inductivism where objective facts mysteriously enter a passive mind and leave their traces there. It was prepared by Kant, by Dingler, and, in a very different manner, by Hume.

Having formulated a problem one tries to *solve* it. Solving a problem means inventing a theory that is relevant, falsifiable (to a larger degree than any alternative solution), but not yet falsified. In the case mentioned above (planets at the time of Plato) the problem was to find circular

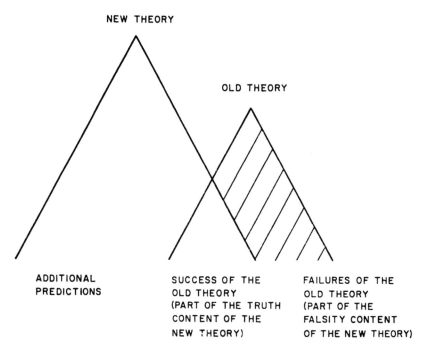

NEW THEORY

OLD THEORY

ADDITIONAL
PREDICTIONS

SUCCESS OF THE
OLD THEORY
(PART OF THE TRUTH
CONTENT OF THE
NEW THEORY)

FAILURES OF THE
OLD THEORY
(PART OF THE
FALSITY CONTENT
OF THE NEW THEORY)

motions of constant angular velocity for the purpose of saving the planetary phenomena. It was solved by Eudoxos.

Next comes the criticism of the theory that has been put forth in the
attempt to solve the problem. Successful criticism removes the theory
once and for all and creates a new problem, viz. to explain (a) why the
theory has been successful so far; (b) why it failed. Trying to solve this
problem we need a new theory that produces the successful consequences
of the older theory, denies its mistakes, and makes additional predictions
not made before. These are some of the *formal conditions* which a *suitable successor of a refuted theory must satisfy.* Adopting the conditions
one proceeds, by conjectures and refutations, from less general theories
to more general theories and expands the content of human knowledge.
More and more facts are discovered (or constructed with the help of expectations) and are then connected in a reasonable manner. There is no
guarantee that man will solve every problem and replace every theory
that has been refuted with a successor satisfying the formal conditions.
The invention of theories depends on our talents and other fortuitous
circumstances, such as a satisfactory sex life. But as long as these talents
hold out the accompanying scheme is a correct account of the growth
of a knowledge that satisfies the rules of critical rationalism.

75

Now, at this point we may raise two questions:

1. Is it *desirable* to live in accordance with the rules of a critical rationalism?

2. Is it *possible* to have both a science as we know it and these rules?

As far as I am concerned the first question is far more important than the second. True—science and other depressing and narrow-minded institutions play an important part in our culture and they occupy the center of interest of most philosophers. Thus the ideas of the Popperian school were obtained by generalizing solutions for methodological and epistemological problems. Critical rationalism arose from the attempt to solve Hume's problem and to understand the Einsteinian revolution, and it was then extended to politics, and even to the conduct of one's private life (Habermas and others therefore seem to be justified in calling Popper a positivist). Such a procedure may satisfy a *school philosopher* who looks at life through the spectacles of his own specific problems and recognizes hatred, love, happiness only to the extent to which they occur in these problems. But if we consider the interests of *man* and, above all, the question of his freedom (freedom from hunger, despair, from the tyranny of constipated systems of thought, *not* the academic "freedom of the will"), then we are proceeding in the worst possible fashion.

For is it not possible that science as we know it today (the science of critical rationalism that has been freed from all inductive elements) or a "search for the truth" in the style of traditional philosophy will create a monster? Is it not possible that it will harm man, turn him into a miserable, unfriendly, self-righteous mechanism without charm and without humor? "Is it not possible," asks Kierkegaard, "that my activity as an objective [or a critico-rational] observer of nature will weaken my strength as a human being?" [186] I suspect the answer to all these questions must be affirmative and I believe that a reform of the sciences that makes it more anarchistic and more subjective (in Kierkegaard's sense) is therefore urgently needed. But this is not what I want to discuss in the present essay. Here I shall restrict myself to the second question and I shall ask: is it possible to have both a science as we know it and the rules of a critical rationalism as just described? And to *this* question the answer seems to be a resounding *no*.

To start with we have seen, though rather briefly,[187] that the actual development of institutions, ideas, practices, and so on often does not start from a problem but rather from some irrelevant activity, such as playing, which, as a side effect, leads to developments which later on can be interpreted as solutions to unrealized problems. Are such developments to be excluded? And if we *do* exclude them, will this not considerably reduce the number of our adaptive reactions and the quality of our learning process?

Secondly, we have seen, in sections 4ff, that a strict principle of falsification, or a "naive falsificationism" as Imre Lakatos calls it, combined with the demand for maximum testability and non-adhocness would wipe out science as we know it, and would never have permitted it to start. This has been realized by Lakatos who has set out to remedy the situation.[188] His remedy is not mine, it is not anarchism. His remedy consists in slight modification of the "critical standards" he adores. (He also tries to show, with the help of amusing numerological considerations, that it is already foreshadowed in Popper.)

According to naive falsificationism, a theory is judged, i.e., either accepted or condemned, as soon as it is introduced into the discussion. Lakatos gives a theory time, he permits it to develop, to show its hidden strength, and he judges it only "in the long run." The "critical standards" he employs provide for an interval of hesitation. They are applied "with hindsight." If the theory gives rise to interesting new developments, if it engenders "progressive problem shifts," then it may be retained despite its initial vices. If on the other hand the theory leads nowhere, if the ad hoc hypotheses it employs are not the starting point but the end of all research, if the theory seems to kill the imagination and to dry up every resource of speculation, if it creates "degenerating problem shifts," i.e., changes which terminate in a dead end, then it is time to give it up and to look for something better.

Now it is easily seen that standards of this kind have practical force only if they are combined with a *time limit*. What looks like a degenerating problem shift may be the beginning of a much longer period of advance, so—how long are we supposed to wait? But if a time limit *is* introduced, then the argument against the more conservative point of view, against "naive falsificationism," reappears with only a minor modification. For if you can wait, then why not wait a little longer? Besides there are theories which for centuries were accompanied by degenerating problem shifts until they found the right defenders and returned to the stage in full bloom. The heliocentric theory is one example. The atomic theory is another. We see that the new standards which Lakatos wants to de-

77

fend either are vacuous—one does not know when and how to apply them —or else can be criticized on grounds very similar to those which led to them in the first place.

In these circumstances one can do one of the following two things. One can stop appealing to permanent standards which remain in force throughout history, and govern every single period of scientific development and every transition from one period to another. Or one can retain such standards as a verbal ornament, as a memorial to happier times when it was still thought possible to run a complex and catastrophic business like science by a few simple and "rational" rules. It seems that Lakatos wants to choose the second alternative.

Choosing the second alternative means abandoning permanent standards in fact, though retaining them in words. In fact Lakatos's position now is identical with the position of Popper as summarized in the marvelous (because self-destructive) Appendix i/15 of the fifth edition of the Open Society.[189] According to Popper, we do not "need any . . . definite frame of reference for our criticism," we may revise even the most fundamental rules and drop the most fundamental demands if the need for a different measure of excellence should arise.[190] Is such a position irrational? Yes and no. Yes, because there no longer exists a single set of rules that will guide us through all the twists and turns of the history of thought (science), either as participants or as historians who want to reconstruct its course. One can of course force history into a pattern, but the results will always be poorer and less interesting than were the actual events. No, because each particular episode is rational in the sense that some of its features can be explained in terms of reasons which were either accepted at the time of its occurrence or invented in the course of its development. Yes, because even these local reasons which change from age to age are never sufficient to explain all the important features of a particular episode. One must add accidents, prejudices, material conditions, e.g., the existence of a particular type of glass in one country and not in another for the explanation of the history of optics, the vicissitudes of married life (Ohm!), superficiality, pride, oversight, and many other things, in order to get a complete picture. No, because, transported into the climate of the period under consideration and endowed with a lively and curious intelligence, we might have had still more to say; we might have tried to overcome accidents, and to "rationalize" even the most whimsical sequence of events. But, and now I come to a decisive point for the discussion of incommensurability, how is the transition from certain standards to other standards to be achieved? More especially, what happens to our standards, as opposed to our theories, during a period of revolution? Are they changed in the manner suggested by Mill, by a critical discussion of alternatives, or are there processes which defy a rational analysis? Well, let us see!

That standards are not always adopted on the basis of argument has been emphasized by Popper himself. Children, he says, "learn to imitate others . . . and so learn to look upon standards of behavior as if they consisted of fixed, 'given' rules . . . and such things as sympathy and imagination may play an important role in this development." [191] Similar considerations apply to those grownups who want to continue learning, and who are intent on expanding both their knowledge and their sensibility. This we have already discussed in section 1. Popper also admits that new standards may be discovered, invented, accepted, imparted to others in a very irrational manner. But, he points out, one can criticize them *after* they have been adopted, and it is *this* possibility which keeps our knowledge rational. "What, then, are we to trust?" he asks after a survey of possible sources for standards.[192] "What are we to accept? The answer is: whatever we accept we should trust only tentatively, always remembering that we are in possession, at best, of partial truth (or rightness), and that we are bound to make at least some mistake or misjudgement somewhere—not only with respect to facts but also with respect to the adopted standards; secondly, we should trust (even tentatively) our intuition only if it has been arrived at as the result of many attempts to use our imagination; of many mistakes, of many tests, of many doubts, *and of searching criticism.*"

Now this reference to tests and to criticism, which is supposed to guarantee the rationality of science, and, perhaps, of our entire life, may be either to *well-defined procedures* without which a criticism or test cannot be said to have taken place, or to a purely *abstract* notion, so that it is left to us to fill it now with this, and now with that concrete content. The first case has just been discussed. In the second case we have again but a verbal ornament. The questions asked in the last paragraph but one remain unanswered in either case.

In a way even this situation has been described by Popper, who says that "rationalism is necessarily far from comprehensive or self-contained." [193]

Paul K. Feyerabend

But our present inquiry is not whether *there are* limits to our reason; the question is *where* these limits are *situated.* Are they outside the sciences so that science itself remains entirely rational; or are irrational changes an essential part even of the most rational enterprise that has been invented by man? Does the historical phenomenon "science" contain ingredients which defy a rational analysis, although they may be described with complete clarity in psychological or sociological terms? Can the abstract aim to come closer to the truth be reached in an entirely rational manner, or is it perhaps inaccessible to those who decide to rely on argument only? These are the problems which were raised, first by Hegel and then, in quite different terms, by Kuhn. They are the problems I wish to discuss.

In discussing these further problems, Popper and Lakatos reject considerations of sociology and psychology, or as Lakatos expresses himself, "mob psychology," and assert the rational character of *all* science. According to Popper, it is possible to arrive at a judgment as to which of two theories is closer to the truth, even if the theories should be separated by a catastrophic upheaval such as a scientific or other revolution. (A theory is closer to the truth than another theory if the class of its true consequences, its truth content, exceeds the truth content of the latter without an increase of falsity content.) According to Lakatos, the apparently unreasonable features of science occur only in the material world and in the world of (psychological) thought; they are absent from the "world of ideas, from Plato's and Popper's 'third world.'" It is in this third world that the growth of knowledge takes place, and that a rational judgment of all aspects of science becomes possible.

Now in regard to this convenient flight into higher regions, it must be pointed out that the scientist is, unfortunately, dealing with the world of matter and of psychological (i.e., subjective) thought also. It is mainly this material world he wants to change and to influence. And the rules which create order in the third world will most likely be entirely inappropriate for creating order in the brains of living human beings (unless these brains and their structural features are put in the third world also, a point that does not become clear from Popper's account[194]). The numerous deviations from the straight and rather boring path of rationality which one can observe in actual science may well be necessary if we want to achieve progress with the brittle and unreliable material (instruments; brains; assistants; etc.) at our disposal.

However, there is no need to pursue this objection further. There is no need to argue that science as we know it may differ from its third-world shadow *in precisely those respects* which make progress possible.[195] For the Popperian model of an approach to the truth breaks down even if we confine ourselves to ideas entirely. It breaks down because there are incommensurable theories.

13. Incommensurability

Scientific investigation, says Popper, *starts* with a problem, and it proceeds by *solving* it.

This characterization does not take into account that problems may be wrongly formulated, that one may inquire about properties of things or processes which later research declares to be nonexistent. Problems of this kind are not *solved*, they are *dissolved* and removed from the domain of legitimate inquiry. Examples are the problem of the absolute velocity of the earth, the problem of the trajectory of an electron in an interference pattern, or the important problem whether incubi are capable of producing offspring or whether they are forced to use the seeds of men for that purpose.[196]

The first problem was dissolved by the theory of relativity which denies the existence of absolute velocities. The second problem was dissolved by the quantum theory which denies the existence of trajectories in interference patterns. The third problem was dissolved, though much less decisively so, by modern (i.e., post-sixteenth century) psychology and physiology as well as by the mechanistic cosmology of Descartes.

Now changes of ontology such as those just described are often accompanied by conceptual changes.

The discovery that certain entities do not exist may force the scientist to redescribe the events, processes, observations which were thought to be manifestations of them and were therefore described in terms assuming their existence. Or, rather, it may force him to use new concepts as the older words will remain in use for a considerable time. Thus the term "possessed" which was once used for giving a causal description of the behavioral peculiarities connected with epilepsy was retained, but it was voided of its devilish connotations.

An interesting development occurs when the faulty ontology is comprehensive, that is, when its elements are thought to be present in every process in a certain domain. In this case every description inside the domain must be changed and must be replaced by a different statement (or by no statement at all). Classical physics is a case in point. It has developed a comprehensive terminology for describing the most fundamental mechanical properties of our universe, such as shapes, speeds, and masses. The conceptual system connected with this terminology assumes that the properties *inhere* in objects and that they change only if one interferes with the objects, not otherwise. The theory of relativity teaches us, at least in one of its interpretations, that there are no such inherent properties in the world, neither observable, nor unobservable, and it produces an entirely new conceptual system for description inside the domain of mechanics. This new conceptual system does not just deny the existence of classical states of affairs, it does not even permit us to *formulate statements* expressing such states of affairs (there is no arrangement in the Minkowski diagram that corresponds to a classical situation). It does not, and cannot, share a single statement with its predecessor. As a result the formal conditions for a suitable successor of a refuted theory (it has to repeat the successful consequences of the older theory, deny its false consequences, and make additional predictions) cannot be satisfied in the case of relativity versus classical physics and the Popperian scheme of progress breaks down. It is not even possible to connect classical statements and relativistic statements by an *empirical hypothesis*.[197] Formulating such a connection would mean formulating statements of the type "whenever there is possession by a demon there is discharge in the brain" which perpetuate rather than eliminate the older ontology. Comprehensive theories of the kind just mentioned are therefore completely disjointed, or *incommensurable*. The existence of incommensurable theories provides another difficulty for critical rationalism (and, a fortiori, for its more positivistic predecessors). We shall discuss this difficulty by discussing and refuting objections against it.

It was pointed out that progress may lead to a complete replacement of statements (and perhaps even of descriptions) in a certain domain. More especially, it may replace certain natural interpretations by others. This case has already been discussed (see above, section 6). Galileo replaces the idea of the operative character of all motion by his relativity principle in order to accommodate the new views of Copernicus. It is entirely natural to proceed in this way. A cosmological theory such as the heliocentric theory, or the theory of relativity, or the quantum theory (though the last one only with certain restrictions) makes assertions about the world as a whole. It applies to observed and to unobserved (unobservable, 'theoretical') processes. It can therefore demand to be used always, and not only on the theoretical level. Now such an adaptation of observation to theory, and this is the gist of the *first objection*, removes conflicting observation reports and saves the theory in an ad hoc manner. Moreover, there arises the *suspicion* that observations which are interpreted in terms of a new theory can no longer be used to refute that theory. It is not difficult to reply to these points.

As regards the objection we point out, in agreement with what has been said before (toward the end of section 4), that an inconsistency between theory and observation may reveal a fault of our *observational terminology* (and even of our sensations) so that it is quite natural to change this terminology, to adapt it to the new theory, and to see what happens. Such a change gives rise, and should give rise, to new auxiliary subjects (hydrodynamics, theory of solid objects, optics in the case of Galileo) which may more than compensate for the empirical content lost by the adaptation. And as regards the suspicion we must remember that the predictions of a theory depend on its postulates, the associated grammatical rules, *as well as* on initial conditions while the meaning of the "primitive" notions depends on the postulates (and the associated grammatical rules) only.[198] In those rare cases, however, where a theory *entails* assertions about possible initial conditions[199] we can refute it with the help of *self-inconsistent observation reports* such as "object A does not move on a geodesic" which, if analyzed in accordance with the Einstein-Infeld-Hoffmann account reads "singularity a which moves on a geodesic does not move on a geodesic."

The *second objection* criticizes the interpretation of science that brings about incommensurability. To deal with it we must realize that the question "are two particular comprehensive theories, such as classical celestial mechanics (CM) and the special theory of relativity (SR) incommensurable?" is not a complete question. Theories can be interpreted in different ways. They will be commensurable in some interpretations, incommensurable in others. Instrumentalism, for example, makes commensurable all those theories which are related to the same observation language and are interpreted on its basis. A realist, on the other hand, wants to give a unified account, both of observable and of unobservable matters, and he will use the most abstract terms of whatever theory he is contemplating for that purpose.[200] This is an entirely natural procedure. SR, so one would be inclined to say, does not just invite us to rethink unobserved length, mass, duration; it would seem to entail the relational character of all lengths, masses, durations, whether observed or unobserved, observable or unobservable.

Now, and here we only repeat what was said not so long ago, extending the concepts of a new theory, T, to all its consequences, observational reports included, may change the interpretation of these consequences to such an extent that they disappear from the consequence classes either of earlier theories or of the available alternatives. These earlier theories and alternatives will then all become incommensurable with T. The relation between SR and CM is a case in point. The concept of length as used in SR and the concept of length as presupposed in CM are different concepts. Both are *relational* concepts, and very complex relational concepts at that (just consider determination of length in terms of the wave length of a specified spectral line). But relativistic length, or relativistic shape, involves an element that is absent from the classical concept and is in principle excluded from it.[201] It involves the relative velocity of the object concerned in some reference system. It is of course true that the relativistic scheme very often yields numbers which are practically identical with the numbers obtained from CM, but this does not make the *concepts* more similar. Even the case $c \to \infty$ (or $v \to O$) which yields identical predictions cannot be used as an argument for showing that the concepts must coincide, at least in this special case. Different magnitudes based on different concepts may give identical values on their respective scales without ceasing to be different magnitudes. The same remark applies to the attempt to identify classical mass with relativistic rest mass.[202] This conceptual disparity, if taken seriously, infects even the most "ordinary" situations. The relativistic concept of a certain *shape*, such as the shape of a table, or of a certain temporal sequence, such as my saying "Yes," will differ from the corresponding classical concept also. It is therefore futile to expect that sufficiently long derivations may eventually return us to the older ideas.[203] The consequence classes of SR and CM are not related in any way. A comparison of content and a judgment of verisimilitude cannot be made.[204]

The situation becomes even clearer when we use the Marzke-Wheeler interpretation of SR. For it can be easily shown that the methods of measurement provided by these authors, while perfectly adequate in a relativistic universe, either collapse or give nonsensical results in a classical world (length, for example, is no longer transitive, and in some coordinate systems it may be impossible to assign a definite length to any object[205]).

We are now ready to discuss the *second* and most popular objection against incommensurability. This objection proceeds from the version of realism described above. "A realist," we said, "will want to give a unified account, both of observable and of unobservable matters, and he will use the most abstract terms of whatever theory he is contemplating for his purpose." He will use such terms in order either to give meaning to observation sentences or else to *replace* their customary interpretation. (For example, he will use the ideas of SR in order to replace the customary CM-interpretation of everyday statements about shapes, temporal sequences, and so on.) Against this, it is pointed out that theoretical terms receive their interpretation by being connected with a preexisting observation language, or with another theory that has already been connected with such an observation language, and that they are devoid of content without such connection. Thus Carnap asserts[206] that "[t]here is no independent interpretation for L_T [the language in terms of which a certain theory, or a certain world view, is formulated]. The system T [the axioms of the theory and the rules of derivation] is in itself an uninterpreted postulate system. [Its] terms . . . obtain only an indirect and incomplete interpretation by the fact that some of them are connected by the [correspondence] rules C with observation terms . . ." Now, if theoretical terms have no "independent interpretation," then surely they cannot be used for correcting the interpretation of the observation statements, which is the one and only source of their meaning. It follows that realism as described here is an impossible doctrine.

The guiding idea behind this very popular objection is that new and abstract languages cannot be introduced in a direct way, but must be first connected with an already existing, and presumably stable, observational idiom.[207]

This guiding idea is refuted at once by noting the way in which children learn to speak and in which anthropologists and linguists learn the unknown language of a newly discovered tribe.

The first example is instructive for other reasons also, for incommensurability plays an important role in the early months of human development. As has been suggested by Piaget and his school[208] the child's perception develops through various stages before it reaches its relatively stable adult

85

form. In one stage objects seem to behave very much like afterimages,[209] and they are treated as such. In this stage the child follows the object with his eyes until it disappears, and he does not make the slightest attempt to recover it, even if this would require but a minimal physical (or intellectual) effort, an effort, moreover, that is already within the child's reach. There is not even a tendency to search; and this is quite appropriate, "conceptually" speaking. For it would indeed be nonsensical to "look for" an afterimage. Its "concept" does not provide for such an operation.

The arrival of the concept and of the perceptual image of material objects changes the situation quite dramatically. There occurs a drastic reorientation of behavioral patterns, and, so one may conjecture, of thought. Afterimages, or things somewhat like them, still exist, but they are now difficult to find and must be discovered by special methods. (The earlier visual world therefore *literally disappears*.) Such special methods proceed from a new conceptual scheme (afterimages occur in *humans*, not in the outer physical world, and are tied to them) and cannot lead back to the exact phenomena of the previous stage (these phenomena should therefore be called by a different name, such as "pseudo-afterimages"). Neither afterimages nor pseudo-afterimages are given a special position in the new world. For example, they are not treated as "evidence" on which the new notion of a material object is supposed to rest. Nor can they be used to *explain* this notion: afterimages arise *together with it*, and are absent from the minds of those who do not yet recognize material objects. And pseudo-afterimages *disappear* as soon as such recognition takes place. It is to be admitted that every stage possesses a kind of observational "basis" to which one pays special attention and from which one receives a multitude of suggestions. However, this basis (i) *changes from stage to stage*; and (ii) *is part of the conceptual apparatus of a given stage*; it is *not* its one and only source of interpretation.

Considering developments such as these, one may suspect that the family of concepts centering upon "material object" and the family of concepts centering upon "pseudo-afterimage" are incommensurable in precisely the sense that is at issue here. Is it reasonable to expect that conceptual and perceptual changes of this kind occur in childhood only? Should we welcome the fact, if it is a fact, that an adult is stuck with a stable perceptual world and an accompanying stable conceptual system which he can modify in many ways, but whose general outlines have forever become immobilized? Or is it not more realistic to assume that fundamental changes, entailing incommensurability, are still possible, and that they should be encouraged lest we remain forever excluded from what might be a higher stage of knowledge and of consciousness? (Cf. on this point again section 1, especially on the role of scientific and other revolutions in bringing about such a higher stage.) Besides, the question of the mobility of the adult stage is at any rate an empirical question, which must be attacked by research and which cannot be settled by methodological fiat. The attempt to break through the boundaries of a given conceptual system and to escape the reach of "Popperian spectacles" (Lakatos) is an essential part of such research (and should be an essential part of any interesting life).[210]

Looking now at the second element of the refutation, anthropological field work, we see that what is anathema here (and for very good reasons) is still a fundamental principle for the contemporary representatives of the philosophy of the Vienna Circle. According to Carnap, Feigl, Nagel, and others, the terms of a theory receive their interpretation in an indirect fashion, by being related to a different conceptual system which is either an older theory or an observation language.[211] This older theory, this observation language, is not adopted because of its theoretical excellence. It cannot possibly be: the older theories are usually refuted. It is adopted because it is "used by a certain language community as a means of communication."[212] According to this method, the phrase "having much larger relativistic mass than . . ." is partially interpreted by first connecting it with some prerelativistic terms (classical terms, common-sense terms), which are "commonly understood" (presumably, as the result of previous teaching in connection with crude weighing methods), and it is used only after such connection has given it a well-defined meaning.

This is even worse than the once quite popular demand to clarify doubtful points by translating them into Latin. For while Latin was chosen because of its precision and clarity, and also because it was conceptually richer than the slowly evolving vulgar idioms,[213] the choice of an observation language or of an older theory as a basis for interpretation is justified by saying that they are "antecedently understood": the choice is based on sheer popularity. Besides, if prerelativistic terms which are pretty far removed from reality (especially in view of the fact that they come from an incorrect theory implying a nonexistent ontology) can be taught ostensively, for example, with the help of crude weighing methods (and one must assume that they can be so taught, or the whole scheme collapses),

87

then why should one not introduce the *relativistic* terms *directly*, and *without* assistance from the terms of some other idiom? Finally, it is but plain common sense that the teaching or the learning of new and unknown languages must not be contaminated by external material. Linguists remind us that a perfect translation is never possible, even if one is prepared to use complex contextual definitions. This is one of the reasons for the importance of *field work* where new languages are learned from scratch, and for the rejection, as inadequate, of any account that relies on 'complete' or 'partial' translation. *Yet just what is anathema in linguistics is taken for granted by logical empiricism*, a mythical "observation language" replacing the English of the translators. Let us commence field work in this domain also, and let us study the language of new theories not in the definition factories of the double language model, but in the company of those metaphysicians, theoreticians, playwrights, courtesans who have constructed new world views! This finishes my discussion of the guiding principle behind the second objection against realism and the possibility of incommensurable theories.

Another point that is often made is that there exist *crucial experiments* which refute one of two allegedly incommensurable theories and confirm the other (example: the Michelson-Morley experiment, the variation of the mass of elementary particles, the transverse Doppler effect, are said to refute CM and confirm SR). The answer to this problem is not difficult either: adopting the point of view of relativity, we find that the experiments, *which of course will now be described in relativistic terms*, using the relativistic notions of length, duration, speed, and so on,[214] are relevant to the theory. And we also find that they support the theory. Adopting CM (with, or without an ether), we again find that the experiments, which are now described in the very different terms of classical physics, i.e., roughly in the manner in which Lorentz described them, are relevant. But we also find that they *undermine* CM, i.e., the conjunction of classical electrodynamics and of CM. Why should it be necessary to possess terminology that allows one to say that it is the *same* experiment which confirms one theory and refutes the other? But did we not ourselves use such terminology? Well, for one thing it should be easy though somewhat laborious to express what was just said *without* asserting identity. Secondly, the identification is of course not contrary to our thesis, for we are now not *using* the terms of either relativity or classical physics, as is done in a test, but are *referring* to them and their relation to the physical world. The language in which *this* discourse is carried out can be classical, or relativistic, or ordinary. It is no good insisting that scientists act as if the situation were much less complicated. If they act that way, then they are either instrumentalists (see above) or mistaken (many scientists are nowadays interested in *formulas*, while the subject here is *interpretations*). It is also possible that being well acquainted with both CM and SR, they change back and forth between these theories with such speed that they seem to remain within a single domain of discourse.

It is also said that by admitting incommensurability into science we can no longer decide whether a new view explains what it is supposed to explain, or whether it does not wander off into different fields.[215] For example, we would not know whether a newly invented physical theory is still dealing with problems of space and time or whether its author has not by mistake made a biological assertion. But there is no need to possess such knowledge. For once the fact of incommensurability has been admitted, the question which underlies the objection does not arise. Conceptual progress often makes it impossible to ask certain questions and to explain certain things; thus we can no longer ask for the absolute velocity of an object, at least as long as we take relativity seriously. Is this a serious loss for science? Not at all! Progress was made by the very same "wandering off into different fields" whose undecidability now so greatly exercises the critic: Aristotle saw the world as a super *organism*, as a *biological* entity, while one essential element of the new science of Descartes, Galileo, and their followers in medicine and in biology is its exclusively *mechanistic* outlook. Are such developments to be forbidden? And if they are not, what, then, is left of the complaint?

A closely connected objection starts from the notion of *explanation* or *reduction* and emphasizes that this notion presupposes continuity of concepts; other notions could be used for starting exactly the same kind of argument. (Relativity is supposed to explain the valid parts of classical physics; hence it cannot be incommensurable with it!) The reply is again obvious. As a matter of fact it is a triviality for anyone who has only the slightest acquaintance with the Hegelian philosophy: why should the relativist be concerned with the fate of classical mechanics except as part of a historical exercise? There is only *one* task we can legitimately demand of a theory, and it is that it should give us a correct account of the world, i.e., of the totality of facts *as seen through its own concepts*. What have the principles of explanation got to do with this demand? Is it not reasonable to assume that a point of view such as the point of view of classical mechanics that has been found wanting in various respects, that gets in difficulty with its own facts (see above, on crucial experiments), and must therefore be regarded as self-inconsistent (another application of Hegelian principles!), cannot have entirely adequate concepts? Is it not equally reasonable to try replacing its concepts with those of a more promising cosmology? Besides, why should the notion of explanation be burdened by the demand for conceptual continuity? This notion has been found to be too narrow before (demand of derivability), and it had to be widened so as to include partial and statistical connections. Nothing prevents us from widening it still further and admitting, say, "explanations by equivocation."

Incommensurable theories, then, can be *refuted* by reference to their own respective kinds of experience, i.e., by discovering the *internal* contradictions from which they are suffering (in the absence of commensurable alternatives these refutations are quite weak, however[216]). Their content cannot be compared, nor is it possible to make a judgment of *verisimilitude* except within the confines of a particular theory. None of the methods which Popper (or Carnap, or Hempel, or Nagel) want to use for rationalizing science can be applied, and the one that can be applied, refutation, is greatly reduced in strength. What remains are esthetic judgments, judgments of taste, and our own subjective wishes.[217] Does this mean that we are ending up in subjectivism? Does this mean that science has become arbitrary, that it has become an element of the general relativism which so much exercises the conscience of some philosophers? Well, let us see.

14. The Choice between Comprehensive Ideologies

To start with, it seems to me that an enterprise whose human character can be seen by all is preferable to one that looks "objective" and impervious to human actions and wishes.[218] The sciences, after all, are our own creation, including all the severe standards they seem to impose on us. It is good to be constantly reminded of this fact. It is good to be constantly reminded of the fact that science as we know it today is not inescapable, and that we can construct a world in which it plays no role whatever. (Such a world, I venture to suggest, would be more pleasant to behold than the world we live in today, both materially and intellectually.) What better reminder is there than the realization that the choice between theories which are sufficiently general to yield a comprehensive world view and which are empirically disconnected may become a matter of taste? *That the choice of a basic cosmology may become a matter of taste?*

Secondly, matters of taste are not completely beyond the reach of argument. Poems, for example, can be compared in grammar, sound structure, imagery, rhythm, and can be evaluated on such a basis (cf. Ezra Pound on progress in poetry[219]). Even the most elusive mood can be analyzed *and should be analyzed* if the purpose is to present it in a manner that either can be enjoyed or increases the emotional, cognitive, perceptual, etc., inventory of the reader. Every poet who is worth his salt compares, improves, argues until he finds the correct formulation of what he wants to say.[220] Would it not be marvelous if this free and entertaining[221] process played a role in the sciences also?

Finally, there are more pedestrian ways of explaining the same matter which may be somewhat less repulsive to the tender ears of a professional philosopher of science. One may consider the *length* of derivations leading from the principles of a theory to its observation language, and one may also draw attention to the number of *approximations* made in the course of the derivation. All derivations must be standardized for this purpose so that unambiguous judgments of length can be made. (This standardization concerns the form of the derivation, it does not concern the content.) Smaller length and smaller number of approximations would seem to be preferable. It is not easy to see how this requirement can be made compatible with the demand for simplicity and generality which, so it seems, would tend to increase both parameters. However that may be, there are many ways open to us once the fact of incommensurability is understood, and taken seriously.

15. Conclusion

The idea that science can and should be run according to some fixed rules, and that its rationality consists in agreement with such rules, is both unrealistic and vicious. It is *unrealistic*, since it takes too simple a view of the talents of men and of the circumstances which encourage, or cause, their development. And it is *vicious*, since the attempt to enforce the rules will undoubtedly erect barriers to what men might have been, and will reduce our humanity by increasing our professional qualifications. We can free ourselves from the idea and from the power it may possess over us (i) by a detailed study of the work of revolutionaries such as Gali-

91

leo, Luther, Marx, or Lenin; (ii) by some acquaintance with the Hegelian philosophy and with the alternative provided by Kierkegaard; (iii) by remembering that the existing separation between the sciences and the arts is artificial, that it is a side effect of an idea of professionalism one should eliminate, that a poem or a play can be intelligent as well as informative (Aristophanes, Hochhuth, Brecht), and a scientific theory pleasant to behold (Galileo, Dirac), and that we can change science and make it agree with our wishes. We can turn science from a stern and demanding mistress into an attractive and yielding courtesan who tries to anticipate every wish of her lover. Of course, it is up to us to choose either a dragon or a pussycat as our companion. So far mankind seems to have preferred the latter alternative: "The more solid, well defined, and splendid the edifice erected by the understanding, the more restless the urge of life . . . to escape from it into freedom." We must take care that we do not lose our ability to make such a choice.

Appendix. Science without Experience

1. One of the most important properties of modern science, at least according to some of its admirers, is its *universality*: any question can be attacked in a scientific way leading either to an unambiguous answer or else to an explanation of why an answer cannot be had. Let us therefore ask whether the *empirical hypothesis* is correct, i.e., whether experience can be regarded as a true source and foundation (testing ground) of knowledge.

2. Asking this question and expecting a scientific answer assumes that a science *without* experience is a *possibility*, that is, it assumes that the idea is neither absurd nor self-contradictory. It must be possible to imagine a natural science without sensory elements, and it should perhaps also be possible to indicate how such a science is going to work.

3. Now experience is said to enter science at three points: testing; assimilation of the results of test; understanding of theories.

A test may involve complex machinery and highly abstract auxiliary assumptions. But its final outcome has to be recognized by a human observer who *looks* at some piece of apparatus and *notices* some observable change. Communicating the results of a test also involves the senses: we *hear* what somebody says to us; we *read* what somebody has written down. Finally, the abstract principles of a theory are just strings of signs, without relation to the external world unless we know how to connect them with experiment and that means, according to the first item on the list, with experience, involving simple and readily identifiable sensations.

4. It is easily seen that experience is needed at none of the three points just mentioned.

To start with, experience does not need to enter the process of test: we can put a theory into a computer, provide the computer with suitable instruments directed by him (her, it) so that relevant measurements are made which return to the computer leading there to an evaluation of the theory. The computer can give a simple yes-no response from which a scientist may learn whether or not a theory has been confirmed without having in any way participated in the test (i.e., without having been subjected to some relevant experience).

5. Learning what a computer says means being informed about some simple occurrence in the macroscopic world. Usually such information travels through the senses giving rise to distinct sensations. But this is not always the case. Subliminal perception leads to reactions directly, and without sensory data. Latent learning leads to memory traces directly, and without sensory data. Posthypnotic suggestion leads to (belated) reactions directly, and without sensory data. In addition there is the whole unexplored field of telepathic phenomena. I am not asserting that the natural sciences as we know them today could be built on these phenomena alone and could be freed from sensations entirely. Considering the peripheral nature of the phenomena and considering also how little attention is given to them in our education (we are not trained to effectively use our ability for latent learning) this would be both unwise and impractical. But the point is made that sensations are not necessary for the business of science and that they occur for practical reasons only.

6. Considering now the objection that we understand our theories, that we can apply them only because we have been told how they are connected with experience, one must point out that experience arises together with theoretical assumptions, not before them, and that an experience without theories is just as uncomprehended as is (allegedly) a theory without experience: eliminate part of the theoretical knowledge of a sensing subject and you have a person who is completely disoriented and incapable of carrying out the simplest action. Eliminate further knowledge and his sensory world (his "observation language") will start disintegrating, even colors and other simple sensations will disappear until he is in a stage even more primitive than a small child. A small child, on the other

hand, does not possess a stable perceptual world which he uses for making sense of the theories put before him. Quite the contrary. He passes through various perceptual stages which are only loosely connected with each other (earlier stages *disappear* when new stages take over) and which embody all the theoretical knowledge achieved at the time. Moreover, the whole process (including the very complex process of learning up to three or four languages) gets started only because the child reacts correctly toward signals, *interprets them correctly*, because he possesses means of interpretation even before he has experienced his first clear sensation. Again we can imagine that this interpretative apparatus acts without being acompanied by sensations (as do all reflexes and all well-learned movements such as typing). The theoretical knowledge it contains certainly can be *applied* correctly, though it is perhaps not *understood*. But what do sensations contribute to our understanding? Taken by themselves, i.e., taken as they would appear to a completely disoriented person, they are of no use, either for understanding or for action. Nor is it sufficient to just *link them* to the existing theories. This would mean extending the theories by further elements so that we obtain longer *expressions*, i.e., longer series of events, not the *understanding* of the shorter expressions which we wanted. No—the sensations must be incorporated into our behavior in a manner that allows us to pass smoothly from them into action. But this returns us to the earlier situation where the theory was applied but allegedly not yet understood. Understanding in the sense demanded here thus turns out to be ineffective and superfluous. Result: sensations can be eliminated from the process of understanding also (though they may of course continue to *accompany* it, just as a headache accompanies deep thought).

NOTES

1. V. I. Lenin, '*Left Wing*' *Communism, an Infantile Disorder* (Peking: Foreign Language Press, 1965), p. 100 (the book was first published in 1919 in order to criticize certain puritanical elements in German communism). Lenin speaks of parties and the revolutionary vanguard rather than of scientists and methodologists. The lesson is, however, the same.

2. H. Butterfield, *The Whig Interpretation of History* (New York: Norton, 1965), p. 66.

3. *Ibid.*, p. 21.

4. *Ibid.*, p. 25.

5. Lenin, '*Left Wing*' *Communism*, p. 100. It is interesting to see how a few substitutions can turn a political lesson into a lesson for methodology which, after all, is part of the process by means of which we move from one historical stage to another. We

also see how an individual who is not intimidated by traditional boundaries can give useful advice to everyone, philosophers of science included. Cf. notes 27 and 33, 35, 38.

6. P. A. Schilpp, ed., *Albert Einstein, Philosopher-Scientist* (Evanston, Ill.: Tudor, 1948), p. 683.

7. D. Hume, *A Treatise of Human Nature* (Oxford: Oxford University Press, 1888), p. 180.

8. Popper and his followers distinguish between the socio-psychological process of science where errors abound and rules are constantly broken and a "third world" where knowledge is changed in a rational manner, and without interference from "mob psychology," as Lakatos expresses himself. For details and a criticism of this poor man's Platonism see the text to note 194 below.

9. E. H. Carr, *Michael Bakunin* (London: Macmillan, 1937), pp. 8–9.

10. Thus external pressure is replaced by bad conscience, and freedom remains restricted as before. Marx describes a similar development in the case of Luther in the following words: ". . . Luther eliminates *external* religiousness and turns religiousness into the *inner* essence of man . . . he negates the raving parish-priest outside the layman, for he puts him right into his heart." *Nationaloekonomie und Philosophie*; quoted from *Marx, die Frühschriften*, ed. S. Landshut (Stuttgart: Kroner, 1953), p. 228.

Whatever remains of irrationality in history is suppressed by the quasi-historical and indeed quite mythological manner in which scientists describe the genesis of their discoveries, or of the discoveries of others. ". . . history is wholly subordinated to the needs of the present, and indeed only survives to such an extent, and in such form, as serves present needs." Among the present needs, however, the propagation of what is thought to be good science is the most important one. Hence, history is replaced by myths "which are to be consonant with what [one thinks] to be good physics, and they are to be internally consistent." Paul Forman, "The Discovery of the Diffraction of X-Rays by Crystals: A Critique of the Myths," *Archive for the History of the Exact Sciences*, 6 (1969), 68–69. Forman's paper presents an interesting example to illustrate this statement. Another example is the myths which have been invented to explain the origin of the special theory of relativity. For an excellent account with plentiful sources see G. Holton, "Einstein, Michelson, and the 'Crucial' Experiment," *Isis*, 60 (1969), 133–197.

11. "This unique prevalence of the *inner* logic of a subject over and above the *outer* influences is not . . . to be found at the beginning of modern science." H. Blumenberg, *Die Kopernikanische Wende* (Frankfurt: Suhrkamp, 1965), p. 8.

12. "Nothing is more dangerous to reason than the flights of the imagination . . ." Hume, *A Treatise of Human Nature*, p. 267.

13. An expert is a man or a woman who has decided to achieve excellence in a narrow field at the expense of a balanced development. He has decided to subject himself to standards which restrict him in many ways, his style of writing and the patterns of his speech included, and he is prepared to conduct most of his waking life in accordance with these standards (this being the case, it is likely that his dreams will be governed by these standards, too). He is not averse to occasionally venturing into different fields, to listen to fashionable music, to adopt fashionable ways of dressing (though the business suit still seems to be his favorite uniform, in this country and abroad), or to seduce his students. However, these activities are aberrations of his *private* life; they have no relation whatever to what he is doing as an expert. A love for Mozart, or for *Hair*, will not make his physics more melodious, or give it a better rhythm. Nor will an affair make his chemistry more colorful.

This separation of domains has very unfortunate consequences. Not only are special subjects voided of ingredients which make a human life beautiful and worth living, but these ingredients are impoverished, too, emotions become crude and thoughtless, just as thought becomes cold and inhumane. Indeed, the private parts of one's existence suffer much more than does one's official capacity. Every aspect of professionalism has its watchdogs; the slightest change, or threat of a change, is examined, broadcast, warnings are issued, and the whole depressing machinery moves at once in order to restore the status quo. Who takes care of the quality of our emotions? Who watches those parts of our language which are supposed to bring people together more closely, which have the function of giving comfort, understanding, and perhaps a little personal criticism and encouragement? There are no such agencies. As a result professionalism takes over even here.

To mention some examples:

In 1610 Galileo reported for the first time his invention of the telescope and the observations he made with it. This was a scientific event of the first magnitude, far more important than anything we have achieved in our megalomaniac twentieth century. Not only was here a new and very mysterious instrument introduced to the learned world (it was introduced to the *learned* world, for the essay was written in Latin), but this instrument was at once put to a very unusual use: it was directed toward the sky; and the results, the astonishing results, quite definitely seemed to support the new theory which Copernicus had suggested over sixty years earlier, and which was still very far from being generally accepted. How does Galileo introduce his subject? Let us hear.

"About 10 months ago a report reached my ears that a certain Dutchman had constructed a spyglass by means of which visible objects, though very distant from the eye of the observer, were distinctly seen, as if nearby. Of this truly remarkable effect several experiences were related, to which some persons gave credence while others denied them. A few days later the report was confirmed to me in a letter from a noble Frenchman in Paris, Jacques Badovere, which caused me to apply myself wholeheartedly to enquire into the means by which I might arrive at the invention of a similar instrument . . ." Quoted from Stillman Drake, ed., *Discoveries and Opinions of Galileo* (New York: Doubleday Anchor Books, 1957), pp. 28–29.

We start with a personal story, a very charming story, which slowly leads us to the discoveries, and these are reported in the same clear, concrete, and colorful way: "There is another thing," writes Galileo, describing the face of the moon, "which I must not omit, for I beheld it not without a certain wonder; this is that almost in the center of the moon there is a cavity larger than all the rest, and perfectly round in shape. I have observed it near both the first and last quarters, and have tried to represent it as correctly as possible in the second of the above figures . . ." Quoted from Drake, ed., *Discoveries and Opinions of Galileo*, p. 36. Galileo's drawing attracts the attention of Kepler who was one of the first to read Galileo's essay. He comments: "I cannot help wondering about the meaning of that large circular cavity in what I usually call the left corner of the mouth. Is it a work of nature, or of a trained hand? Suppose that there are living beings on the moon (following the footsteps of Pythagoras and Plutarch I enjoyed toying with this idea, long ago . . .). It surely stands to reason that the inhabitants express the character of their dwelling place, which has much bigger mountains and valleys than our earth has. Consequently, being endowed with very massive bodies, they also construct gigantic projects . . ." Quoted from *Kepler's Conversations with Galileo's Sidereal Messenger*, trans. Edward Rosen (New York: Johnson Reprint Corporation, 1965), pp. 27–28.

"I have observed"; "I have seen"; "I have been surprised"; "I cannot help wondering"; "I was delighted"—this is how one speaks to a friend or, at any rate, to a live human being.

The awful Newton who more than anyone else is responsible for the plague of professionalism from which we suffer today starts his first paper on colors in a very similar style: ". . . in the beginning of the year 1666 . . . I procured me a triangular glass prisme, to try therewith the celebrated phenomena of colours. And in order thereto having darkened my chamber, and made a small hole in my window shuts, to let in a convenient quantity of the sun's light, I placed my prisme at its entrance, that it might be thereby refracted to the opposite wall. It was at first a very pleasing divertisement, to view the vivid and intense colours produced thereby; but after a while applying myself to consider them more circumspectly, I became surprised to see them in an oblonge form . . ." Quoted from *The Correspondence of Isaac Newton*, vol. I (Cambridge: Cambridge University Press, 1959), p. 92.

Remember that all these reports are about cold, objective, "inhuman" *inanimate* nature, they are about stars, prisms, lenses, the moon, and yet these are described in a most lively and fascinating manner, communicating to the reader an interest and an excitement which the discoverer felt when first venturing into strange new worlds.

Now compare with this the introduction to a recent book, a best seller even, *Human Sexual Response* by W. H. Masters and V. E. Johnson (Boston: Little, Brown, 1966). I have chosen the book for two reasons. First, because it is of general interest. It removes prejudices which influence not only the members of some profession, but the everyday behavior of a good many apparently "normal" people. Second, because it deals with a subject that is new and without special terminology. Also, it is about man rather than about stones and prisms. So one would expect a beginning even more lively and interesting than that of Galileo, or Kepler, or Newton. What do we read instead? Behold, oh patient reader: "In view of the pervicacious gonadal urge in human beings, it is not a little curious that science develops its sole timidity about the pivotal point of the physiology of sex. Perhaps this avoidance . . ." and so on. This is human speech no more. This is the language of the expert.

Note that the subject has completely left the picture. Not "*I* was very surprised to find" or, since there are two authors, "*We* were very surprised to find," but "*It is* surprising to find"—only not expressed in these simple terms. Note also to what extent irrelevant technical terms intrude and fill the sentences with antediluvian barks, grunts, squeaks, belches. A wall is erected between the writers and their readers not because of some lack of knowledge, not because the writers do not know their readers, but in order to make utterances conform to some curious professional ideal of objectivity. And this ugly, inarticulate, and inhuman idiom turns up everywhere, and takes over the function of the most simple and the most straightforward description.

Thus on page 65 of the book we hear that the female, being capable of multiple orgasm, must often masturbate after her partner has withdrawn in order to complete the physiological process that is characteristic for her. And, so the authors want to say, she will stop only when she gets tired. This is what they *want* to say. What they actually say is: "usually physical exhaustion alone terminates such an active masturbatory session." You don't just masturbate; you have an "active masturbatory session." On the next page the male is advised to ask the female what she wants or does not want rather than try to guess it on his own. "He should ask her"—this is what our authors want to convey. What is the sentence that actually lies there in the book? Listen: "The male will be infinitely more effective if he encourages vocalization on her part." "Encourages vocalization" instead of "asks her"—well, one might want to say, the authors want to be *precise*, and they want to address their *fellow professionals* rather than the general public and, naturally, they have to use a special lingo in order to make themselves understood. Now as regards the first point, precision, remember that they also say that the male will be "*infinitely* more effective" which, considering the circumstances, is not a very precise statement of the facts. And as regards the second point we must say that we are not dealing with the structure of organs, or with special physiological processes which might have a special name in medicine, but with an ordinary affair such as *asking*. Besides, Galileo and Newton could do *without* a special lingo although the physics of their time was highly specialized and contained many technical terms. They could do without a special lingo because they wanted to start afresh and because they were sufficiently free and inventive not to be dominated by words, but to be able to dominate them. Masters and Johnson find themselves in the same position, but they cannot speak straight any more, their linguistic talents and sensibilities have been distorted to such an extent that one asks oneself whether they will ever be able to speak normal English again.

The answer to *this* question is contained in a little pamphlet which came into my hands and which contains the report of an ad hoc committee formed for the purpose

97

of examining rumors of police brutality during some rather restless weeks in Berkeley (winter 1968–69). The members of the committee were all people of good will. They were interested not only in the *academic* quality of life on campus; they were even more interested in bringing about an atmosphere of understanding and of compassion. Most of them came from sociology and from related fields, that is, they came from fields which deal not with lenses, stones, stars, as did Galileo in his beautiful little book, but with humans. There was a mathematician among them who had devoted considerable time to setting up and defending student-run courses and who finally gave up in disgust —he could not change the "established academic procedures." How do these nice and decent people write? How do they address those to whose cause they have devoted their spare time and whose lives they want to improve? Are they able to overcome the boundaries of professionalism at least on this occasion? Are they able to *speak*? They are not.

The authors want to say that policemen often make arrests in circumstances when people are bound to get angry. They say: "When *arousal* of those present is the inevitable consequence." "Arousal"; "inevitable consequence"—this is the lingo of the laboratory, this is the language of people who habitually mistreat rats, mice, dogs, rabbits, and carefully notice the effects of their mistreatment, but the language they use is now applied to humans, too, to humans, moreover, with whom one sympathizes, or says one sympathizes, and whose aims one supports. They want to say that policemen and strikers hardly talk to each other. They say: "*Communication* between strikers and policemen is nonexistent." Not the strikers, not the police, not people are at the center of attention, but an abstract process, "communication," about which one has learned a thing or two and with which one feels more at ease than with living human beings. They want to say that more than 80 people took part in the venture, and that the report contains the common elements of what about 30 of them have written. They say: "This report tries to reflect a consensus from the 30 reports submitted by the 80 plus faculty observers who participated." Need I continue? Or is it not already clear that the effects, the miserable effects, of professionalism are much deeper and much more vicious than one would expect at first sight? That some professionals have even lost the ability to *speak* in a civilized manner, that they have returned to a state of mind more primitive than that of an eighteen-year-old who is still able to adapt his language to the situation in which he finds himself, talking the lingo of physics in his physics class and quite a different language with his friends in the street (or in bed)?

Many colleagues who agree with my general criticism of science find this emphasis on language farfetched and exaggerated. Language, they say, is an *instrument* of thought that does not influence it to the extent I surmise. This is true as long as a person has different languages at his disposal, and as long as he is still able to switch from one to another as the situation demands. But this is not the case here. Here a single and rather impoverished idiom takes over all functions and is used under all circumstances. Does one want to insist that the thought that hides behind *this* ugly exterior has remained nimble and humane? Or must be not rather agree with V. Klemperer and others who have analyzed the deterioration of language in fascistic societies that "words are like small doses of arsenic: they are swallowed unawares, they do not seem to have any noticeable effect, and yet the poisonous influence will be there after some time. If someone frequently enough replaces words such as 'heroic' and 'virtuous' by 'fanatical' he will believe in the end that without fanaticism there is no heroism and no virtue." *Die Unbewaeltigte Sprache* (Munich: Deutscher Taschenbuch Verlag, 1969), p. 23. Similarly the frequent use of abstract terms from abstract disciplines ("communication"; "arousal") in subjects dealing with humans is bound to make people believe that a human being can be dissolved into a few bland processes and that things such as emotion and understanding are just disturbing elements, or, rather, misconceptions belonging to a more primitive stage of knowledge.

In their search for a bland and standardized language with uniform spelling, punctuation, standardized references, and so on experts receive increasing support from publishers. Idiosyncrasies of style and expression that have been overlooked by a referee will certainly be noticed by printers or editors, and much energy is wasted in quarrels over a phrase, or the position of a comma. It seems that language has ceased to be the property of writers and readers and has been purchased by publishing houses, so that authors are no longer allowed to express themselves as they see fit and to make their contribution to the growth of English.

14. John Stuart Mill, *On Liberty*, quoted from *The Philosophy of John Stuart Mill*, ed. Marshall Cohen (New York: Modern Library, 1961), p. 258.

15. *Ibid.*, p. 265.

16. Even in undetermined and ambiguous situations uniformity of action is soon achieved, and adhered to tenaciously. Cf. M. Sherif, *The Psychology of Social Norms* (New York: Harper Torchbooks, 1964).

17. *A Treatise of Human Nature*, p. xxii. The word "reason" has been replaced by "understanding" in order to establish coherence with the terminology of the German idealists.

18. The first part of the quotation, up to "appearing as," is taken from *Differenz des Fichte'schen und Schelling'schen Systems der Philosophie*, ed. G. Lasson (Hamburg: Felix Meiner, 1962), p. 13. The second part is from the *Wissenschaft der Logik*, vol. I (Hamburg: Felix Meiner, 1965), p. 6.

19. Letter to Gert Micha Simon of October 11, 1949. Quoted from *Gottfried Benn, Lyrik und Prosa, Briefe und Dokumente* (Wiesbaden: Limes Verlag, 1962), p. 235.

20. For details and further literature see "Problems of Empiricism, Part II," in *The Nature and Function of Scientific Theory*, ed. R. G. Colodny (Pittsburgh: University of Pittsburgh Press, 1970).

21. One of the few physicists to see and to understand this feature of the development of scientific knowledge was Niels Bohr: ". . . he would never try to outline any finished picture, but would patiently go through all the phases of the development of a problem, starting from some apparent paradox, and gradually leading to its elucidation. In fact, he never regarded achieved results in any other light than as starting points for further exploration. In speculating about the prospects of some line of investigation, he would dismiss the usual considerations of simplicity, elegance or even consistency with the remark that such qualities can only be properly judged *after* [my italics] the event . . ." L. Rosenfeld in S. Rozental, ed., *Niels Bohr, His Life and Work as Seen by His Friends and Colleagues* (New York: Interscience, 1967), p. 117.

One must of course realize that science does not achieve final results and that it is therefore always "before" the event, never "after" it. Simplicity, elegance, consistency are therefore never a *conditio sine qua non* of scientific knowledge.

Considerations like these are usually criticized by the childish remark that a contradiction entails every statement and that self-inconsistent views are therefore useless for science. I call the remark childish because it assumes that a self-consistent science is a realistic possibility, that the rule which leads to the result just mentioned is the only possible rule, and that the scientist is obliged to play the thinking games of the logician. There is of course no such obligation. Quite the contrary, the scientist can criticize the logician for providing him with inadequate instruments that make nonsense of the complex, delicate, and often self-inconsistent theories he uses.

For further information concerning Bohr's philosophy see my essay "On a Recent Critique of Complementarity," *Philosophy of Science*, 35 (1968), 309–331, and 36 (1969), 82–105. The essay also cites relevant literature.

22. Children "learn to imitate others . . . and so learn to look upon standards of behavior as if they consisted of fixed, 'given' rules . . . and such things as sympathy and imagination may play an important role in this development . . ." K. R. Popper, *The Open Society and Its Enemies* (New York: Harper Torchbooks, 1967), II, 390.

99

Paul K. Feyerabend

One should also compare the remainder of appendix i/15 which gives a clear account of the irrational elements in our knowledge.

23. In one of his numerous lucubrations in praise of Ordinary English ("Moore and Ordinary Language," in *The Philosophy of G. E. Moore*, ed. P. A. Schilpp, New York: Tudor, 1952, pp. 354ff) Malcolm makes the following comment: ". . . if a child who was learning the language were to say, in a situation where we were sitting in a room with chairs about, that it was 'highly probable' that there were chairs there, we should smile, *and correct his language*" (italics in the original). One can only hope that the children whom Malcolm addresses in this manner are not as gullible as are most of his students and that they will retain their intelligence, their imagination, and especially their sense of humor in the face of this and other "methods" of education.

24. Cf. below, text to note 208.

25. Commenting on his early education by his father, and especially on the explanations he received on matters of logic, J. S. Mill made the following observations: "The explanations did not make the matter at all clear to me at the time; but they were not therefore useless; they remained as a nucleus for my observations and reflections to crystallize upon; the import of his general remarks being interpreted to me, by the particular instances which came under my notice afterwards." *Autobiography* (London: Oxford University Press, 1963), p. 16. In "Problems of Empiricism, Part II" I have argued that the development of science exhibits phase differences of precisely this kind. A strange and incomprehensible new principle often serves as a "nucleus for observations and reflections to crystallize upon" until we obtain a theory that is understood even by the most uneducated empiricist. For a general discussion of the problem touched upon in this remark, see Hegel, *Wissenschaft der Logik*, I, 51–64. See also St. Augustine, *De doctrina Christiana*, 11/9: "The first . . . case is to know these books [i.e., the books of the old and of the new testament]. Altogether, we may not yet understand them, but by reading we can either memorise them, or become somehow acquainted with them." The way in which apparently aimless talk may lead to new ideas and to a new state of consciousness has been described, briefly, but exquisitely, by Heinrich von Kleist, "Ueber die allmaehliche Verfertigung der Gedanken beim Reden," in Hans Meyer, ed., *Meisterwerke Deutscher Literaturkritik* (Stuttgart: Goverts, Neue Bibliotek der Weltliteratur, 1962), 741–747.

26. "Recourse to direct action changed the whole tenor of the struggle, for the workers' self-confidence is enormously increased (and their knowledge transformed) once they act without delegating any of their power to political parties or trade unions. 'The factory is ours so do we need to start working for the bosses again?' This idea arose quite spontaneously, not by command, or under the aegis of the so-called vanguard of the proletariat [with its special methods, rules, prescriptions, and its special idea of rationality], but simply as a *natural response to a concrete situation*." D. Cohn-Bendit, *Obsolete Communism: The Left Wing Alternative*, trans. A. Pomerans (London: André Deutsch, 1968), p. 67. Cohn-Bendit's emphasis on "spontaneity . . . The chief enemy of all bureaucrats" (p. 154) agrees with the tenor of the present paper which wants to eliminate excessive bureaucracy not only from government, but also from the administration of knowledge (where it appears as an appeal to rationality). For the formation of natural responses to ambiguous situations, see also Sherif, *The Psychology of Social Norms*.

27. (A) K. R. Popper, whose views I have in mind when criticizing the omnipresence of argument, has admitted that "rationalism is necessarily far from comprehensive or self-contained." *The Open Society and Its Enemies*, II, 231. But the question I am asking is not whether there are limits to our reason. The question is where these limits are situated. Are they outside the sciences, so that science itself remains entirely rational (though the decision to become scientific may be an irrational decision); or are irrational changes an essential part of even the most rational enterprise that has been invented by man? Does the historical phenomenon 'science' contain ingredients which defy a rational analysis? Can the abstract goal of coming closer to the

truth be reached in an entirely rational fashion, or is it perhaps inaccessible to those who decide to rely on argument only? These are the questions to which I want to address myself in the present essay.

(B) Surprising insights into the limitations of methodological rules as well as into their dependence on a certain developmental stage of mankind are found in Lenin's and Mao's political writings and, of course, in Hegel's philosophy. It needs only a little imagination to turn the positive advice contained in these writings into advice for the scientist, or the philosopher of science.

Thus, we read on pp. 40ff of Lenin's 'Left Wing' Communism (a book that is very useful as a theoretical basis for the criticism of contemporary left radicalism, campus radicals, leftist puritans, and other leftovers from the undialectical political stone age) : "we can (and must) begin to build socialism, not with imaginary human material [as does the doctrine of critical rationalism], nor with human material specially prepared by us [as do all Stalinists, in politics as well as in the philosophy of science], but with the [quite specific] human material bequeathed to us by capitalism. True, that is very 'difficult,' but no other approach to this task is serious enough to warrant discussion." Replace "socialism" by "rationality of the future," "capitalism" by "critical rationalism," and our case is stated with perfect clarity.

It seems to me that such attention to the wider political context will free the philosopher of science from the Nagel-Carnap-Popper-Kuhn carrousel. The only philosopher who secretly imbibes the forbidden brew of Leninism is Lakatos—and the results are evident in his magnificent work. All that is required now is that he confess his vices openly so that others may learn to delight and enlighten us in a similar way.

(C) An excellent example of the need for moving forces in addition to argument is provided by the history of witchcraft in the thirteenth to seventeenth centuries. "No mere skepticism, no mere 'rationalism,' could have driven out the old cosmology," writes H. Trevor-Roper in his analysis (The European Witch Craze, New York: Harper Torchbooks, 1969, p. 181). "A rival faith had been needed . . ." Despite all the arguments against it "the intellectual basis of the witch craze remained firm all through the seventeenth century. No critic had improved on the arguments of Weyer; none had attacked the substance of the myth . . ." (pp. 160-161). Such attacks did not occur, and they could not have been effective. They could not have been effective, because the science of the schools was "empirically confirmed" (p. 191), because it "created its own evidence" (p. 166), because it was firmly rooted in common belief (p. 124), leading to strong experiences, to "illusions" which were "centralised . . . around" the main characters of the dominating myth such as for example "the devil" (p. 125), and because strong emotional forces were expressed by the myth as well. The existence of the empirical evidence made it difficult to argue against witchcraft in a "scientific" manner. The existence of the emotional force would have neutralized even an effective scientific counterargument. What was needed was not simply a formal criticism, or an empirical criticism; what was needed was a change of consciousness, a "rival faith" as Trevor-Roper expresses himself, and this rival faith had to be introduced against tremendous odds, and even in the face of reason. Now it is of course correct that a general and forceful education in the rules of rationalism, dogmatic, critical, or otherwise, will make it more easy for arguments to win the day—well-trained dogs heel more promptly than do their anarchistic counterparts—but the discussion of the value of argument will now be considerably more difficult, and perhaps entirely impossible. Besides, man was not meant to be just a rational animal. At any rate he was not meant to be castrated and cut apart. But whatever our position on that head, we shall have to admit that rational argument works with rational people only and that an appeal to rational argument is therefore discriminatory. Rational people are specially prepared, they have been conditioned in special ways, their freedom of action and of thought has been considerably restricted. If we oppose discrimination and mental restriction, then the omnipresence of reason can no longer be guaranteed and our assertion in the text holds. Cf. also Burr's letter to A. D. White, quoted from George Lincoln Burr, His Life and

Paul K. Feyerabend

Selected Writings (Ithaca, N.Y.: Cornell University Press, 1943), p. 56, my italics: "To my thought—and here I differ widely from both Buckle and Lecky . . . —it was not science, not reason that put an end to inhumanity in so many fields: *the pedants were as cruel as the bigots.* Reason came in only to sanction here reforms which had been wrought in spite of her. *The real antagonist of theology and of rationalism alike* [and it does not make any difference whether we speak here of dogmatic rationalists, or of skeptics, or of critical rationalists as is shown by the example of Glanville] *was the unreasoning impulse of human kindliness.*"

(D) The example of witchcraft shows that the wider context we need in order to see science, or the "search for truth," in perspective need not be politics. It can be religion, metaphysics, theology, or what have you. In "Classical Empiricism" (in R. E. Butts, ed., *The Methodological Heritage of Newton*, Toronto: University of Toronto Press, 1970), I have linked developments of science with developments in theology and I have commented on the wider perspective of the theologians when compared with that of the scientists. Today, of course, politics are much more popular. Besides, Professor Imre Lakatos, the secretary general of the slowly disintegrating Popperian party, is a politician first, and a theologian only much much later, and he knows Lenin better than he knows St. Thomas. This is why I have taken my extrascientific quotations from revolutionary politics, and not from revolutionary theology (besides, everyone has by now forgotten that St. Thomas was a revolutionary, too).

28. According to Popper we do not "need any . . . definite frame of reference for our criticism"; we may revise even the most fundamental rules and drop the most fundamental demands if the need for different measures of excellence should arise. *The Open Society and Its Enemies*, II, 390.

29. "No new progressive epoch has ever defined itself by its own limitations . . . In our case, however, watching the boundaries is regarded as more virtuous than transcending them." Speech of Milan Kundera at the IVth Congress of Czech Authors, Prague, June 1967. Quoted from *Reden Zum IV. Kongress des Tschchoslowakischen Schriftstellerverbandes* (Frankfurt: Suhrkamp, 1968), p. 17. "Our case" is of course also the case of revolutionary developments in science and methodology. In his introduction to the German translation of Burke's writings on the French Revolution, Gentz comments in a similar vein (quoted from P. G. Gooch, *Germany and the French Revolution*, London: Longmans, 1920, p. 95): ". . . the eulogist of new systems always finds opinion on his side [optimist!], while the defender of the old must [read: will] appeal to reason." The "opinion" of today is, of course, the "reason" of tomorrow which is already present in a naive, immediate, undeveloped form.

30. Leon Trotsky, *The Revolution Betrayed*, trans. M. Eastman (Garden City, N.Y.: Doubleday, 1937), pp. 86–87.

31. The priority of idea over behavior, problem over physical adaptation, brain over body—these are other versions of the ideology I am criticizing, and all of them have been refuted by more recent research. Thus the discovery of the australopithecines confronted us with a being that combines the brain of an ape with nearly human dentition, posture, and (possibly), behavior. Such a combination "was not anticipated in earlier speculation" (George G. Simpson et al., *Life: An Introduction to College Biology*, New York: Harcourt, Brace, 1957, p. 793) where it was assumed that it is the brain that is responsible for the remaining human features and not the other way around: man became erect, he started using his hands, because his brain told him so. Today, we must admit that a new posture leading to new tasks may "create" the brain needed for these tasks (this, essentially, was also Engels's conjecture in his little essay about the function of the hand in the humanization of our apelike ancestors).

It also seems that certain comprehensive features of early civilization such as domestication or agriculture did not arise as attempts to solve problems. Rather "man at play inadvertently discovered their practical use." F. Alexander, *Fundamentals of Psychoanalysis* (New York: International Universities Press, 1948), p. 113; cf. also G. Róheim, *The Origin and Function of Culture* (New York: Nervous and Mental Disease Monographs, 1943), pp. 40–47, on the origin of the economic activity of mankind, and *Psychoanalysis and Anthropology* (New York: International Universities Press, 1950), p. 437, on the reasons why parents take care of their children. This is most easily proved by the fact that wool in sheep, a surplus of milk in cows, an abundant amount of eggs laid by fowls are all a consequence of domestication and cannot have acted as a reason for it. Hahn (*Die Haustiere in ihrer Beziehung zur Gesellschaft des Menschen*, Leipzig: Johann Ambrosius Barth, 1896, pp. 79, 154, 300, paraphrased after R. H. Lowie, *The History of Ethnological Theory*, New York: Farrar and Reinhart, 1937, pp. 112ff) suggests that people kept poultry originally as alarm clocks, or for cockfights—both non-economic motives. He also suggests that early man was an idler, doing useful labor as a pastime, rather than with serious and problem-conscious intent. Q. H. Schultz ("Some Factors Influencing the Social Life of Primates in General and of Early Man in Particular," in S. L. Washburn, ed., *Social Life of Early Man*, Chicago: Aldine, 1961, p. 63) says: "It was no radical innovation for Dawn man to use their hands for picking up rocks or clubs as ready defence to overcome the lack of large teeth. Nearly every captive macaque delights in carrying new objects around its cage, and apes are entertained for hours by a blanket, or a bucket which they will not let out of their hands without a fight" (my italics).

Wherever we look we see a happy and playful activity leading to accidental solutions of unrealized problems. We do *not* see serious problem-conscious thinkers engaged in the attempt to intellectually discuss and then properly solve the problems they have set up. Later on the sequence is of course inverted by postulating either a divine inventor or a problem situation to which the minds of the contemporaries are supposed to have found the appropriate solution. Such an intellectualistic account is neither correct nor helpful for it prevents us from improving unknown faults of our situation in a spontaneous way and it also prevents us from recognizing such faults in retrospect, after their removal has made their substance clear. By all means, let us be rational: But let us not make the mistake of believing that man can and should improve his lot by reasoned planning only.

32. Cf. notes 22 and 25.

33. I cannot believe that a revolution such as the French Revolution occurred "*in the full consciousness of* [the] rights [which people possess] as men and citizens" as Wilhelm von Humboldt expresses himself (quoted from Gooch, *Germany and the French Revolution*, p. 109), or that a revolution such as the Copernican Revolution proceeded in the full consciousness of the ideas and methods, and with a full understanding of the instruments about (i.e., within the next 300 years) to be invented. In all these cases the element of *action*—unreasonable, nonsensical, mad, immoral action when seen from the point of view of a contemporary—is a necessary presupposition of whatever clarity one would like to *possess*, but can achieve only *after the event*, as the *result of the actions performed*. For material from the history of science see my "Problems of Empiricism, Part II," especially sections 7, 8, 11.

In *politics* and *religion* the point just made implies the need for (mass) action in addition to (party) doctrine, even if the doctrine should happen to contain definite and absolutely clear rules of procedure. *For such rules, while clear and complete when compared with other rules, are always woefully inadequate vis-à-vis the ever changing multitude of social conditions.* (In physics the situation is exactly the same: the formalism of the elementary quantum theory is a monster of beauty and precision. But it is very difficult to exactly specify experimental arrangements capable of *measuring* even the simplest observable. Here we must still rely on the correspondence principle.) But it is just to such conditions that their content must be referred and in the process 'anarchistic' action, i.e., action that is directly related neither to theory nor to the existing institutions, plays an essential part: "We cannot tell . . . what immediate cause will most serve to rouse [a revolution], kindle it, and impel very wide masses [of scientists, for example] who are at present dormant into the struggle . . . History generally, and the history of revolutions in particular, is always richer in content, more varied, more many-sided, more lively and 'subtle' than even the best parties and the most class conscious vanguards of the most advanced classes imagine . . . From this follow two very important practical conclusions: first, that in order to fulfil its task the revolutionary class must be able to master all forms, or sides of social activity without exception . . . second, that the revolutionary class must be ready to pass from one form to another in the quickest and most unexpected manner." Lenin, *'Left Wing' Communism*, p. 100. Cf. also the text to note 5. The application to science is quite straightforward if we keep the proper rules of translation (note 27(B)) in mind. Cohn-Bendit, *Obsolete Communism*, gives a vivid account of an anarchism of the kind. "Problems of Empiricism, Part II" applies the lesson to science. Cf. also notes 35 and 38.

[Addition in fall 1969: I now prefer the label of *Dadaism* to that of *anarchism*. There is not much difference between the two procedures *theoretically* (for partial argument see my essay "The Theatre as an Instrument of the Criticism of Ideologies," *Inquiry*, 10 (1967), 298–310, especially footnote 12 and text). *But an anarchist is prepared to kill while a Dadaist would not hurt a fly. The only thing he does hurt is the "professional conscience" of the defenders of the status quo which at any rate must be exposed to discomfort if one wants to find its limits and if one wants to move beyond them. The necessity for mass action (interruption of "professional meetings," for example) is not denied—but it must be restricted by a dogmatic respect for human lives and by a somewhat less dogmatic respect for the views of the opposition.*]

In *philosophy* this point implies the dependence of theoretical structure on individual action and individual decision: Kierkegaard's analysis of the Ethical applies to the sciences as well. See note 35.

34. The phrase "magical" is quite appropriate, for the inclusion of well-formed observational reports was demanded in books on magic, down to Agrippa's *De occulta philosophia*.

35. Our understanding of ideas and concepts, says Hegel (*Gymnasialreden*; quoted from K. Loewith and J. Riedel, eds., *Hegel, Studienausgabe*, vol. I, Frankfurt: Fischer Bücherei, 1968, p. 54), starts with "an uncomprehended knowledge of them" ("Es ist damit derselbe Fall wie mit anderen Vorstellungen und Begriffen, deren Verstehen gleichfalls mit einer unverstandenen Kenntnis anfaengt . . ."). Cf. also *Logik*, I, 39–40. "It sometimes happens that at a new turning point of a movement, theoretical absurdities cover up some practical truth." Lenin, diary note at the Stuttgart Conference of the Second International, quoted from Bertram D. Wolfe, *Three Who Made Revolution* (Boston: Beacon, 1948), p. 599.

The ideas which are needed in order to explain and to justify a certain procedure in the sciences are often created only by the procedure itself and remain unavailable if the procedure is not carried out. This shows that the element of action and faith which some believe has been eliminated from the sciences is absolutely essential for it: "Even intellectual history, we now admit, is relative, and cannot be dissociated from the wider social context with which it is in constant interaction." Trevor-Roper, *The European Witch Craze*, p. 100. "We are here up against an extremely interesting historical and philosophical phenomenon," writes Ronchi in his discussion of Galileo and the telescope ("Complexities, Advances, and Misconceptions in the Development of the Science of Vision: What Is Being Discovered?" in A. C. Crombie, ed., *Scientific Change*, London: Heinemann, 1963, p. 552), "which illustrates the possible harm that can be caused by logic and reason [i.e., by the exclusive use of well-established ideas and rational methods] while pure faith—for all its unreasonableness—may bring about the most fruitful results."

It is also interesting to note to what extent Kierkegaard's ideas about the role of faith, passion, subjectivity apply to our scientific life (provided, of course, we are interested in fundamental discoveries, and not just in the preservation of the status quo, in methodology, and elsewhere). Cf. *Concluding Unscientific Postscript*, trans. David

F. Swensen and Walter Lowrie (Princeton, N.J.: Princeton University Press, 1941), especially chapter II: "Truth as Subjectivity." Kierkegaard emphasizes the process over the result. "While objective thought translates everything into results and helps all mankind to cheat, by copying these off and reciting them by rote, subjective thought puts everything in process and omits the result; partly, because this belongs to him who has the way, partly because as an existing individual he is constantly in process of coming to be which holds true of every human being who has not permitted himself to be deceived into becoming objective, inhumanly identifying himself with speculative philosophy in the abstract [for example, with the rules of critical rationalism]" (p. 68). One might add that the results of objective thought which are supposed to give reason to everything emerge only at the end of a long process which therefore will have to occur without reason and will have to be passed through on faith only: The "rationality" of the early Royal Society, to take but one example, was entirely a matter of faith.

Kierkegaard's thought has had a decisive influence on Bohr (for material see M. Jammer, *The Conceptual Development of Quantum Mechanics*, New York: McGraw-Hill, 1966, pp. 172ff). It could be used, in conjunction with material from the history of science, to help us construct a new methodology which takes into consideration the role of the individual thinker, not just because he is there, and because his fate should be of interest to us, but because even the most dehumanized and "objective" form of science could not exist without his unreasonable and humorless passionate efforts. Cf. also note 27.

36. H. Marcuse, *Reason and Revolution* (London: Oxford University Press, 1941), p. 130. The quotation is about Hegel's logic.

37. Cf. note 18.

38. "It would be absurd to formulate a recipe or general rule . . . to serve all cases. One must use one's own brains and be able to find one's bearings in each separate case." Lenin, *'Left Wing' Communism*, p. 64. Cf. also note 27(B).

The reader should remember that despite all my praise for Marxism and its various proponents I am defending its *anarchistic* elements only and that I am defending those elements only insofar as they can be used for a criticism of epistemological and moral rules. I quote Lenin because of his insight into the complexity of historical conditions (which is incomparably superior to the insight of scientists and of philosophers of science) and because he recommends an appropriately complex method. I recommend Luxemburg because in elaborating her method she has always the individual before her eyes (one cannot say the same about Sir Karl Popper). I quote Mao because he is prepared to abandon doctrine, to experiment, even in quite fundamental matters. However, I do *not* quote these authors because of their defense of a uniform society of the future, or because of their belief in inexorable laws of history (in the case of Lenin the latter belief is present in a more critical form, for it is connected with *potentialities* rather than with actual developments). Such a society, such laws, it seems to me, would be even less attractive than the "system" of today whose dogmatism has the advantage of being tempered by dishonesty, doubt, cowardice, and indolence.

Some of my friends have chided me for elevating a statement such as "anything goes" into a fundamental principle of epistemology. They did not notice that I was joking. Theories of knowledge as I conceive them *develop*, like everything else. We find new principles, we abandon old ones. Now there are some people who will accept an epistemology only if it has some stability, or "rationality" as they are pleased to express themselves. Well, they can have such an epistemology, and "anything goes" will be its only principle.

39. "Problems of Empiricism," in *Beyond the Edge of Certainty*, ed. R. G. Colodny (Englewood Cliffs, N.J.: Prentice-Hall, 1965), sections IVff, especially section VI. (The relevant material has been reprinted in P. H. Nidditch, ed., *The Philosophy of Science*, London: Oxford University Press, 1969, pp. 12ff, especially pp. 25–33.) "Realism and Instrumentalism," in *The Critical Approach to Science and Philosophy*, ed. M. Bunge (Glencoe, Ill.: Free Press, 1964). "Reply to Criticism," in *Boston Studies*

Paul K. Feyerabend

in the *Philosophy of Science*, vol. II, ed. R. S. Cohen and M. W. Wartofsky (New York: Humanities, 1965).

40. Looking back into history we see that progress, or what is regarded as progress today, has almost always been achieved by counterinduction. Thales' principle according to which there is unity behind the variety of appearances lies at the bottom of all science, ancient and modern. Yet it is contradicted by observations of the most primitive kind (change; the difference between air and iron, for example). The same applies, and to an even larger extent, to Parmenides' principle of the impossibility of all motion. (Even a rationalist like Popper now feels inclined to attack Parmenides on empirical grounds.) The modern interpretation of mental illness as being due not to the action of some external spiritual principle but to autonomous disturbances of the sick organism ran counter to numerous instances where the action of such a principle was both *felt* (split personality, hearing voices, forced movement, objective appearance of emotions and dreams, nightmares, etc.) and *objectively observed* (phantom pregnancy, disintegration of speech patterns). Denying the power of the devil in these times was almost as foolish as (or, considering the threat of hellfire, much more foolish than) denying the existence of material objects is regarded today. Then, Copernicus put forth his magnificent hypothesis and upheld it in the face of plain and indubitable experience (for literature see the reference in note 20). Even Newton, who explicitly advises against the use of alternatives for hypotheses which are not yet contradicted by experience and who invites the scientist not merely to *guess*, but to *deduce* his laws from "phenomena" (cf. his famous rule IV), can do so only by using as "phenomena" *laws which are inconsistent with the observations at his disposal.* (As he says himself: "In laying down . . . Phenomena, I neglect those small and inconsiderable errors." *Sir Isaac Newton's Mathematical Principles of Natural Philosophy and His System of the World*, trans. A. Motte, rev. F. Cajori, Berkeley: University of California Press, 1953, p. 405.) For a more detailed analysis of Newton's dogmatic philosophy and of his dialectical method see my paper "Classical Empiricism."

Yet all these lessons are in vain. Now as ever counterinduction is ruled out by methodology. "The Counterinductive rule," says W. Salmon in his essay "The Foundation of Scientific Inference" (*Mind and Cosmos*, ed. R. G. Colodny, Pittsburgh: University of Pittsburgh Press, 1966, p. 185), is "demonstrably unsatisfactory." He fails to explain how the application of a "demonstrably unsatisfactory" rule can lead to so many satisfactory results which could not have been obtained in any other way.

41. "Fantasy as encountered in many people today is split off from what the person regards as his mature, sane, rational, adult experience. We do not then see fantasy in its true function, but experienced merely as an intrusive, sabotaging, infantile nuisance." R. D. Laing, *The Politics of Experience* (New York: Ballantine, 1967), p. 31.

Laing restricts his discussion of experience and of fantasy to their effect upon interpersonal relations (p. 23: "here, however, I am concentrating upon what we do to ourselves and to each other"). Fantasy, for him, is "a particular way of relating to the [social] world" (p. 31), telling us of problems, abilities, wishes which have become suppressed. The domain of natural science, the physical universe, remains unaffected.

But why should we restrict ourselves to rebuilding man's perception of *society* and of his *fellow men*? Why should we be interested in social reform alone and consider only new pictures of *society*? Is the structure of our *physical world* to be taken for granted? Are we expected to meekly accept the fact that we are living in a lousy material universe, that we are alone in a great ocean of lifeless matter? Or should we not try to change our vision of this universe, too, by leaving the domain of orthodox physics and considering more charming cosmologies? (The only alternative is to become mechanical oneself—this is the path chosen by some scientists, astronauts, and other strange beings.) Proliferation (revival of astrology, witchcraft, magic, alchemy, elaboration of Leibnitz's *Monadology*, and so on) will be a powerful guide in these matters. Psychiatrists and sociologists, however, must not rest content with changing perception and society. They must interfere with the *physical* world and contemplate *its* reform in terms of our fantasies.

42. Those who want to consider the psychological consequences of proliferation will have to distinguish between intraindividual proliferation (plurality of world views within one and the same individual) and interindividual proliferation (plurality of world views in society, each individual holding only a single view and developing it according to his talents and his drive).

Intraindividual proliferation may in extreme cases lead to multiple personality. If we believe the teaching of psychoanalysis then there are always at least two elements present, the ego and the ego ideal, and the latter is ambivalent, being the result of the Oedipus complex. Freud, *Das Ich und das Es* (Leipzig-Vienna-Zurich: Internationaler Psychoanalytischer Verlag, 1923), p. 40. It is this ambivalence which turns the elements *against each other*, contributes to the *development* of both, and creates the dynamics of the individual. (In an animal which is also guided by different principles, for example by different instincts, the principles are not in competition but *work peacefully side by side*, each becoming active in specific circumstances only: G. Róheim, *Psychoanalysis and Anthropology*, p. 430.) This participation of various elements *in any particular human action* explains the "increase of flexibility, as compared to the animal world"; it explains why man "is the only organism normally and inevitably subjected to psychological conflict" (J. Huxley, *The Uniqueness of Man*, London: The Mall, 1941, p. 22); but it also explains why human behavior always presents "a mild case of insanity" (Róheim, *Psychoanalysis and Anthropology*, p. 442). The situation is further complicated by teachers, deans, bosses, and other authorities "who perpetuate the role of the father, whose demands and restrictions have remained active in the ego-ideal, and now act as moral censors in the form of our conscience" (Freud, *Das Ich und das Es*, p. 44). Imposing such a multiplicity of demands with merciless insistence, with a great amount of moralistic grumbling, threatening, headshaking, is bound to lead to crises in the life of the individual so treated and to extreme actions. "There are . . . disastrous choices [such] as those which confronted young people who felt that the service of God demanded forswearing the world forever, as in the Middle Ages, or cutting off one's finger as a religious offering, as among the Plains Indians." M. Mead, *Coming of Age in Samoa* (New York: Morrow, 1961), p. 200. Are we forced to renounce pluralism in favor of happiness and a balanced development?

I do not think we are driven to those extremes. Proliferation produces crises only if the chosen alternatives are played against each other with a vengeance. "The organization of science," writes R. K. Merton ("Behavior Patterns of Scientists," *American Scholar*, 38 (Spring 1969), 220), "operates as a system of institutionalized vigilance, involving competitive cooperation. It affords both commitment and reward for finding where others have erred or have stopped before tracking down the implications of their results, or have passed over in their work what is there to be seen by the fresh eye of another. In such a system, scientists are at the ready to pick apart and appraise each new claim to knowledge. This unending exchange of critical judgment, of praise and punishment, is developed in science to a degree that makes the monitoring of children's behavior by their parents seem little more than child's play." In a warlike community of this kind proliferation will certainly lead to tension and nastiness (and there exists a good deal of nastiness in science, as well as in other critically rationalistic enterprises) but there is no need to combine proliferation with a war of all against all. All that is needed is less moralism, *less seriousness*, less concern for the truth, a vastly deflated "professional conscience," a more playful attitude, conventionalization of "a lack of deep feeling" (Mead, *Coming of Age in Samoa*, p. 7; cf. also p. 35: "and with this goes the continual demand that [one] should not be too efficient, too outstanding, too precocious. [One] *must never excel his fellows by more than a little*," my italics)— and a good deal of laziness—and we shall be able to have our cake: to have freedom of choice in practical as well as in intellectual matters—and to eat it: to have this freedom without too much mental or emotional strain. This is one of the reasons why I regard

Paul K. Feyerabend

the moralism of today, whether it is now found on the right, with the defenders of "The System," or on the left, with the "New Revolutionaries," whether it carries with itself the invitation to "search for the truth," or the admonition to pursue some practical aim, as one of the most vicious ideologies invented by man.

43. *Autobiography* (London: Oxford University Press, 1963), p. 215. Many people are inclined to call Mill a liberal and to dismiss him because of the weaknesses of the liberal creed they have perceived. This is somewhat unjust, for Mill is very different indeed from much that is called "liberalism" today. He is a radical in many ways. Even as a radical, however, he excels by his rationality and his humanity. Cf. R. Lichtman, "The Façade of Equality in Liberal Democratic Theory," *Inquiry*, 12 (1969), 170–208.

44. For one particular element of this plurality, see K. R. Popper, "Back to the Presocratics," *Conjectures and Refutations* (New York: Basic Books, 1962), p. 136.

45. "Coleridge," in Cohen, ed., *The Philosophy of John Stuart Mill*, p. 62. (Numbers in parentheses in the text are pages in this edition.) ". . . I had to learn that I would recognize the value of health even in sickness, the value of rest through exertion, the spiritual through deprivation of material things . . . through evil the value of good . . . I suppose all that I ever tried to teach is expressed in these words." Sybil Leek, *Diary of a Witch* (New York: Quadrangle, 1969), pp. 49, 122.

46. Cf. also my essay "Outline of a Pluralistic Theory of Knowledge and Action," in *Planning for Diversity and Choice*, ed. S. Anderson (Cambridge, Mass.: MIT Press, 1968), which establishes the connection with scientific method alluded to toward the end of the last section.

For the relation between idea and action see the text to note 31. Emphasis on action within a libertarian framework plays an important role in Cohn-Bendit, *Obsolete Communism*, especially chapter V, p. 254: "Every small action committee [in the customary political language of the West: every institution, however small], no less than every mass movement [every large institution, including government bodies, etc.] which seeks to improve the lives of all men must resolve: (i) to respect and guarantee the plurality and diversity of political currents [in the widest sense of including scientific theories and other ideologies] . . . It must accordingly grant minority groups [such as witches, to mention only one example] the right of independent action—only if the plurality of ideas is allowed *to express itself in social practice* does this idea have any real meaning." In addition Cohn-Bendit demands flexibility and a democratic base for all institutions: "all delegates are accountable to, and subject to immediate recall by those who have elected them . . ." For example, one must "oppose the introduction of specialists and specialization" and one must "struggle against the formation of any kind of hierarchy" including the hierarchies of our educational institutions, universities, schools of technology, and so on. As regards knowledge the task is "to ensure a continuous exchange of ideas, and to oppose any control of information and knowledge." It seems to me that the best starting point in our attempt to remove the still existing fetters to thought and action is a combination of Mill's general ideas and of a practical anarchism such as that of Cohn-Bendit. Such a combination produces an ideology and a people that refuses to be intimidated, or restricted, by specialist knowledge (including the specialist knowledge disseminated by our contemporary critical rationalists), that tries to reform the corresponding institutions, especially those graceless safe-deposit boxes of wisdom, our universities, and that encourages the free flow of individuals from position to position ("[N]o function must be allowed to petrify or become fixed . . . the commander of yesterday can become a subordinate tomorrow"—Bakunin, quoted after James Joll, *The Anarchists*, London: Eyre and Spottiswoode, 1964, p. 109), assuring at the same time *that every position in society is equally attractive, and is treated with equal respect.* Let no one say that science, being purely theoretical, has nothing to do with action and politics. The scientist whose results are received with respect and even with fear by the rest of the community and whose "methods" are eagerly imitated lives in a peculiar and often quite constipated environment. It has its own style (cf. note 13), its own rules, its own silly jokes, its own standards of 'integrity' which are likely to poison the whole republic unless special preventive measures (elimination of specialists from positions of power: careful supervision of the educational process so that personal or group idiosyncrasies do not become a national malaise; and *absolute distrust of expert testimony and of expert morality*) are taken. The connection between theory and politics must *always* be considered.

47. For the propagandistic function of medieval art, see Rosario Assunto, *Die Theorie des Schoenen im Mittelalter* (Cologne: DuMont Schauberg, 1963), especially pp. 21–22.

48. "Ideological struggle," says Mao Tse-Tung ("On the Correct Handling of Contradictions among the People," quoted from *Four Essays on Philosophy*, Peking: Foreign Language Press, 1966, p. 116), "is not like other forms of struggle. The only method to be used in this struggle is that of painstaking reasoning and not crude coercion." ". . . the growth of new things may be hindered in the absence of deliberate suppression simply through lack of discernment. It is therefore necessary to be careful about questions of right and wrong in the arts and sciences, to encourage free discussion and avoid hasty conclusions. We believe that such an attitude can help to ensure a relatively smooth development of the arts and sciences" (p. 114). "People may ask, since Marxism is accepted as the guiding ideology by the majority of the people in our country, can it be criticised? Certainly it can . . . Marxists should not be afraid of criticism from any quarter. Quite the contrary, they need to temper and develop themselves and win new positions in the teeth of criticism and in the storm and stress of struggle . . . What should our policy be towards non-Marxist ideas? . . . Will it do to ban such ideas and deny them any opportunity for expression? Certainly not. It is not only futile but very harmful to use summary methods in dealing with ideological questions among the people . . . You may ban the expression of wrong ideas, but the ideas will still be there. On the other hand, if correct ideas are pampered in hothouses without being exposed to the elements or immunized from disease, they will not win out against erroneous ones. Therefore, it is only by employing the method of discussion, criticism and reasoning that we can really foster correct ideas and overcome wrong ones, and that we can really settle issues" (pp. 111–118). The similarity to Mill, whom Mao read in his youth, is remarkable.

It is to be noted that this advice is not put forth generally, but "in the light of China's specific conditions, on the basis of the recognition that various kinds of contradictions still exist in socialist society, and in response to the country's urgent need to speed up its economic and cultural development" (p. 113; see also p. 69, i.e., "On contradiction": ". . . we must make a concrete study of the circumstances of each specific struggle of opposites, and should not arbitrarily apply the formula to everything. Contradiction and struggle are universal and absolute, but the methods of resolving contradictions, that is, the forms of struggle, differ according to the differences in the nature of the contradictions"). Cf. also note 89.

Nor is freedom of discussion granted to everyone: "As far as unmistakable counter-revolutionaries and saboteurs of the socialist cause are concerned, the matter is easy: we simply deprive them of their freedom of speech." *Four Essays on Philosophy*, p. 117. (Cf. H. Marcuse, "Repressive Tolerance," in R. P. Wolff, B. Moore, Jr., H. Marcuse, *A Critique of Pure Tolerance*, Boston: Beacon Press, 1965, p. 100. Marcuse's case is quite interesting. He demands that certain powerful elements be excluded from the democratic debate. This assumes that he has the *power* to suppress them and to prevent them from speaking out and making themselves heard. Now, if he has *this* power, then he certainly has the power to make his own views better known, and he has also the power to educate people in the art of critical thinking. One wonders why he prefers to use an imaginary power which he does not yet possess but which he (or his wife) would certainly like to have, for *suppressing opponents* rather than for *education* and a more balanced discussion of views. Does he perhaps realize that well-educated people would never follow him, no matter how omnipresent his slogans and how seductive his presentation?)

109

The restriction occurs already in Mill, though with different reasons, and expressed in different terminology: "It is, perhaps, hardly necessary to say that this doctrine is meant to apply only to human beings in the maturity of their faculties . . . The early difficulties in the way of spontaneous progress are so great that there is seldom any choice of means for overcoming them; and a ruler full of the spirit of improvement is warranted in the use of any expedients that will attain an end perhaps otherwise unattainable. Despotism is a legitimate mode of government in dealing with barbarians, provided the end be their improvement and the means justified by actually effecting that end. Liberty, as a principle, has no application to any state of things anterior to the time when mankind have become capable of being improved by free and equal discussion. . . ." *On Liberty*, pp. 197–198; cf. Lenin, '*Left Wing*' *Communism*, p. 40: "We can (and must) begin to build socialism not with imaginary human material . . . but with the human material bequeathed to us . . ." The *difference between Mill and Popper*, however, seems to lie in this. For Mill the (material and spiritual) welfare of the individual, the full development of his capabilities, is the primary aim. The fact that the methods used for achieving this aim also yield a scientific philosophy, a book of rules concerning the 'search for the truth,' is a side effect, though a pleasant one. For Popper the search for the truth seems to be much more important and it seems occasionally to even outrank the interests of the individual. In this issue my sympathies are firmly with Mill.

49. This and similar remarks make it clear that Mill (and Popper, who follows Mill in all the respects so far enumerated) is not "dedicated to a national religion of skepticism, to the suspension of judgement" and that he does not "den[y] the existence . . . not only of a public truth, but of any truth whatever," as we can read in Willmore Kendall's bombastic but uninformed essay "The 'Open Society' and Its Fallacies," *American Political Science Review*, 54 (1960), 972ff, quoted from P. Radcliff, ed., *Limits of Liberty* (Belmont, Calif.: Wadsworth, 1966), pp. 38 and 32. To refute the charge of suspension of judgment we should also consider this passage: "No wise man ever acquired his wisdom in any mode but this; nor is it in the nature of human intellect to become wise in any other manner. The steady habit of correcting and completing his own opinion by collating it with those of others, *so far from causing doubt and hesitation* in carrying it into practice, is the only stable foundation for a just reliance on it; for, being cognizant of all that can, at least obviously, be said against him, and having taken up his position against all gainsayers—knowing that he has sought for objections and difficulties instead of avoiding them, and has shut out no limit which can be thrown upon the subject from any quarter—he has a right to think his judgment better than that of any person, or any multitude, who have not gone through a similar process" (p. 209; my italics). Nor is the insinuation correct that Mill's society is, "so to speak, a *debating club*" (p. 36, italics in the original). Just think of Mill's insistence on different "*experiments of living*" (p. 249). Of course, such attention to detail is not to be expected from a self-righteous conservative for whom any discussion of freedom, and any attempt to achieve it, is but "evil teaching" (p. 35).

The possibilities of Mill's liberalism can be seen from the fact that it provides room for any human desire, and for any human vice. There are no general principles apart from the principle of minimal interference with the life of individuals, or groups of individuals who have decided to pursue a common aim. For example, *there is no attempt to make the sanctity of human life a principle that would be binding for all*. Those among us who can realize themselves only by killing humans and who feel fully alive only when in mortal danger are permitted to form a subsociety of their own where human targets are selected for the hunt, and are hunted down mercilessly, either by a single individual or by specially trained groups (for a vivid account of such forms of life see the film *The Tenth Victim*). So whoever wants to live a dangerous life, whoever wants to taste human blood, will be permitted to do so within the domain of his own subsociety. But *he will not be permitted to implicate others*; for example, he will not be permitted to force others to participate in a "war of national honor," or what have you. He will not be permitted to cover up whatever guilt he may feel by making a potential murderer out of everyone. It is very strange to see how the general idea of the sanctity of human life that frowns upon simple, innocent, and rational murders such as the murder of a nagging wife by a henpecked husband does not object to the general murder of people one has not seen and with whom one has no quarrel. Let us admit that we have different tastes, let those who want to wallow in blood receive the opportunity to do so without giving them the power to make "heroes" of the rest of society. As far as I am concerned a world in which a louse can live happily is a better world, a more instructive world, a more mature world than a world in which a louse must be wiped out. (For this point of view see the work of Carl Sternheim; for a brief account of Sternheim's philosophy, see Wilhelm Emrich's Preface to C. Sternheim, *Aus dem Buergerlichen Heldenleben*, Neuwied: Hermann Luchterhand, 1969, pp. 5–19.) Mill's essay is a first step in the direction of constructing such a world.

It also seems to me the United States is very close to a cultural laboratory in the sense of Mill where different forms of life are developed and different modes of human existence tested. There are still many cruel and irrelevant restrictions, and excesses of so-called lawfulness threaten the possibilities which this country contains. However, these restrictions, these excesses, these brutalities occur in the brains of human beings; they are not all found in the Constitution. Accordingly, they can be removed by propaganda, enlightenment, special bills, personal effort (Ralph Nader!), and numerous other legal means. Of course, if such enlightenment is regarded as superfluous, if one thinks it irrelevant, if one assumes from the very beginning that the existing possibilities for change are either insufficient or condemned to failure, if one is determined to use "revolutionary" methods (methods, incidentally, which real revolutionaries, such as Lenin, would have regarded as utterly infantile, and which must increase the resistance of the opposition rather than removing it), then, of course, the "system" will appear much harder than it really is. It will appear harder *because one has hardened it oneself*, and the blame falls back on the bigmouth who calls himself a critic of society. It is depressing to see how a system that has much inherent elasticity is increasingly made less responsive by fascists on the Right and extremists on the Left until democracy disappears without ever having had a chance. My criticism, and my plea for anarchism, is therefore directed *both* against the traditional puritanism in science and society *and* against the "new," but actually age-old, antediluvian, primitive puritanism of the "new" Left which is always based on anger, on frustration, on the urge for revenge, but never on imagination. Restrictions, demands, moral arias, generalized violence everywhere. A plague on both your houses!

50. For a different argument which is entirely in Mill's spirit, see my "Problems of Empiricism," p. 185. Today increase of testability can be added to the list of epistemological benefits presented by Mill ("Problems of Empiricism," section vi). This is not a real addition, however, but only a more detailed and more technical presentation of ideas already developed by him.

51. This quotation has been added mainly for the benefit of Professor Herbert Feigl who keeps making fun of me for adopting extreme positions. Extreme positions are of extreme value. They induce the reader to think along different lines. They break his conformist habits. They are strong instruments for the criticism of what is established and well received. On the other hand, the current infatuation with "syntheses" and "dialogues" which are defended in the spirit of tolerance and of understanding can only lead to an end of all tolerance and of all understanding. To defend a "synthesis" by reference to tolerance means that one is not prepared to tolerate a view that does not show an admixture of one's own pet prejudices. To invite to a "dialogue" by reference to tolerance means inviting one to state one's views in a less radical and therefore mostly less clear way. An author who can write, in the spirit of "dialogue," that "Christianity and Marxism are not contrary to each other" (Guenther Nenning, quoted from the *Newsletter of the American Institute for Marxist Studies*, vol. 6, no. 1 (January–Febru-

111

ary 1969), first page bottom) will hardly be prepared to accept the doctrines of a tough-minded Marxist who is interested in progress, not in peace of mind.

52. In a singularly pretentious, ignorant, and narrow-minded book, *The Poverty of Liberalism* (Boston: Beacon, 1968), R. P. Wolff objects to proliferation on the grounds that it does not follow from the happiness principle. This criticism is certainly irrelevant to the thesis of *On Liberty*. The purpose of *On Liberty* is not to *establish a proposition*, be it now by reference to happiness, or in any other way; the purpose is *to set an example*, to present, explain, defend a certain form of life and to show its consequences in special cases (this becomes crystal clear from the relevant pages of the *Autobiography*). True—Mill also wrote on the happiness principle; but he was free and inventive enough not to restrict himself to a single philosophy, but to pursue different lines of thought. As a result maximum happiness plays no role in *On Liberty*. What *does* play a role is the free and unrestricted development of an individual. One can understand, however, why the author concentrates on happiness. This gives him the opportunity to display his knowledge (if one can call it that) of some of the tools which analytic "philosophers" have constructed for the endless discussion of hedonism.

In addition to the complaint just mentioned—for one can hardly call it an argument—Wolff offers what amounts to a series of rhetorical questions. "It is hard to believe," he says (p. 17), "that even the most dedicated liberal will call for the establishing of chairs of astrology in our astronomy departments or insist that medical schools allow a portion of their curriculum to the exposition of chiropractice in order to strengthen our faith in the germ theory of disease." This is hard to believe indeed, for our "most dedicated liberals" are often moral and intellectual cowards who would not dream of attacking that prize exhibit of the twentieth century—science. Besides, who would think that increasing the number of university chairs is going to lead to a more critical point of view? Are university chairs the only things a contemporary "radical philosopher" (text on front flap of the book) can think on when considering the possibilities of intellectual improvement? Are the limits of a university also the limits of the imagination of our academic radicals? If so, then the attack against Mill collapses at once, for how can a person with such a restricted point of view hope to even *comprehend* the simple nonacademic message of Mill's philosophy?

"Does anyone suppose," Wolff continues his inquiry (p. 16), "that a bright young physicist must keep his belief in quantum mechanics alive by periodically rehearsing the crucial experiments which gave rise to it?" Yes sir, there are lots of people who suppose just that, among them the founders of the quantum theory. There are lots of people who point out that science was often advanced with the help of some *historical* piece of knowledge and who explain the boorishness of much of contemporary physics by the very same lack of perspective which our radical author takes as the basis of his criticism. Of course, "no material harm" (p. 16) will come from the suppression of history and of alternatives just as brothels do not suffer from the philosophical ignorance of the whores—they flourish, and continue flourishing. But a philosophical courtesan certainly is preferable to a common broad because of the added techniques she can develop; and a science with alternatives is preferable to the orthodoxy of today for exactly the same reasons.

It is interesting to see how conservative so-called "radicals" become when confronted with the apparently more solid and more difficult parts of the establishment, such as for example science. Which again shows that they are moral cowards who dare to sing their arias only when there is absolutely no danger of a serious intellectual fight and when they can be absolutely sure of the support of what they think are the "progressive" elements of society.

53. Later in the nineteenth century proliferation was defended by *evolutionary arguments*: Just as animal species improve by producing variations and weeding out the less competitive variants, science was thought to improve by proliferation and criticism. Conversely, "well-established" results of science and even the "laws of thought" were regarded as temporary results of adaptation; they were not given absolute validity. Ac-

cording to Boltzmann (*Populaere Schriften*, Leipzig: Johann Ambrosius Barth, 1905, pp. 398, 318, 258–259), the latter "error finds its complete explanation in Darwin's theory. Only what was adequate was also inherited. . . . In this way the laws of thought obtained an impression of infallibility that was strong enough to regard them as supreme judges, even of experience . . . One believed them to be irrefutable and perfect. In the same way our eyes and ears were once assumed to be perfect, too, for they are indeed most remarkable. Today we know that we were mistaken—our senses are not perfect." Considering the hypothetical status of the laws of thought, we must "oppose the tendency to apply them indiscriminately, and in all domains" (p. 40). This means, of course, that there are circumstances, not factually circumscribed or determined in any other way, in which we must introduce ideas that contradict them. We must be prepared to introduce ideas inconsistent with the most fundamental assumptions of our science even before these assumptions have exhibited any weakness. Even "the facts" are incapable of restricting proliferation, for "there is not a single statement that is pure experience" (pp. 286, 222). Proliferation is important not only in science but in other domains too: "We often regard as ridiculous the activity of the conservatives, of those pedantic, constipated, and stiff judges of morality and good taste who anxiously insist on the observance of every and any ancient custom and rule of behavior; but this activity is beneficent and it must be carried out in order to prevent us from falling back into barbarism. Yet petrification does not set in, for there are also those who are emancipated, relaxed, the hommes sans gêne. Both classes of people fight each other and together they achieve a well-balanced society" (p. 322).

But Boltzmann does not always carry his ideas through to the end. Occasionally he relies on a more simplistic empiricism such as when he says that "a well-determined fact remains unchanged forever" (p. 343), or when he regards "my waking sensations [as] the only elements of my thought" (p. 173) so that "we infer the existence of objects from the impressions made on our senses" (p. 19), or when he declares, more than once, that the task of science is "to adapt our thoughts, ideas, and concepts to the given rather than subjecting the given to the judgment of the laws of thought" (p. 354; cf. with this the assertion, on p. 286, that "the simplest words such as yellow, sweet, sour, etc. which seem to represent mere sensations already stand for concepts which have been obtained by abstracting from numerous facts of experience"). He also warns us not to "go too far beyond experience." This vacillation between a sound scientific philosophy and a bad positivistic conscience is characteristic of almost all so-called "realists" from Boltzmann up to, and including, Herbert Feigl. Reasons for this phenomenon are found in Lenin's *Materialism and Empirio-Criticism* (New York: International Publishers, 1927). Popper's theory of falsification which tells us why we can and should go as far beyond experience as possible has considerably improved the situation. All that is needed now is a little dialectics and attention to specific historical conditions (cf., for example, note 27(B)).

54. Popper, for example, takes it for granted that the subject cannot enter the domain of science, and he also uses a rather simple form of mechanical materialism in his attack on Bohr. For details see part II of "On a Recent Critique of Complementarity," especially section 9. All these principles are used by him dogmatically, and without the shred of an argument. No Hegelian would ever proceed in such a simpleminded manner.

55. Cf. below, sections 12 and 13.

56. "Verhaeltnis des Skeptizismus zur Philosophie," quoted from *Hegel, Studienausgabe*, I, 113; cf. also p. 112.

57. *Differenz des Fichte'schen und Schelling'schen Systems*, p. 13.

58. "Process becomes converted back to praxis, the patient becomes an agent." Laing, *The Politics of Experience*, p. 35. There is a good deal of similarity between Hegel's attempt to set concepts in motion and the attempts of some contemporary psychiatrists to return to the individual the control of some of the defense and projection mechanisms he has himself invented.

59. *Logik*, II, 61.

60. "Reflective reason . . . is nothing but the understanding which uses abstraction, separates, and insists that the separation be maintained and taken seriously." *Logik*, I, 26.

61. *Logik*, I, 82.

62. Cf. *Differenz*, p. 14.

63. Cf. the Carnap quotation, text to note 206.

64. *Logik*, I, 25.

65. *Encyclopaedie der Philosophischen Wissenschaften*, ed. G. Lasson (Leipzig: Teubner, 1920), pp. 72–73. In the original the reference is to Kant, not to scientific empiricism.

66. *Logik*, I, 25.

67. *Logik*, II, 211.

68. *Differenz*, p. 14. Cf. Lenin's comments on a similar passage in his notes on Hegel's *Logic*, quoted in V. I. Lenin, *Aus Dem Philosophischen Nachlass* (Berlin, 1949), pp. 136ff, especially p. 142.

69. Cf. also "Skeptizismus," *Hegel, Studienausgabe*, p. 117: "that scepticism is intrinsically connected with every true philosophy." Also p. 118: "Where can we find a more perfect and independent document and system of true scepticism than in Plato's . . . *Parmenides*? Which embraces and destroys the whole domain of a knowledge achieved by the concepts of our understanding."

70. *Differenz*, p. 25.

71. "It is my aim to read Hegel in a materialistic fashion . . ." Lenin, *Nachlass*, p. 20. The same is true of Professor D. Bohm.

72. Cf. the note on the limit and the ought, *Logik*, I, 121–122: "Even a stone, being something, is differentiated into its being for itself and its Being and so it, too, transcends its limit . . . If it is a basis for acidification, then it can be oxidized, neutralized, and so on. In the process of oxidation, neutralization, etc. its limit, i.e., only to be a basis, is removed . . . and it contains the *ought* to such an extent that only force can prevent it from ceasing to be a basis . . ."

73. *Logik*, I, 71.

74. "Everything that exists is linked in this way to everything else: to the *total* process of the universe. This linkage is either direct, by means of a single quantum, or else indirect, through a series of such linkages." This is how Bohm describes (*Scientific Change*, ed. Crombie, p. 478) the situation created by the quantum theory. The similarity to Hegel is no accident. Bohm has studied Hegel in detail, and he has taken the *Logic* especially as the point of departure for some of his scientific views: ". . . may we not try to understand the world as a total process, in which all parts (for example, the system under observation, observing apparatus, man, etc.) are aspects or sides whose relationships are determined by the way in which they are generated in the process? Of course, in physics, man can, in an adequate approximation, probably be left out of the totality, because he obtains his information from a piece of apparatus on the large-scale level, which is influenced in a negligible way by his looking at it. But at a quantum mechanical level of accuracy, the apparatus and the system under observation must be recognized to be linked indivisibly. Should not the theory be formulated so as to say that this is so . . . ? In a total process of the kind that I am talking about, an observation is regarded as a particular kind of movement, in which some aspects of the process are, as it were, 'projected' into certain large-scale results . . . This process of projection is . . . an integral part of the total process that is being projected" (p. 482).

75. *Logik*, II, 53.

76. *Logik*, I, 67. Cf. also the physical model for this identity in I, 78–79, according to which neither "pure light" nor "pure darkness" gives rise to (the perception of) objects which are recognized and "distinguished only in the determined light . . . which is turbid light."

77. Bohm will therefore not be able to keep contradiction out of his ideas as he occasionally seems to believe (e.g., in *Scientific Change*, p. 482, second paragraph). He agrees in other places but tries to circumvent particular contradictions by moving to a different level of reality. Cf. his *Causality and Chance in Modern Physics* (New York: Harper Torchbooks, 1961).

78. Lenin, *Nachlass*, p. 27.

79. *Logik*, I, 115.

80. *Jenenser Logik, Metaphysik und Naturphilosophie*, ed. G. Lasson (Hamburg: Felix Meiner, 1967), p. 31.

81. In German the statement is more impressive: "Die Wahrheit [des] Seins der endlichen Dinge ist ihr *Ende*."

82. *Logik*, I, 117.

83. *Ibid.*

84. *Ibid.*, p. 36.

85. *Ibid.*

86. Cf. below, section 13, as well as footnote 116 of "Problems of Empiricism, Part II."

87. *Logik*, I, 36; cf. also II, 54, 58ff.

88. *Logik*, I, 117.

89. F. Engels, *Anti-Duehring* (Chicago: Charles H. Kerr, 1935), pp. 144–145; my italics. I am quoting Engels, Lenin, Mao, and similar thinkers rather than the usual bunch of Hegelian or anti-Hegelian scholars as they have still kept the freshness of mind that is necessary to interpret *and* to concretely *apply* the Hegelian philosophy. The same is true of such physicists as Bohm, Vigier, and even Bohr who may occasionally be regarded as an unconscious Hegelian. Cf. the remarks on subject and object below. Cf. also note 38.

90. *Logik*, I, 107.

91. Mathematics was for a long time regarded as lying outside the domain of dialectics. The examples used by Hegel and Engels and especially the example of the differential calculus, so it was thought, only showed the immaturity of the mathematics of the time and the limitations of even the greatest philosophers. One should not have been quite so generous, however. What Hegel says of mathematics applies to *informal* mathematics and, insofar as informal mathematics is the source of the rest, to all of mathematics. That a dialectical study of mathematics can lead to splendid discoveries, even today, is shown by Lakatos's *Proofs and Refutations* (first published in the *British Journal for the Philosophy of Science*, 1963–64). One must praise Lakatos for having made such excellent use of his Hegelian upbringing. On the other hand one must perhaps also criticize him for not revealing his source of inspiration in a more straightforward manner but giving the impression that he is indebted to a much less comprehensive and much more mechanical school of thought. Or has his temporary membership in this school made him lose his sense of perspective? So that he prefers being mistaken for a Wittgensteinian to being classified with the dialectical tradition to which he belongs? Cf. also note 27(B).

92. *Anti-Duehring*, pp. 143–144.

93. *Ibid.*, pp. 138–139.

94. *Ibid.*, p. 144; my italics. Epistemologically these laws belong to the Aristotelian rather than to the Newtonian tradition.

95. *Encyclopaedie der Philosophischen Wissenschaften, ergaenzt durch Vortraege und Kollegienhefte*, ed. L. Henning et al. (Berlin, 1840), pp. 395–396; cf. also Lenin, *Nachlass*, p. 102. Or, to use Bohm's terminology: "as long as, by our customary habits of thinking, we try to say that in an experiment, some part of the world is observed [and described], with the aid of some other part, we introduce an element of confusion into our thought process. Indeed, even the very word 'observation' is misleading, as it generally implies a separation between the observing apparatus and the object under observation, of a kind that does not actually exist." *Scientific Change*, pp. 482–483.

Paul K. Feyerabend

The reader should go on and consider the beautiful example of the observation of a mirror image.

96. *Logik*, II, 224.
97. *Ibid.*, p. 227.
98. *Ibid.*, p. 408.
99. *Ibid.*, p. 225.
100. *Ibid.*, p. 408.
101. Lenin, *Nachlass*, p. 114.
102. *Logik*, II, 410.
103. *Ibid.*, pp. 408–409.
104. *Ibid.*, p. 228. "Knowledge is the eternal infinite approach of thought and object. The *mirroring* of nature in human thought is not 'dead,' it is not 'abstract,' *it is not without motion*, not without its contradictions but is to be conceived as an eternally moving process that gives rise to contradictions and removes them." Lenin, *Nachlass*, p. 115.
105. *Logik*, II, 228. The whole introduction to the *Subjective Logik*, i.e., II, 213–234, can be used for a criticism of what has become known as Tarski's theory of truth. If I remember correctly, this criticism is similar to a criticism voiced by the late Professor Austin in his lectures in Berkeley in 1959. Which shows that even an Oxford philosopher occasionally stumbles upon The Truth.
106. *The Assayer*, quoted from S. Drake and C. D. O'Malley, eds., *The Controversy on the Comets of 1618* (London: Oxford University Press, 1960), pp. 184–185.
107. *Dialogue concerning the Two Chief World Systems*, trans. S. Drake (Berkeley: University of California Press, 1953), p. 328.
108. D. Brouwer and G. M. Clemence, *Methods of Celestial Mechanics* (New York: Academic, 1961), p. v. Cf. also R. H. Dicke, "Remarks on the Observational Basis of General Relativity," in Hong-Yee Chiu and W. F. Hoffmann, eds., *Gravitation and Relativity* (New York: Benjamin, 1964), pp. 1–16. For a more detailed discussion of some of the difficulties of classical celestial mechanics see chapters IV and V of J. Chazy, *La Théorie de la Relativité et la Méchanique Céleste*, vol. I (Paris: Gauthier-Villars, 1928).
109. Cf. section 22 of Jammer, *The Conceptual Development of Quantum Mechanics*. For an analysis see the paper by Lakatos referred to in note 188 of the present essay.
110. H. A. Lorentz studied Miller's work for many years and could not find the trouble. It was only in 1955, 25 years after Miller had finished his experiments, that a satisfactory account of his results was found. See R. S. Shankland, "Conversations with Einstein," *American Journal of Physics*, 31 (1963), 47–57, especially p. 51, as well as footnotes 19 and 34. See also the inconclusive discussion at the "Conference on the Michelson-Morley Experiment," *Astrophysical Journal*, 68 (1928), 341ff. For general relativity see Chazy, *La Théorie de la Relativité*, I, 228ff.
111. For arguments see my essay "In Defence of Classical Physics" in the first issue of the *Studies in the History and Philosophy of Science*, Spring 1970, pp. 59–85.
112. This has been pointed out by K. R. Popper, for example in his paper "Rationality and the Search for Invariants" (Opening Address to the International Colloquium for the Philosophy of Science, London, 1965).
113. W. Heisenberg, "Der gegenwaertige Stand der Theorie der Elementarteilchen," *Naturwissenschaften*, 42 (1955), 640ff. For a comprehensive account of Heisenberg's philosophy, see Herbert Hörz, *Werner Heisenberg und die Philosophie* (Berlin: Deutscher Verlag der Wissenschaften, 1966).
114. *Physics*, book VI; *De coelo*, 303a3ff; *De generatione et corruptione*, 316a. Aristotle's theory of the continuum seems to be closely connected with his empiricism. In Aristotle the empirical doctrine is not just a philosophical dogma, it is a cosmological hypothesis that is clearly formulated (one hears, for a change, what kind of process experience is supposed to be) and leads to a solution of problems which arose in other,

116

and more 'metaphysical,' traditions. The problem of the continuum seems to be one of these problems.

115. Cf. A. Grünbaum, "A Consistent Conception of the Extended Linear Continuum as an Aggregate of Unextended Elements," *Philosophy of Science*, 19 (1952), 288ff.

116. Sir Isaac Newton, *Opticks* (New York: Dover, 1952), p. 266.

117. The rule is enunciated in Kepler's *Ad Vitellionem Paralipomena, Johannes Kepler, Gesammelte Werke herausgegeben im Auftrage der Deutschen Forschungsgemeinschaft und der Bayrischen Akademie der Wissenschaften*, vol. II (Munich: C. H. Beck'sche Verlagsbuchhandlung, 1939), p. 72. For a detailed discussion of Kepler's rule and its influence see Vasco Ronchi, *Optics: The Science of Vision* (New York: New York University Press, 1957), sections 43ff.

118. *Lectiones XVIII Cantabrigiae in Scholis publicis habitae in quibus Opticorum Phenomenon genuinae Rationes investigantur ac exponentur* (London, 1669), pp. 125–126. The passage is used by Berkeley in his attack on the traditional, 'objectivistic' optics. *An Essay towards a New Theory of Vision*, vol. I, *Works*, ed. A. C. Fraser (London, 1901), pp. 137ff.

119. Assuming M to be the observed mass of the charged particle, we obtain for its acceleration at time t the value

$$b(t) = b(O) \cdot \exp[3/2 \cdot Mc^3/e^2] \cdot t.$$

Cf. D. K. Sen, *Fields and/or Particles* (New York: McGraw-Hill, 1968), p. 10.

120. G. Källén, *Helvetica Physica Acta*, 25 (1952), 417, as well as Sen, *Fields and/or Particles*, pp. ix and 73. ". . . this treatment illustrates how we can extract sensible numbers that can be compared with observation despite the divergence difficulties inherent in the present form of field theory." J. J. Sakurai, *Advanced Quantum Mechanics* (Reading, N.Y.: Addison-Wesley, 1967), p. 72.

121. The difficulty was realized by Bohr in his thesis. Bohr also pointed out that the velocity changes due to the change of the external field would equalize after the field is established so that no magnetic effect could arise. Cf. J. L. Heilbron and T. S. Kuhn, "The Genesis of the Bohr Atom," *Historical Studies in the Physical Sciences*, 1 (1969), 221.

The argument in the text is taken from vol. II of *The Feynman Lectures* (Reading, N.Y.: Addison-Wesley, 1965), chapter 34.6. For a somewhat clearer account see R. Becker, *Theorie der Elektrizitaet*, vol. II (Leipzig: Teubner, 1949), p. 132.

122. Cf. my translation of Ehrenhaft's lectures on singular magnetic poles which can be obtained from me at the drop of a postcard.

123. Example: the theory of Eudoxus was misunderstood for a considerable time until Schiaparelli made it comprehensible through calculations of his own. For details see N. Herz, *Geschichte der Bahnbestimmung von Planeten und Kometen*, vol. I, *Die Theorien des Altertums* (Leipzig: Teubner, 1887), pp. 18ff. This is one of the reasons why even obscure or refuted theories should not be abandoned but be made available to all, so that some sympathetic and intelligent guy may pick them up and demonstrate their hidden virtue.

124. "The ephemerides are calculated in accordance with the Newtonian law of gravitation, modified by the theory of general relativity." *Explanatory Supplement to the Astronomical Ephemeris and the American Ephemeris and Nautical Almanack* (London: Her Majesty's Stationery Office, 1961), p. 11. "In the theory of relativity the law of attraction can be formulated rigorously only for the movement of an infinitely small mass under the influence of a fixed spherical mass; this movement is determined by the geodesics of the ds^2 of Schwarzschild . . . if we now want to pass in the study of planetary movements from the Newtonian theory to the theory of relativity, then it suffices . . . to add to the Newtonian perturbations the advancements of the perihelia . . . obtained from the ds^2 of Schwarzschild." J. Chazy, *La Théorie de la Relativité et la Méchanique Céleste*, I, 228. "This mixture of the theories of

Newton and Einstein is intellectually repellent, since the two theories are based upon such different fundamental concepts. The situation will be made clear only when the many-body problem has been handled relativistically in a rational and mathematically satisfactory way." J. L. Synge, *Relativity, the General Theory* (Amsterdam: North-Holland, 1964), pp. 296–297.

125. One might be inclined to deny this statement by referring to the numerous "derivations" of classical mechanics from the general theory of relativity, some of them dealing quite explicitly with the n-body problem. Now, such derivations are but formal exercises *unless* it is shown that not only *momentary* effects but also *long-term* effects are excluded, and this for the whole period for which useful astronomical observations are available (more than 3000 years!): one would have to show that the minute deviations neglected in the usual approximations *have no cumulative effect* which might endanger the stability of the planetary system. This is precisely what is missing in the derivation of the ds² of Schwarzschild that is given in J. Chazy, *La Théorie de la Relativité et la Méchanique Céleste*, vol. II (Paris: Gauthier-Villard, 1930), chapters IX to XI. Planets are here quite properly embedded in the solar system, and the basic equations of relativity are used to show that the combination, referred to in note 124, of Newtonian perturbation theory and the ds² of Schwarzschild is valid to the degree of approximation used. However, Chazy's statement (p. 182) that "this method has thereby been shown to be legitimate" cannot be accepted, for cumulative effects have been omitted from the calculation. Considering the difficulties of the relativistic many-body problem it is not likely that they will soon be taken into account. And even if they are some day, we must still concede the existence of periods in the history of science which, from a sternly methodological point of view, are close to madness, but whose elimination is bound to wipe out science.

126. "The complete, or almost complete mistakes and failures are usually forgotten, by the prophets as well as by the faithful," says a "modern man" and decided opponent about astrology (Franz Boll and Carl Bezold, *Sternglaube und Sterndeutung*, Leipzig: Teubner, 1931, pp. 74, 72). It is clear that the judgment applies to the so-called "sciences" as well.

127. For details see again "Problems of Empiricism, Part II." The fact that science, or any historically grown subject, contains components of different age and different sophistication which hinder each other has been seen and described in a political context by Lenin, Trotsky, and others: "The gist of the matter lies in this," writes Trotsky ("The School of Revolutionary Strategy," Speech at a General Party Membership Meeting of the Moscow Organization, July 1921, quoted from *The First Four Years of the Communist International*, vol. II, New York: Pioneer Publishers, 1923, p. 5): "that the different aspects of the historical process—economics, politics, the growth of the working class—do not develop simultaneously along parallel lines." Cf. also Lenin, '*Left Wing*' *Communism*, p. 59, as well as the quotation in note 38 of the present essay.

The same is true of the relation between observation, auxiliary sciences, theories, and so on.

An excellent example of the phase difference between different parts of the historical process is provided by the history of witchcraft. Witchcraft persecutions were at their peak at the beginning of the seventeenth century and later on, when Galileo reported his telescopic discoveries, Kepler found the laws of planetary motion (and had to defend his own mother against the accusation of witchcraft), when Descartes developed his rationalism and his materialistic physics, 80 years after Copernicus, 40 to 50 years after Montaigne, and they continued into the age of Newton. And the belief was very often held by people who were otherwise perfect examples of the new "scientific spirit." In these times "in which science and art were reborn . . . when people were painting and sculpting anew and once more turned towards investigation and writing, the making of new discoveries and new inventions, when the old classical world and bookprinting seemed to recast the face of Western civilization—in those very days humanity stood in one respect on a lower level of mental development than do some of the primitive races of today." C. Binz, *Doctor Johann Weyer* (Bonn: Landesverlag, 1895), p. 3.

128. In what follows the reader is advised to always consult his Hegel and to compare my statements with Hegel's own dialectical formulations. The reader will also realize that my analysis at once invalidates the direct and naively empirical "refutations" of Marxism by Bernstein, Popper, and others. Things are not quite that simple! Cf. also the next section.

129. *Dialogue concerning the Two Chief World Systems*, p. 126.

130. *Ibid.*, p. 125.

131. *Ibid.*, p. 256.

132. "Problems of Empiricism," pp. 204ff.

133. Bacon, *The New Organon*, Introduction.

134. *Dialogue concerning the Two Chief World Systems*, p. 255. My italics.

135. *Ibid.*, p. 256.

136. *Ibid.*, p. 248.

137. *Ibid.*, p. 171. Only one example to support this thesis: In the Middle Ages there existed two theories of planetary motion, one asserting a motion in consequence, with Saturn the slowest planet and the moon the fastest, the other asserting a motion

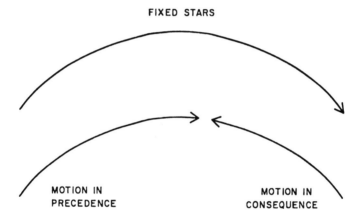

FIXED STARS

MOTION IN PRECEDENCE

MOTION IN CONSEQUENCE

THEORY ii   THEORY i

in precedence, i.e., from east to west, with Saturn the fastest (but not as fast as the celestial sphere) and the moon the slowest. The first theory is held by Plato (*Laws*, 822a), by all the followers of Ptolemy; it occurs in the Sphere of Sacrobosco (Lynn Thorndike, *The "Sphere" of Sacrobosco and Its Commentators*, Chicago: University of Chicago Press, 1949, p. 120, Latin text p. 79), in the German Sphere of Conrad von Megenberg, and in many encyclopedias and textbooks (Vitruvius, Isidore, Bede, Hrabanus Maurus, and others). The second theory occurs in Democritus, in Wolfram von Eschenbach's *Parzival* (cf. G. K. Bauer, *Sternenkunde und Sterndeutung der Deutschen im 9.–14. Jahrhundert*, Berlin, 1937, pp. 27–28). Both theories are compared in the book *De solis affectibus* (Jacques Paul Migne, *Patrologia Latina*, vol. 172, p. 108): "Utrique sententiae, sive contra firmamentum vadunt planetae, seu cum firmamentum potest opponi." Yet we have here a perfect example of relative motion. An even better

119

example of the operative interpretation of motion is provided by the habit of interpreting Bible passages concerning motion as dealing with absolute motion. Altogether the interpreters of the Bible disregard *appearances* and regard terms such as "move," "to be at rest," as absolute terms referring to *objective situations* having unique consequences. This in turn is the result of a naive realism of fantastic proportions. Thus St. Augustine (*De Genesi ad litteram*, II, chapter XVI; Migne, *Patrologia Latina*, vol. 134, p. 277) rejects the idea of fixed stars bigger than the sun on the basis of the duo luminaria magna of Genesis 1:16. The persistence of the belief in witchcraft is at least partly due to this instinctive naive realism that was reluctant to declare as illusory what one had experienced so plainly. Cf. Gregory Zilboorg, *The Medical Man and the Witch during the Renaissance* (Baltimore: Johns Hopkins Press, 1935). Cf. also note 40.

Finally, it must not be overlooked that the *impetus theory* which Galileo accepts in his early writings on mechanics (*De motu; De motu dialogus*) and which had been the *opinio communis* since the fifteenth century demands an absolute view of motion. For if the motive force *resides* in a moving object in the same way in which heat resides in a piece of iron, or sound in a bell that has just been struck (for these examples see *De motu* as translated by I. E. Drabkin in Galileo Galilei, *On Motion and On Mechanics*, ed. S. Drake and I. E. Drabkin, Madison: University of Wisconsin Press, 1960, p. 77, and memorabilia on motion as translated by I. E. Drabkin in Drake and Drabkin, eds., *Mechanics in 16th Century Italy*, Madison: University of Wisconsin Press, 1969, p. 379), then the necessary effect of such a force, i.e., motion, cannot depend on the *relation* of the object to an arbitrarily chosen coordinate system: the impetus theory entails the absolute, or operative, view of *all* motion.

138. Cf. "Problems of Empiricism," pp. 204ff.

139. Cf. Hegel, *Vorlesungen uber die Geschichte der Philosophie*, part I, ed. C. L. Michelet (Berlin: Duncker und Humblot, 1840), p. 289.

140. *Dialogue concerning the Two Chief World Systems*, p. 171.

Galileo's relativism with respect to motion is far from being satisfactory, or even consistent. He proposes the view, (i), expressed in the quotation in the text, that shared motion has no effect whatever. "Motion," he says (*Dialogue*, p. 116), "insofar as it is and acts as motion, to that extent exists relatively to things that lack it; and among things which all share equally in any motion, it does not act and is as if it did not exist." "Whatever motion comes to be attributed to the earth must necessarily remain im-

RIGID CONNECTION

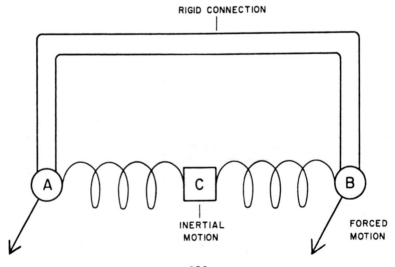

INERTIAL
MOTION

FORCED
MOTION

perceptible . . . so long as we look only at terrestrial objects" (p. 114). ". . . motion that is common to many moving things is idle and inconsequential to the relation of these movables among themselves . . ." (p. 116).

On the other hand, (ii), there is the assertion (cf. *Dialogue,* p. 19) that "nothing . . . moves in a straight line by nature. The motion of all celestial objects is in a circle; ships, coaches, horses, birds, all move in a circle around the earth; the motions of the parts of animals are all circular; in sum—we are forced to assume that only *gravia deorsum* and *levia sursum* move apparently in a straight line; but even that is not certain as long as it has not been proven that the earth is at rest."

Now, if (ii) is adopted, then (i) cannot be correct. For assume that two objects, A and B, being rigidly connected, move in a straight line and that a third object, C, is fastened to them by a spring. Clearly C, being moved forcibly, will tend to assume its natural circular motion and will therefore change its relation to A and B, thus contradicting the assertion, inherent in (i), that common motion does not affect the relation between things. It is this inconsistency which has forced me to split the argument in the text into two steps, one dealing with the relativity of motion (only relative motion it *noticed*), the other dealing with inertial laws (and only inertial motion *leaves* the relation between *the parts of a system unaffected*—assuming, of course, that neighboring inertial motions are approximately parallel). For the two steps of the argument see the beginning of section 8.

It is also important to realize that accepting the relativity of motion even for inertial paths means giving up the impetus theory (cf. the last part of note 137). This Galileo seems to have done by now, for his argument for the existence of "boundless" or "perpetual" motions which he outlines on pp. 147ff of the *Dialogue* appeals to motions which are neutral, i.e., neither natural nor forced and which may therefore(?) be assumed to go on forever.

141. J. L. Austin, *Sense and Sensibilia* (New York: Oxford University Press, 1964), p. 74.

142. For details see the Appendix at the end of this paper.

143. *Dialogue concerning the Two Chief World Systems,* pp. 171–172.

144. *Ibid.,* pp. 249–250.

145. *Ibid.,* pp. 172–173.

146. *Ibid.,* p. 250.

147. Ptolemy, *Syntaxis,* i.7.

148. *Dialogue concerning the Two Chief World Systems,* p. 416. Cf. the *Dialogues concerning Two New Sciences,* trans. Henry Crew and Alfonso de Salvio (London, 1914; New York: Dover, 1958), p. 164: "The same experiment which at first glance seemed to show one thing, when more carefully examined, assures us of the contrary."

149. *Dialogue concerning the Two Chief World Systems,* p. 131.

150. *Ibid.,* p. 327.

151. *Ibid.,* p. 330.

The idea that there is an absolute direction in the universe has a very interesting history. It rests on the structure of the gravitational field on the surface of the earth or of that part of the earth which the observer knows, and generalizes the experiences made there. The generalization is only rarely regarded as a separate hypothesis; it rather enters the "grammar" of common sense and gives the terms "up" and "down" an absolute meaning. (This is a natural interpretation, in precisely the sense that was explained in the text above.) Lactantius, a church father of the fourth century, appeals to this meaning when he asks (*Divinae Institutiones,* III, de falsa sapientia): "Is one really going to be so confused as to assume the existence of humans whose feet are above their heads? Or of regions where the objects which fall with us rise instead? Where trees and fruit grow not upward, but downwards?" The same use of language is presupposed by that "mass of untutored men" who raise the question why the antipodes are not falling off the earth (Pliny, *Natural History,* II, 161–166; cf. also Ptolemy,

Syntaxis, i.7). The attempts of the Presocratics, Thales, Anaximenes, Xenophanes, to find support for the earth which prevents it from falling "down" (Aristotle, *De coelo*, 294a12ff) show that almost all early philosophers with only the exception of Anaximander have shared in this way of thinking. (For the Atomists who assume that the atoms originally fall "down," see M. Jammer, *Concepts of Space*, Cambridge, Mass.: Harvard University Press, 1953, p. 11). Even Galileo, who thoroughly ridicules the idea of the falling antipodes (*Dialogue concerning the Two Chief World Systems*, p. 331) occasionally speaks of the "upper half of the moon" (p. 65), meaning that part of the moon "which is invisible to us." And let us not forget that some linguistic philosophers of today "who are too stupid to recognize their own limitations" (p. 327) want to revive the absolute meaning of "up-down" at least *locally*. Thus the power over the minds of his contemporaries of a primitive conceptual frame assuming an anisotropic world, which Galileo had also to fight, must not be underestimated. For an examination of some aspects of common sense at the time of Galileo, including astronomical common sense, the reader is invited to consult E. M. W. Tillyard, *The Elizabethan World Picture* (London: Penguin, 1963). The agreement between popular opinion and the central-symmetrical universe is frequently asserted by Aristotle. See, for example, *De coelo*, 308a23f.

152. *Dialogue concerning the Two Chief World Systems*, p. 327.

153. *Ibid.*, p. 327; italics added.

154. *Ibid.*, pp. 132, 416.

155. Cf. footnote 137 of "Problems of Empiricism, Part II."

156. *Dialogue concerning the Two Chief World Systems*, p. 341. Galileo here quotes part of Copernicus's address to Pope Paul III in *De revolutionibus*. Cf. also the Narratio Prima (quoted from E. Rosen, *Three Copernican Treatises*, New York: Dover, 1959, p. 165): "For all these phenomena appear to be linked most nobly together, as by a golden chain; and each of the planets, by its position, and order, and every inequality of its motion, bears witness that the earth moves and that we who dwell upon the globe of the earth, instead of accepting its changes of position, believe that the planets wander in all sorts of motions of their own." Note that empirical reasons are absent from the argument, and they have to be, for Copernicus himself admits (Commentariolus, Rosen, *Three Copernican Treatises*, p. 57) that the Ptolemaic theory is "consistent with the numerical data."

157. *Dialogue concerning the Two Chief World Systems*, p. 120. In their book *Geschichte der Hexenprozesse*, vol. I (Stuttgart: Cotta, 1880), p. 64, W. G. Soldan and H. Heppe comment on the fluidity of concepts such as striga, empusa, Lamia, and they continue: "it must not be forgotten that no physiology has been written for the domain of superstition and that there remained, despite the existence of certain essential elements, sufficient leeway for variety in the particulars, according to age, locality, or the fantasy of the individual poet." Cf. also the material assembled by J. Frank, "Geschichte des Wortes Hexe," in J. Hansen, *Quellen und Untersuchungen zur Geschichte des Hexenwahns und der Hexenverfolgungen im Mittelalter* (Bonn: Olbers, 1901), chapter VII. My analysis of Galileo shows that such fluidity is a characteristic of science also and that it takes possession not only of the accidental elements of a concept, but of its very essence. Moreover, it is a precondition of scientific progress. The stability of concepts is not the differentia specifica that separates science from witchcraft (magic, poetry, and so on).

158. Cf. "Classical Empiricism."

159. Cf. note 140.

160. *Dialogue concerning the Two Chief World Systems*, p. 145.

161. *Ibid.*, p. 147.

162. Cf. note 140.

163. Charles B. Schmitt, in an interesting and very important article ("Experience and Experiment: A Comparison of Zabarella's View with Galileo's *De motu*," *Studies*

in the Renaissance, 16 (1969), 80–138) discusses the various notions of experience which were active in the sixteenth and seventeenth centuries and tries to determine Galileo's own position during his years in Pisa. Galileo then regarded experience as a "useful device to resolve a particular dispute. By merely observing the world around us we can sometimes decide either for or against a particular opinion which has been brought forth. Therefore, Aristotle can sometimes be criticized for holding positions which are not in conformity with experience. On the other hand, Aristotle sometimes relies too much upon experience, to the extent that he does not allow a sufficient role to *rationes*; but according to Galileo it is through *rationes* that demonstration takes place. That is to say: demonstration and proof depend upon 'objects of thought' rather than 'objects of experience' (pp. 111–112). Accordingly, "with the young Galileo . . . experience is not always so carefully selected and, more often than not, proves to be deceptive or, at least, not capable of resolving the problem at hand" (p. 124). In addition Galileo seems to distrust experience because of its occult overtones (p. 135): there was a tradition, in the sixteenth and seventeenth centuries, when experience went hand in hand with the study of magic and of the occult, being a source of knowledge in cases which could not be reached by reason: "There are hidden forces," writes Cornelius Agrippa in his *Occult Philosophy* (I, 10), "whose causes are inaccessible because reason cannot thoroughly explore them. Therefore philosophers have studied the greater number of these causes by experience rather than by thought" (cf. Schmitt, pp. 86ff, and the literature there). Now our hypothesis is (a) that later on Galileo neither rejected experience nor relied on it to the exclusion of everything else, *but that he changed it so that 'objects of thought' became perceptible,* and (b) that his belief in the Copernican theory played an essential part in the transformation. This hypothesis will of course have to be supported by a more detailed study than I have given here. Schmitt's article shows that the situation is less settled than is usually assumed and that many popular views concerning Galileo (including those put forth more recently by Geymonat—see Schmitt, footnote 133) are oversimplified, to say the least. "What are some of the broader implications of our investigation?" asks Schmitt at the end of his paper (p. 136). "Although it seems obvious that the 'experimental method' which emerged in the 17th century was in some way or another an outgrowth of the observationalist and experientialist tradition of the preceding centuries, it is not at all clear precisely *how*. To see it as merely an outgrowth of technological practice on the one hand, or of Aristotelian empiricism on the other, seems to oversimplify the situation to the point of distortion. At the same time, one must be careful not to dismiss entirely the significance of observation and experience and to make the 'scientific revolution' merely a conceptual revolution in which a Platonic view of the universe replaced an Aristotelian one. It seems clearly to have been a more complex process than either of these interpretations would seem to suggest. It is the belief of the present writer," Schmitt continues, "that more light could be shed on this subject through a detailed study of sixteenth century writings on natural philosophy, both scholastic, and nonscholastic." I would add that one must also study the manner in which the wish to make Copernicus true has influenced the procedures of some thinkers, and especially of Galileo.

164. Cf. *Physics,* 208b10ff.

165. Galileo seems to have been aware of this situation. He silently abandons the idea of the nonoperative character of circular motion in his attempt to prove the motion of the earth from the tides. Cf. H. L. Burstyn, "Galileo's Attempt to Prove That the Earth Moves," *Isis,* 53 (1962), 161–185, and the literature cited there.

166. Such a stone, says Galileo (*Dialogue concerning the Two Chief World Systems,* p. 233), would arrive ahead of the tower.

167. One might be inclined to assume that the phenomenon of projectile motion which in the fourteenth century had led to various criticisms of the Aristotelian point of view was a clear empirical basis for a law such as Galileo wanted to establish. A

little consideration shows that this cannot have been the case. All one knew about projectile motion was that it proceeded for some time and then came to a halt. Galileo's circular law is in no way determined by this phenomenon. It is determined by his wish to preserve the Copernican view. Cf. also notes 137, 140, and the text below. Buridan, incidentally, *rejected* the rotation of the earth.

168. Cf. note 140.

169. For an enumeration of such experiments see A. Armitage, "The Deviation of Falling Bodies," *Annals of Science*, 5 (1941–47), 342–351. For further material and discussion see A. Koyré, *Metaphysics and Measurement* (Cambridge, Mass.: Harvard University Press, 1968). For a comprehensive survey see G. Hagen, *La Rotation de la Terre* (Rome, 1911). It is interesting to see how the experiments ceased after their first inconclusive results, and how they were resumed when Newton made a new prediction concerning their outcome. Cf. Armitage, "The Deviation of Falling Bodies," p. 346.

170. *De motu*, trans. Drabkin, p. 73.

171. *Ibid.*, p. 78.

172. Drabkin translation, p. 338. Drake in footnote 10 of the same page comments that "Galileo was not a Copernican when he wrote this."

173. Quoted from Drake and Drabkin, eds., *Mechanics in 16th Century Italy*, p. 228.

174. *De motu*, pp. 73–74.

175. *Ibid.*, p. 74.

176. *Physics*, VII, 1; 241b34–36.

177. *De motu*, p. 79. Cf. also notes 137 and 140.

178. *De motu*, chapter XVIII in Drabkin's subdivision.

179. Cf. notes 137 and 140.

180. *De revolutionibus*, i.8.

181. *Dialogue concerning Two New Sciences*, pp. 215, 250.

182. Pp. 147ff. Cf., however, the inconsistency described in note 137 of the present essay.

183. According to Anneliese Maier, *Die Vorlaeufer Galileis im 14. Jahrhundert* (Rome: Edizioni di Storia e Litteratura, 1949), pp. 151ff, Galileo replaces impetus by inertia in order to explain the "fact" that "neutral" motions go on forever. Now, to start with, there is no such fact. Secondly, Galileo initially does not believe, and rightly so, that there is such a fact. This we have just seen. There is therefore no need for him "to explain certain newly detected *phenomena*" (p. 151). The need was purely theoretical: to accommodate, to "save," not a phenomenon, *but a new world view*.

184. The so-called scientific revolution led to astounding discoveries and it considerably extended our knowledge of physics, physiology, and astronomy. This was achieved by pushing aside and regarding as irrelevant, *and often as nonexistent*, those facts which had supported the older philosophy. Thus all the evidence for witchcraft, demonic possession, all the empirical phenomena one had been able to cite in favor of the existence of the devil, were pushed aside *together with the* "superstitions" they once confirmed. The result was that "towards the close of the Middle Ages science was forced away from human psychology, so that even the great endeavour of Erasmus and his friend Vives, as the best representatives of humanism, did not suffice to bring about a rapprochement, and psychopathology had to trail centuries behind the developmental trend of general medicine and surgery ["The hatred and jealousy of the doctors," says von Gleichen, "when they persecute, [is] as dangerous as that of the priests"]. As a matter of fact . . . the divorcement of medical science from psychopathology was so definite that the latter was always totally relegated to the domain of theology and ecclesiastic and civil law—two fields which naturally became further and further removed from medicine . . ." Zilboorg, *The Medical Man and the Witch*, pp. 3–4 as well as 70ff. ("Dr. Zilboorg," says H. Sigerist in his introduction to the book, "recognised that witchcraft is the central problem in the development of occidental psychiatry. In the changing attitude towards witchcraft modern psychiatry was born as a medical disci-

pline.") Astronomy advanced, but our knowledge of man slipped back into an earlier, more primitive stage. Cf. note 127.

Another example is *astrology*. "In the early stages of the human mind," writes A. Comte (*Philosophie Positive*, Paris: Littré, 1836, III, 273–280), "these connecting links between astronomy and biology were studied from a very different point of view, but at least they were studied and not left out of sight, as is the common tendency in our own time, under the restricting influence of a nascent and incomplete positivism. Beneath the chimaerical belief of the old philosophy in the physiological influence of the stars, there lay a strong though confused recognition of the truth that the facts of life were in some way dependent on the solar system. Like all primitive inspirations of man's intelligence this feeling needed rectification by positive science, but not destruction; though unhappily in science, as in politics, it is often hard to reorganise without some brief period of overthrow."

185. "Neurath fails to give . . . rules [which distinguish empirical statements from others] and thus unwittingly throws empiricism overboard." K. R. Popper, *The Logic of Scientific Discovery* (New York: Basic Books, 1959), p. 97.

186. *Papirer*, ed. P. A. Heiberg (Copenhagen, 1909), VII, part I, see A, Nr. 182. Cf. also sections 7ff of my forthcoming paper "Abriss einer anarchistischen Erkenntnislehre."

187. Cf. note 31 and text.

188. "Criticism and the Methodology of Scientific Research Programs," in *Criticism and the Growth of Knowledge*, ed. I. Lakatos and A. Musgrave (Amsterdam: North-Holland, 1969). Quotations are from the typescript of the paper which Lakatos distributed liberally before its publication. In this typescript the reference is mostly to Popper. Had Lakatos been as careful with acknowledgments as he is when the Spiritual Property of the Popperian Church is concerned, he would have pointed out that his liberalization which sees knowledge as a *process* is indebted to Hegel.

189. Popper, *The Open Society and Its Enemies*, pp. 388ff.

190. *Ibid.*, p. 390. Cf. also note 28.

191. *Ibid.* Cf. note 22 and the corresponding text.

192. *Ibid.*, p. 391.

193. *Ibid.*, p. 231.

194. I am referring here to the following two papers: "Epistemology without a Knowing Subject," in Bob Van Rootselaar and J. F. Staal, eds., *Logic, Methodology and a Knowledge of Science*, vol. III (Amsterdam: North-Holland, 1968), as well as "On the Theory of the Objective Mind." In the first paper, birdnests are assigned to the "third world" (p. 341) and an interaction is assumed between them and the remaining worlds. They are assigned to the third world *because of their function*. But then stones and rivers can be found in this third world too, for a bird may sit on a stone, or take a bath in a river. As a matter of fact, everything that is noticed by some organism will be found in the third world, which will therefore contain the whole material world and all the mistakes mankind has made. It will also contain "mob psychology."

195. Cf. again "Problems of Empiricism, Part II."

196. Cf. *Malleus Maleficarum*, trans. Montague Summers (London: Pushkin Press, 1928), part II, question 1, chapter IV: "Here follows the way whereby witches copulate with those Devils known as Incubi," second item, as to the acts, "whether it is always accompanied with the injection of semen received from some other man." The theory goes back to St. Thomas Aquinas.

197. It is of course possible to establish correlations between the *sentences* of the two theories, but one must realize that the elements of the correlation, when interpreted, cannot be both meaningful, or both true: if relativity is true, then classical descriptions are either always false or are always nonsensical. Continued use of classical sentences must therefore be regarded as an abbreviation for sentences of the following kind: "Given conditions C, the classical sentence S was uttered by a classical physicist whose sense organs are in order, and who understands his physics"—and sentences of this kind,

if taken together with certain psychological assumptions, can be used for a test of relativity. However, the *statements* which are expressed by these sentences are part of the *relativistic* framework, for they use relativistic terms. This situation is overlooked by Lakatos who argues as if classical terms and relativistic terms can be combined at will and who infers from this assumption the nonexistence of incommensurability.

198. This became clear to me in a discussion with Mr. L. Briskman, in Professor Watkins's seminar at the London School of Economics.

199. This seems to occur in certain versions of the general theory of relativity. Cf. A. Einstein, L. Infeld, and B. Hoffmann, "The Gravitational Equations and the Problem of Motion," *Annals of Mathematics*, 39 (1938), 65, and Sen, *Fields and/or Particles*, pp. 19ff.

200. This consideration has been raised into a principle by Bohr and Rosenfeld, *Kgl. Danske Videnskab. Selskab, Mat.-Fys. Medd.*, 12, no. 8 (1933), and, more recently, by Robert F. Marzke and John A. Wheeler, "Gravitation as Geometry I," in Chiu and Hoffmann, eds., *Gravitation and Relativity*, p. 48: "every proper theory should provide in and by itself its own means for defining the quantities with which it deals. According to this principle, classical general relativity should admit to calibrations of space and time that are altogether free of any reference to [objects which are external] to it such as rigid rods, inertial clocks, or atomic clocks [which involve] the quantum of action."

201. It is possible to base space-time frames on this new element entirely, and to avoid contamination by earlier modes of thought. All one has to do is to replace distances by light times and to treat time intervals in the relativistic fashion, for example, by using the k-calculus (Cf. chapter II of J. L. Synge, "Introduction to General Relativity," in *Relativity, Groups, and Topology*, ed. C. M. DeWitt and B. B. DeWitt, New York: Gordon, 1964. For the k-calculus see H. Bondi, *Assumption and Myth in Physical Theory*, London: Cambridge University Press, 1967, pp. 28ff, as well as D. Bohm, *The Special Theory of Relativity*, New York: Benjamin, 1965, chapter XXVI.) The resulting concepts (of distance, velocity, time, etc.) are a necessary part of relativity, in the sense that all further ideas, such as the idea of length as defined by the transport of rigid rods, must be changed and adapted to them. They therefore suffice for explaining relativity. Following their own principle as described in note 200 Marzke and Wheeler have given an account of relativistic terms that does not involve any objects external to the theory (this account goes back to Robert F. Marzke, "The Theory of Measurement in General Relativity," A.B. senior thesis, Princeton University, 1959; the article by Marzke and Wheeler is the only published report available so far). All intervals, whether spatial or temporal, are expressed in terms of some (spatial or temporal) standard interval. There is no difference between the units used for intervals of distance and intervals of time. The construction which leads to measurement in terms of the standard interval is carried out with the help of light and mass points only and involves neither rigid rods nor clocks whose construction would have to be explained in nonrelativistic terms. "The importance of light rays and the light cone in the intrinsic geometry of physics comes more directly to the surface. The true function of the speed of light is no longer confused with the trivial task of relating two separate units of interval, the meter and the second, of purely historic and accidental origin" (Marzke and Wheeler, "Gravitation as Geometry I," p. 56). The difference between such terms and classical terms is now very obvious and the assertion of incommensurability is made much more precise. Cf. also note 205.

202. For this point and further arguments see A. S. Eddington, *The Mathematical Theory of Relativity* (Cambridge: Cambridge University Press, 1963), p. 33. The more general problem of concepts and numbers has been treated by Hegel, *Logik*, I, Das Mass.

203. This takes care of an objection which Professor J. W. N. Watkins has raised on various occasions.

204. For further details, especially concerning the concept of mass, the function of "bridge laws" or "correspondence rules," and the two-language model, see section iv of "Problems of Empiricism." It is clear that, given the situation described in

the text, we cannot derive classical mechanics from relativity, not even approximately. For example, we cannot derive the classical law of mass conservation from relativistic laws. The possibility of connecting the *formulas* of the two disciplines in a manner that might satisfy a pure mathematician, or an instrumentalist, is, however, not excluded. For an analogous situation in the case of quantum mechanics see section 3 of my paper "On a Recent Critique of Complementarity." See also section 2 of the same article for more general considerations.

205. Marzke and Wheeler measure length in the following way (for details see "Gravitation as Geometry I"): First, flatness of space is ascertained to the degree of precision desired. Next, a method is devised for constructing a parallel to any straight line in space-time (the method uses inertial trajectories and light rays only, thus eliminating all nonclassical space-time notions). Third, a "geodesic clock" is constructed by

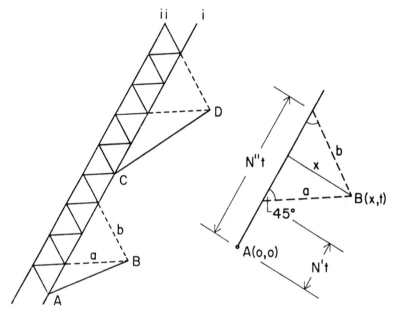

letting a pulse of light be reflected back and forth between two parallels: the intersections of the pulse with one of the lines are the "ticks" of the clock. Finally, two arbitrary intervals, AB and CD, are compared, thus: A trajectory, i, is found that connects A and C. A parallel, ii, is drawn to i. A geodesic clock is constructed between i and ii. A light ray, a, is sent from i toward B and reflected from it back to i (b). $N'\tau$ and $N''\tau$ are the times of departure and arrival of this light ray, counted from A. Assuming c equal to 1 we see that

$N'\tau = t - x$

$N''\tau = t + x$, so that

$N'N''\tau^2 = t^2 - x^2 = (t - O)^2 - (x - O)^2 = (t_B - t_A)^2 - (x_B - x_A)^2 = (\Delta s_{AB})^2$,

hence

$(\Delta s_{AB}) = AB = \tau\sqrt{N'N''}$

$(\Delta s_{CD}) = CD = \tau\sqrt{N'N''}$, so that

$$\frac{CD}{AB} = \sqrt{\frac{N'''N''''}{N'N''}}$$

which gives the numerical values resulting from the comparison.

Now this method clearly works only if we can assume that it gives the same results in all inertial systems, for example, if we can assume that c has the same value in all inertial systems. If there exists a preferred system, or if the classical principle of velocity addition is still assumed to be valid, then the method no longer leads to useful numbers, and transitivity of length cannot be guaranteed for all inertial systems (just assume three systems, A, B, and C, B moving with c/ 2 relatively to A and C with c/ 2 relatively to B, then length measured in C will always be O, assuming A is the rest system).

206. R. Carnap, "The Methodological Character of Theoretical Concepts," *Minnesota Studies in the Philosophy of Science*, vol. I, ed. H. Feigl and M. Scriven (Minneapolis: University of Minnesota Press, 1956), p. 47.

207. An even more conservative principle is sometimes used when discussing the possibility of languages with a logic different from our own: "Any allegedly new possibility must be capable of being fitted into, or understood in terms of, our present conceptual or linguistic apparatus." B. Stroud, "Conventionalism and the Indeterminacy of Translations," *Synthese*, 1968, p. 173.

208. As an example the reader is invited to consult J. Piaget, *The Construction of Reality in the Child* (New York: Basic Books, 1954).

209. *Ibid.*, pp. 5ff.

210. For the condition of research formulated in the last sentence see section 8 of "Reply to Criticism," *Boston Studies in the Philosophy of Science*, vol. II, ed. Cohen and Wartofsky. For the role of observation see section 7 of the same article. For the application of Piaget's work to physics and, more especially, to the theory of relativity see the appendix of Bohm, *The Special Theory of Relativity*. Bohm and Schumacher have also carried out an analysis of the various informal structures which underlie our theories. One of the main results of their work is that Bohr and Einstein argued from incommensurable points of view. Seen in this way the case of Einstein, Podolsky, and Rosen cannot refute the Copenhagen Interpretation and it cannot be refuted by it either. The situation is, rather, that we have two theories, one permitting us to formulate EPR, the other not providing the machinery necessary for such a formulation. We must find independent means for deciding which one to adopt. For further comments on this problem see section 9 of my "On a Recent Critique of Complementarity."

211. For what follows cf. also my review of Nagel's *Structure of Science* on pp. 237–249 of the *British Journal for the Philosophy of Science*, 6 (1966), 237–249.

212. Carnap, "The Methodological Character of Theoretical Concepts," p. 40. Cf. also C. G. Hempel, *Philosophy of Natural Science* (Englewood Cliffs, N.J.: Prentice-Hall, 1966), pp. 74ff.

213. It was for this reason that Leibniz regarded the German of his time and especially the German of the artisans as a perfect observation language, while Latin, for him, was already too much contaminated by theoretical notions. See his "Unvorgreifliche Gedancken, betreffend die Ausübung und Verbesserung der Teutschen Sprache," published in *Wissenschaftliche Beihefte zur Zeitschrift des allgemeinen deutschen Sprachvereins*, IV, 29 (Berlin: F. Berggold, 1907), pp. 292ff.

214. For examples of such descriptions see the article of Synge referred to in note 201.

215. This objection was raised at a conference by Prof. Roger Buck.

216. For this point see section I of "Reply to Criticism," as well as the corresponding sections in "Problems of Empiricism."

217. That the choice between comprehensive theories rests on one's interests entirely and reveals the innermost character of the one who chooses has been emphasized by Fichte in his "Erste Einleitung in die Wissenschaftslehre." Fichte discusses the opposition between idealism and materialism which he calls dogmatism. He points out that there are no facts and no considerations of logic which can force us to adopt either the one or the other position. ". . . we are here faced," he says (*Erste und Zweite Einleitung in die Wissenschaftslehre*, Hamburg: Felix Meiner, 1961, p. 19), "with an

absolutely first act that depends on the freedom of thought entirely. It is therefore determined in an arbitrary manner [durch Willkur] and, as an arbitrary decision must have a reason nevertheless, by our inclination and our interest. The final reason for the difference between the idealist and the dogmatist is therefore the difference in their interests."

218. Here once more the familiar problem of alienation arises: what is the result of our own activity becomes separated from it, and assumes an existence of its own. The connection with our intentions and our wishes becomes more and more opaque so that in the end we, instead of leading, follow slavishly the dim outlines of our shadow whether this shadow manifests itself objectively, in certain institutions, or subjectively, in what some people are pleased to call their "intellectual honesty," or their "scientific integrity." (". . . Luther eliminates external religiousness and turns religiousness into the inner essence of man . . . he negates the raving parish-priest outside the layman because he puts him into the very heart of the layman." Marx, Nationaloekonomie und Philosophie; quoted from Marx, die Frühschriften, ed. Landshut, p. 228.

In the economic field the development is very clear: "In antiquity and in the Middle Ages exploitation was regarded as an obvious, indisputable, and unchangeable fact by both sides, by the free as well as by the slaves, by the feudal lords as well as by their bondsmen. It was precisely because of this knowledge on the part of both parties that the class structure was so transparent; and it was precisely because of the dominance of agriculture that the exploitation of the lower classes could be seen in the strict sense of the word. In the Middle Ages the serf worked, say, four days and a half per week on his own plot of land and one day and a half on the land of his master. The place of work for himself was distinctly separated from the place of serfdom . . . Even the language was clear, it spoke of 'bondsmen' ["Leibeigene," i.e., those whose bodies are owned by someone else] . . . of 'compulsory service' ["Fronarbeit"] and so on. Thus the class distinctions could not only be seen, they could also be heard. Language did not conceal the class structure, it expressed it in all desirable clarity. That was true in Egypt, Greece, the European Middle Ages, in Asiatic as well as in European languages. It is no longer true in our present epoch . . . Workers in early capitalism spent their whole time in the factory. There was neither a spatial nor a temporal separation between the period they worked for their own livelihood and the period they slaved for the capitalist. This led to the phenomenon I have called . . . the 'sociology of repression.' The fact of exploitation was no longer admitted and the repression was facilitated because exploitation could no longer be seen." Fritz Sternberg, Der Dichter und die Ratio; Erinnerungen an Bertolt Brecht (Gottingen: Sachse und Pohl, 1963), pp. 47ff. Exactly the same development occurred between Galileo and, say, Laplace. Science ceased to be a variable human instrument for exploring and changing the world and became a solid block of "knowledge," impervious to human dreams, wishes, expectations. At the same time the scientists themselves became more and more remote, "serious," greedy for recognition, and incapable and unwilling to express themselves in a way that could be understood and enjoyed by all. Einstein and Bohr, and Boltzmann before them, were notable exceptions. But they did not change the general trend. There are only a few physicists now who share the humor, the modesty, the sense of perspective, and the philosophical interests of these extraordinary people. All of them have taken over their physics, but they have thoroughly ruined it.

It is even worse in the philosophy of science. For some details, see my papers "Classical Empiricism" and "On the Improvement of the Sciences and the Arts, and the Possible Identity of the Two," in Boston Studies in the Philosophy of Science, vol. III, ed. R. S. Cohen and M. W. Wartofsky (Dordrecht: Reidel, 1968).

219. Popper has repeatedly asserted, both in his lectures and in his writings, that while there is progress in the sciences there is no progress in the arts. He bases his assertion on the belief that the content of succeeding theories can be compared and that a judgment of verisimilitude can be made. The refutation of this belief eliminates an important difference, and perhaps the only important difference, between science and

the arts, and makes it possible to speak of styles and preferences in the first, and of progress in the second.

220. Cf. B. Brecht, "Ueber das Zerpfluecken von Gedichten," *Uber Lyrik* (Frankfurt: Suhrkamp, 1964). In my lectures on the theory of knowledge I usually present and discuss the thesis that finding a new theory for given facts is exactly like finding a new production for a well-known play. For painting see also E. Gombrich, *Art and Illusion* (New York: Pantheon, 1960).

221. "The picture of society which we construct we construct for the river-engineers, for the gardeners . . . and for the revolutionaries. All of them we invite into our theater, and we ask them not to forget their interest in *entertainment* when they are with us, for we want to turn over the world to their brains and hearts so that they may change it according to their wishes." Brecht, "Kleines Organon für das Theater," *Schriften zum Theater* (Frankfurt: Suhrkamp, 1964), p. 20; my italics.

A Picture Theory of Theory Meaning

The standard view of scientific theories that Herbert Feigl has pictorially set out here is very helpful in some respects. It is a representation of a scientific theory and the role that it plays with respect to the over-all scientific activity in terms of the syntactical calculus of a theory, the observation statements, and the correspondence rules (the coordinating definitions). What has sometimes been characterized as the ideal language conception of the function of a scientific theory and, on the other hand, what has sometimes been characterized as the black box conception of a scientific theory—both stand out in very clear logical relief in terms of this sort of representation. Historically, there are two extremely good cases, one of which would correspond to the ideal language representation and the other to the black box representation.

The ideal language case is the ideal fluid mechanics of Euler and Bernoulli. Fluid mechanics as represented in the post-Newtonian work of these men was a remarkable theory. They saw to it that at the theoretical level the definitions were so sharp and so idealized that the remainder of the algorithm which was generated was very elegant indeed. From the mathematical point of view it is quite a beautiful theory. All that is required is that the subject matter, in this case fluids, be ideal. That is to say, inviscid, irrotational, incompressible, and obeying something called the sine square law. Of course, it turns out that when one makes the algorithm go as beautifully as Euler and Bernoulli did—and to achieve this they had to set out some very clear axioms or definitions by which it could be made a virtual Euclidean instantiation—then a slightly awkward business comes up at the level of the correspondence rules, or the coordinating activities. As it turns out, there are no ideal fluids. In point of fact, the theory doesn't apply to a damned thing, unless one is prepared to bring in lots of additional qualifications at this level. It becomes, then, a sort of general approximation, in terms of which one can have a kind of gestalt for what types of subject matter fluids really are. One must be prepared to muddy up the calculations with the actual values that have to be plugged into the variable terms if one wants to be any good in civil engineering, plumbing, or the designing of boats. If fluid mechanics is going to make any difference to people of that sort, then a great many adjustments have to be effected right at this practical level. In other words, the objective in the ideal language approach is that certain criteria, formal or at least logical, be met in order that the theory should qualify as being well made. I chose fluid mechanics because it seems to me that this is a historical example of a theory which does meet these criteria up to the hilt.

Contrast the ideal language view with a quite different conception, the black box view. This latter can be represented as the type of theory that is very much concerned with the observed values which one is going to plug into the variable terms of the theory and which coordinate with something that actually constitutes experience. Consider some of the developments of the quantum theory in the late 1930's and certainly in the 1940's. I have in mind now quantum electrodynamics. When that theory concerns itself with weak interactions at high energies, it turns out that certain algorithmic techniques, such as those we owe to thinkers like Dyson and Tomonaga, in particular the technique of renormalization, are from a formal or intra-algorithmic point of view quite objectionable. When one undertakes the technique of renormalization to cope with what are sometimes referred to as the divergencies in the fully developed quantum theory of radiation, it turns out that there are an infinite number of solutions to a given wave equation. One selects, from this infinite set, on terms that are completely extraneous mathematically, a finite number for further examination. Moreover, the renormalization technique, as has been pointed out by Heisenberg, Kallen, Hamilton, and many others, has further disastrous consequences for our understanding of what the quantum theory of radiation is all about. The scattering matrices become non-Hermitian; ghost states, negative probabilities, and all sorts of strange things come up. One has to adopt a fairly tough-minded physicist's attitude toward these. It follows from the fact that the algorithm is so badly designed that one never undertakes to suppose that the argument in favor of the acceptance of a black box theory could be anything other than success. This is one of the strong arguments behind what used to be called the Copenhagen position, to wit, that the thing works pretty well. From the very fact that it's such a mess at the algorithmic level, it seems to follow immediately that the subject matter itself couldn't possibly be like that. Therefore, the whole function of the theory is to provide us with a kind of computational or calculational machine by means of which, when one plugs the values in for the variable terms and turns the handle, the answers come out right. That is the argument in favor of it. Thus, it is clear that both the ideal language representation and the black box representation can be articulated in terms of the standard contrast between the calculus, the observation statement level, and the coordinating definitions or correspondence rules.

This way of putting it does help us to understand precisely what is at stake in some important scientific theories. Sooner or later we would have come to understand quantum theory in terms of a representation like this. Indeed, ideal fluid mechanics has often been characterized in the way in which I have just done it. I want to subscribe with Adolf Grünbaum to the view that the business of hunting for an all or nothing general account of the relationship between theories, their algorithmic structure, correspondence rules, and the observation statements does seem difficult. It's not simply difficult in the sense that it's a great challenge. It seems to me that there certainly are considerations in view of which one might wonder whether this is ever a feasible undertaking.

The language chosen by several of the contributors to this volume concerns correspondence rules, and concerns them in such a way as to suggest that there is a sort of ordering. I don't want this matter to turn simply on an issue in the context of discovery, but it does seem to me that when Professor Hempel refers to the antecedent vocabulary, and when one speaks of finding coordinating definitions for a previously existing theory, there is a commitment here which transcends strictly logical or formal considerations. If one is going to coordinate a certain term in the theory with certain sorts of measurements or observations, there must be a theory which is there and upon which one is going to work. It looks as if the procedure is always a directed series of steps proceeding from having ascertained that the theory is well designed to finding coordinating rules, or activities, or procedures by means of which one succeeds in splicing the algorithms together with experience. A standard example is Euclidean geometry as an algorithm, with coordinating definitions by which one takes a technical term like 'point' or 'straight line' and makes it correspond to something that can be observed, a taut string, or a ray of light, or optically flat glass viewed on edge, or any such thing, and transforms it slowly into a physical theory, i.e., geometrical optics, or physical optics. That seems to be the usual order. I think that most discussions of these matters have rested on an implicit assumption that that is the standard order. While this is helpful in the context where we're concerned with the appropriate representation of theories like ideal fluid mechanics or quantum electrodynamics because of what we can do in terms of this representation, it isn't adequate in all cases. Peter Achinstein has remarked that when he requests a general account of the problem concerning meaning of observation terms or theory, he's always confronted, on the part of Paul Feyerabend and myself, with an example. This is a remarkable thing to complain about. The long tradition of philosophy suffered precisely because anytime one was trying to get some sort of understanding of particular subject matter, the account got too general much too quickly. I am quite prepared to say that a general account is not my objective at all. I would be quite happy if I could take particular cases and get some understanding concerning the relationship in that theory or in that science between the observational terms, the so-called coordinating activities, the theoretical terms, and the theoretical structure itself.

This way of representing the matter, with what I have assumed to be an additional progressive arrow built into it, might indeed apply very, very well to certain theories. I think that it's totally inadequate for other theories, and the discipline I am going to discuss here has concerned itself with one of these other sorts of theories. If one were to try to press some of the standard analyses of the structure of scientific theories on this particular scientific discipline, the result would be a very distorted and a very confused analysis of exactly what the conceptual structure of this physical theory actually consists in.

Certainly the contrast between the context of discovery and the context of justification has been pointed up many, many times. Let us grant that what I am about to do doesn't turn primarily on matters of learning theory, or experimental psychology, or sociology, or navel contemplation, or pink pancreatic responses, or whether or not I salivate when a bell is rung. This is not the objective. It is honestly my objective to try to avoid that sort of thing. I would like to consider what might be called the context of strategies, the sorts of reasonable considerations in virtue of which a man's argument might be designed one way rather than another. I want to consider the possibility that in the full accounting of some important scientific disciplines one might find oneself committed to coordinating the observations and the observation statements in a very significant manner long before there is a full-blown theory in the books. Indeed, with respect to the discipline I have in mind, even at this time there isn't such a full-blown theory. This may be said to be simply a factual matter, a matter for the sociologist to consider. Nonetheless, there seem to be important conceptual considerations which follow with respect to our understanding of the actual activity which counts as scientific reasoning, the dealing with perplexing scientific difficulties.

In important cases, formulating a correspondence rule or a coordinating definition is not always representable as finding those terms in the theory which can be coordinated with "terms in experience" or terms in the observation language. A term like 'is coordinated with' refers to a relationship which operates over two terms. It makes perfectly good sense to suppose that this kind of coordinating activity could operate in either direction. The notion of the coordinating activity does not imply that it must always proceed with one end as prior and stable and the other end picking up its semantical content strictly in terms of the first. It might indeed go in the other direction. Please let me say that I have granted that expositions of quantum electrodynamics and other disciplines as well are helped by analyses along the lines of the standard view, but it seems to me that it fails in some cases. All I'm asking for is that philosophers of science take up the examples of the analysis of scientific theory one at a time, and not try to settle the issue all at once.

The discipline with which I am going to concern myself is fluid mechanics, as understood in the last one hundred and sixty-six years. Fluid mechanics, as we know today, had its origins in Book Two of Newton's *Principia*. Newton's strategy there was simply to reduce the concerns of Book Two to the analyses set out in Book One. In Book One of the *Principia* a fairly good geometer's or logician's catechism of punctiform mechanics is set out with great power and great elegance. To a considerable extent, Newton was committed to the belief that any problem within the subject of matter and motion which was a genuine problem was ultimately going to be resolvable in terms of the analysis appropriate to punctiform masses and the "forces" acting on them. Consequently, when in Book Two the problem of a very perplexing, turbulent subject matter, namely fluids, came up for analysis, it seemed quite a natural move for Newton to reduce this complex subject matter to a macro-observational manifestation of what at the fine level could be construed simply as ensembles of point masses in relationship to each other and in relationship also to resultant forces and aggregate motions. Therefore, for Newton, the behavior of fluids was ultimately going to be analyzable as if a fluid was like a hail of very, very fine sand. If one had enough information about the complete state descriptions of each of the constituent particles, one could sum over this and end up with a totally intelligible algorithm applicable to fluids. This was the strategy.

In order for Newton to make this particular move, he had had to make certain assumptions about the nature of the fluid itself, in order that the analytical apparatus of Book One should apply to the domain of Book Two. This is simply the way in which all theoretical scientists undertake to make an apparently untractable subject matter fit the analysis to which they are committed at a given moment. The definitions are the point at which one undertakes to link the subject matter with this particular inferential apparatus. I have already mentioned four of Newton's assumptions about fluids. If one is going to make the move that he undertakes in Book Two, fluids have to be treated as inviscid, irrotational, incompressible, unlike any fluids we ever encounter. Furthermore, they must have the property that if one were to take such a fluid with laminar flow impinging on a plate, then the force exerted by the corpuscles in the fluid will be proportional to the density of the fluid. If the fluid were liquid mercury, there would be a much greater value for this variable than for ordinary water or air. It would be proportional also to the area of the particular resisting plate. A postcard is going to have a different deflectional effect than a barn door. Similarly, the velocity of these particles as they impinge on the side of the plate is of the order of V^2. Finally, the joker in the Newtonian pack is that the force will vary as the sine squared of the particular angle, a. This has come down to us as the infamous sine square law, by virtue of which it follows that birds can't fly.

This type of analysis, as beautifully precisified in the work of Euler and Bernoulli, this fundamental set of convictions, determined for Newton and for the Newtonians precisely what the nature of the deflection of this plate would be as a result of an impact on the underside. This was the same sort of analysis that Leonardo da Vinci applied to the flight of birds when he wrote in the fifteenth century. For him, the entire capacity of a bird to lift itself through this rather evanescent and immaterial medium was what I will call a waterski effect. It is simply a question of the degree to which the medium can be compressed on the underside of the bird's wing; that was what sustained the animal. The conception of a force on the underside of a lifting or liftable plane was then analyzed in terms of punctiform interactions by Newton. Let me just give two examples. If the degree to which an object was deflected in that direction were proportional to the sine squared of the angle a, the angle of inclination, then a Boeing 320 would actually succeed in getting airborne, given its gross weight of a hundred and sixty tons, if and only if each of its wings were the size of 2.3 football fields. A hummingbird would need both of his wings, given the nature of his sustentation, with a wing plan about the size of a table. Now, what follows from this?

The price that one pays for making this algorithm so beautiful, as it was in the Newtonian framework, is simply that it applies to absolutely nothing. This is the case with ideal fluid mechanics. In point of fact, those who in the middle of the nineteenth century and later were concerned with practical problems of fluid mechanics discovered that they could no longer appeal to physics, theoretical or experimental, for any help whatever.

Paul Feyerabend and I have been previously linked together as apparently being in agreement at Pittsburgh in 1965 on the fact that the Wright brothers didn't know what an airplane was. That was not the point. The point was that the Wright brothers, in a very serious sense, didn't know what flight was. Understanding precisely what it is that takes place when a winged object, whether animate or inanimate, gets itself up into this rather thin medium was not something that was possible before 1908. This is just a calendrical fact. Our understanding of what flight consists in has its roots in the work of Lanchester in 1908 on aerodynamics. Whatever made the achievement of practical flight possible was quite independent of anything that could be offered at this particular time by the theory of fluid mechanics. Let me make one slight digression in order to develop a theoretical point which is important for me at this moment. Long before 1908, Stringfellow, Henson, Horatio Phillips, and others undertook to make measurements with primitive wind tunnels on airfoil sections. They did this in a true Keplerian manner with columns and columns of figures concerning the degree of curvature of the plane itself, the velocity, the area, the density of the fluid in question, the planform of the wing, and the angle of inclination. All of these things were set out and made tabloid in the way in which Boyle's data in 1661 would have had to have

137

been presented before the actual generalization which goes under his name was really formulated. One had here a direct encounter at the experiential level; the observation statements were such that they provided these men with the capacity to set out charts of data points.

Suppose that we digress now from airfoil sections to talk about the internal combustion engine. It will very often be useful for the engineer to represent many parameters on precisely the same chart. He might at one time, for example, want to plot the relative brake mean effective pressure delivered by the airscrew, with a given r.p.m., a given mixture setting, and so forth. He can plot other curves as they intersect with it, concerning things like altitude and angles of tack. From the structural relationship between the slopes of these lines, one can get a very acute and intimate understanding of the dynamical processes which operate inside the engine. There is something very important about the relationship of these sloping lines to the processes that one is actually going to find in the subject matter. One knows what it would be like to make a mistake in any one of these data-point representations. There is a mistake when this representation of the structure, concerning this particular parameter, is not in accord with what one actually observes in the complex subject matter. Consequently, it makes sense to say of a representation of data points of this kind that it is veridical. In other words, it is informative concerning the dynamics of the subject matter. In some important sense the structure of this representation is identical with the structure of the dynamical subject matter, or there has got to be a very intimate correlation at the very least.

For example, what one learns is that as time goes on, the relative viscosity of the lubricant will change one way or another, or as the r.p.m. is increased, the relative inefficiency of the valve clearances might make itself more evident. All these things are dynamical influxes and effluxes which can be plotted in terms of data points on this chart. In a very real sense these give one the only insight that matters into what the engine really is—that is, a structural insight. If one were to go to the Pratt and Whitney or the Allison factories and look at a long array of power plants, all of the same design, say the R2800-30W engine, which designates a series of specific performance expectations, it would be idle to distinguish any two of these on the basis that one was rusty and the other was polished, or that one had 'Kilroy' marked on the side, and the other was quite nascent. The important distinction that one makes is that an engine either does measure up to the fundamental design strategy that was built into the thing, or fails altogether in having the structure which could be represented in this form. If it fails altogether in having the structure, quality control might reject it as not being an adequate example of the R2800-30W engine. But what makes any two or any twenty of these engines the same engine, and that is the way they will be sold, is that they conform to the specifications set out on a chart of this kind. If there were a chart which recorded the data points of a given engine, and if there were a corresponding chart which plotted the characteristics which the designer had in mind, these lines would be more or less congruent, within some degree of error. In some sense of 'structure,' the data-point representation and the subject matter itself have the same structure. When one has a line sloping in X-Y space, there will be some corresponding algebraic expression which will do in the A, B, C variable space precisely what the line in the X-Y space does without any loss of power whatsoever.

The distinction to be drawn between the geometry of a data representation and the corresponding algebraic representation is not a distinction which matters at all, from a logical point of view. If there were an argument in virtue of which one could say that the subject matter and the data-point representation had the same structure, then this point must carry over for the algebraic rendition of precisely the same information. One would then have, with respect to any particular subject matter, a large cluster of expressions, all algebraic in form, which do for one's understanding of the subject matter what the data-point graph did.

According to the general theory of functions, given any two algebraic expressions it will always be possible to find some algebraic expression from which both of them can be generated, within certain boundary conditions. At this stage, the operation is largely one of mathematical creativity. That is to say, one will undertake to find fewer and fewer higher order algebraic commitments from the statement of which whole series of lower order algebraic commitments can be shown to follow deductively. These lower order statements will, as I stressed before, in some very important sense, have the same structure as the subject matter. So far we are still moving in the "up" direction with respect to our diagram of the standard view of scientific theories. The great question comes when the individual moves to higher and higher levels of abstraction, which it is his objective to achieve. (Is it the case that what one could have

139

said about the structural relation of charts to engines is the same as what could be said about the relation of high-level theory to engines? If this is the case, then we might conclude that the decisions made at the level of coordinating activity determine the theory.)

My point in saying, as I did at Pittsburgh, that the Wright brothers didn't understand what flight was, at which moment Paul Feyerabend dramatically agreed, was simply that the nature of the series of experimental approximations, the trial and error that was undertaken until 1903, in no way gave anyone any understanding of precisely what the nature of flight was. There were still many attempts to deal with the problem of flight in terms of underside pressure on the wing. It was in 1908, after the practical achievement of flight, that Lanchester came forward with the essential idea.

Newton himself very clearly described an effect which follows the placing of a cylindrical rod transversely in a flowing fluid. Newton noticed that if the rod is given a sense of rotation, and if the flow lines are laminar and impinge in the appropriate way, this cylindrical rod will be deflected. This is the Magnus effect which was experimentally well known by the nineteenth century. This is the only thing which could really account for the remarkable fact that, Newtonian fluid mechanics to the contrary, birds do fly, and the Wright brothers got off the ground. The theory of Lanchester was the following. Consider any one of those shapes, which had been shown to be more effective by the experimental work of Wendham and Horatio Phillips. It was not possible to understand fully what this deflection consisted in, in terms of the waterski effect, or the underside pressure. The actual effect always seemed to be very much greater than what could be calculated in terms of the underside effect. But Lanchester, since he knew about the Magnus effect, argued that it was one of the fundamental functions of a shape of this sort to initiate a movement of this kind in the fluid through which the shape was moving. In other words, it was the function of this shape to set up a circulation around the wing. This is the beginning of what we still call the circulation theory of airfoil lift, or the lifting line. To explain this fully requires a considerable amount of boundary-layer theory, which I can't go into here. But the nature of this interaction of matter molecules on the surface of the cylindrical rod and immediately adjacent to it in the fluid is very important. It required work in the 1920's by Prandtl to explicate it fully. If one can suppose that the wing-air interaction generates something like the Magnus effect, then many other things follow immediately, such things as coming to understand how the Bernoulli effect gets into the picture. How one builds up the starting vortex of Prandtl and von Karman to explain flight follows immediately from Lanchester's fundamental conception. The marriage of what might be called the physicist's conception of a slightly generalized vortex theory with the actual problem of flight was brought about by Lanchester's fundamental conception. This union has controlled the development of theoretical aerodynamics to the present. At the moment we are a long way from having anything like a well-designed Euclidean algorithm, in any sense. Nonetheless, if there's any motion here, it's in the upward direction. It seems to me that this is not simply something to be identified as lying strictly inside the context of discovery. There are conceptual implications of this for the understanding of how it is that in some sciences the coordinating definitions or the correspondence rules may be related to experience in ways slightly different from what we find in other sciences.

On the "Standard Conception" of Scientific Theories

1. Theories: A Preliminary Characterization

Theories, it is generally agreed, are the keys to the scientific understanding of empirical phenomena: to claim that a given kind of phenomenon is scientifically understood is tantamount to saying that science can offer a satisfactory theoretical account of it.[1]

Theories are normally constructed only when prior research in a given field has yielded a body of knowledge that includes empirical generalizations or putative laws concerning the phenomena under study. A theory then aims at providing a deeper understanding by construing those phenomena as manifestations of certain underlying processes governed by laws which account for the uniformities previously studied, and which, as a rule, yield corrections and refinements of the putative laws by means of which those uniformities had been previously characterized.

Prima facie, therefore, the formulation of a theory may be thought of as calling for statements of two kinds; let us call them *internal principles* and *bridge principles* for short. The internal principles serve to characterize the theoretical setting or the "theoretical scenario": they specify the basic entities and processes posited by the theory, as well as the theoretical laws that are assumed to govern them. The bridge principles, on the other hand, indicate the ways in which the scenario is linked to the previously examined phenomena which the theory is intended to explain. This general conception applies equally, I think, to the two types of theory which Nagel, following Rankine, distinguishes in his thorough study of the sub-

---

[1] This essay develops further, and modifies in certain respects, some ideas set forth in an earlier paper, "On the Structure of Scientific Theories," published in *The Isenberg Memorial Lecture Series, 1965–66* (East Lansing: Michigan State University Press, 1969), pp. 11–38. I am indebted to the Michigan State University Press for permission to include some passages from that essay in the present one.

ject,[2] namely, "abstractive" theories, such as the Newtonian theory of gravitation and motion, and "hypothetical" theories, such as the kinetic theory of heat or the undulatory and corpuscular theories of light.

If I and B are the sets of internal and bridge principles by which a theory T is characterized, then T may be represented as the ordered couple of those sets:

(1a) $\quad\quad T = (I, B).$

Or alternatively, and with greater intuitive appeal, T may be construed as the set of logical consequences of the sum of the two sets:

(1b) $\quad\quad T = c(IUB)$

The formulation of the internal principles will typically make use of a *theoretical vocabulary* V_T, i.e., a set of terms not employed in the earlier descriptions of, and generalizations about, the empirical phenomena which T is to explain, but rather introduced specifically to characterize the theoretical scenario and its laws. The bridge principles will evidently contain both the terms of V_T and those of the vocabulary used in formulating the original descriptions of, and generalizations about, the phenomena for which the theory is to account. This vocabulary will thus be available and understood before the introduction of the theory, and its use will be governed by principles which, at least initially, are independent of the theory. Let us refer to it as the *pre-theoretical, or antecedent, vocabulary*, V_A, relative to the theory in question.

The antecedently examined phenomena for which a theory is to account have often been conceived as being described, or at least describable, by means of an observational vocabulary, i.e., a set of terms standing for particular individuals or for general attributes which, under suitable conditions, are accessible to "direct observation" by human observers. But this conception has been found inadequate on several important counts.[3]

The distinction I have suggested between theoretical and antecedent

---

[2] E. Nagel, *The Structure of Science* (New York: Harcourt, Brace and World, 1961), pp. 125–129.

[3] See the following discussions of the subject, which also give further references to the literature: H. Putnam, "What Theories Are Not," in E. Nagel, P. Suppes, A. Tarski, eds., *Logic, Methodology and Philosophy of Science* (Stanford, Calif.: Stanford University Press, 1962), pp. 240–251; R. Jeffrey's comments on this essay in *Journal of Philosophy*, 61 (1964), 80–84; G. Maxwell, "The Ontological Status of Theoretical Entities," in H. Feigl and G. Maxwell, eds., *Minnesota Studies in the Philosophy of Science*, vol. III (Minneapolis: University of Minnesota Press, 1962), pp. 3–27; P. Achinstein, *Concepts of Science* (Baltimore: Johns Hopkins Press, 1968), chapter 5.

vocabularies hinges on no such assumption. The terms of the antecedent vocabulary need not, and indeed should not, generally be conceived as observational in the narrow sense just adumbrated, for the antecedent vocabulary of a given theory will often contain terms which were originally introduced in the context of an earlier theory, and which are not observational in a narrow intuitive sense. Let us look at some examples.

In the classical kinetic theory of gases, the internal principles are assumptions about the gas molecules; they concern their size, their mass, their large number; and they include also various laws, partly taken over from classical mechanics, partly statistical in nature, pertaining to the motions and collisions of the molecules, and to the resulting changes in their momenta and energies. The bridge principles include statements such as that the temperature of a gas is proportional to the mean kinetic energy of its molecules, and that the rates at which different gases diffuse through the walls of a container are proportional to the numbers of molecules of the gases in question and to their average speeds. By means of such bridge principles, certain micro-characteristics of a gas, which belong to the scenario of the kinetic theory, are linked to macroscopic features such as temperature, pressure, and diffusion rate; these can be described, and generalizations concerning them can be formulated, in terms of an antecedently available vocabulary, namely, that of classical thermodynamics. And some of the features in question might well be regarded as rather directly observable or measurable.

Take, on the other hand, the theoretical account that Bohr's early theory of the hydrogen atom provided for certain previously established empirical laws, such as these: the light emitted by glowing hydrogen gas is limited to certain characteristic discrete wavelengths, which correspond to a set of distinct lines in the emission spectrum of hydrogen; these wavelengths conform to certain general mathematical formulas, the first and most famous of which was Balmer's

$$\lambda = b \frac{n^2}{n^2 - 4}$$

Here, b is a numerical constant; and when n is given the values 3, 4, 5, . . . , the formula yields the wavelengths of the lines that form the so-called Balmer series in the spectrum of hydrogen.

Now let us look briefly at the *internal* principles and the *bridge prin-ciples* of the theory by which Bohr explained these and other empirical laws concerning the hydrogen spectrum.

The *internal principles* formulate Bohr's conception that a hydrogen atom consists of a nucleus about which an electron circles in one or another of a set of discrete orbits with radii r_1, r_2, r_3, . . . , where r_i is proportional to i^2; that when the electron is in the ith orbit, it has an energy E_i characteristic of that orbit and proportional to $(-1/r_i)$; that the electron can jump from a narrower to a wider orbit, or vice versa, and that in this process it absorbs or emits an amount of energy that equals the absolute difference between the energies associated with those or-bits.

The *bridge principles*, which connect these goings-on with the optical phenomena to be explained, include statements such as these: (a) the light given off by glowing hydrogen gas results from the emission of energy by those atoms whose electrons happen to be jumping from outer to inner orbits; (b) the energy released by an electron jump from the ith to the jth orbit $(i > j)$ is given off in the form of monochromatic electromag-netic waves with the wavelength $\lambda = (h \cdot c)/(E_i - E_j)$, where h is Planck's constant and c the velocity of light.

As is to be expected, these bridge principles contain, on the one hand, certain theoretical terms such as 'electronic orbit' and 'electron jump,' which were specifically introduced to describe the theoretical scenario; on the other hand, they contain also certain antecedently available terms, such as 'hydrogen gas,' 'spectrum,' 'wavelength of light,' 'velocity of light,' and 'energy.' And clearly at least some of these terms—for example, 'wave-length of light' and 'hydrogen gas'—are not observational terms in the in-tuitive sense mentioned earlier. Nonetheless, the terms are antecedently understood in the sense indicated above; for when Bohr proposed his theory of the hydrogen atom, principles for their use, including principles for the measurement of optical wavelengths, were already available; they were based on antecedent theories, including wave optics.

2. The Construal of Theories as Interpreted Calculi

In the analytic philosophy of science, theories have usually been characterized in a manner rather different from the one just outlined; and, at least until recently, this characterization was so widely accepted that it could count as the "standard," or the "received," philosophical construal of scientific theories.[4] On this construal, too, a theory is characterized by two constituents, which, moreover, have certain clear affinities to what were called above its internal principles and its bridge principles.

The first constituent is an axiomatized deductive system—sometimes referred to as a calculus—of uninterpreted formulas, the postulates of the system corresponding to the basic principles of the theory. Thus, roughly speaking, the postulates of the calculus may be thought of as formulas obtained by axiomatizing the internal principles of the theory and then replacing the primitive theoretical terms in the axioms by variables or by dummy constants.

The second component is a set of sentences that give empirical import or applicability to the calculus by interpreting some of its formulas in empirical terms—namely, in terms of the vocabulary that serves to describe the phenomena which the theory is to explain. These sentences, which evidently are akin to the bridge principles mentioned above, were characterized by Campbell and by Ramsey as forming a "dictionary" that relates the theoretical terms to pre-theoretical ones;[5] other writers have referred to them as "operational definitions" or "coordinative definitions" for the theoretical terms, as "rules of correspondence," or as "interpretative principles."

The standard conception, then, may be schematized by representing a theory as an ordered couple of sets of sentences:

(2) $T = (C, R)$

<hr>

[4] The appellation "the 'received view' on the role of theories" is Putnam's ("What Theories Are Not," p. 240). Some characteristic stages in the evolution of this construal of scientific theories are represented by the following works: N. R. Campbell, *Physics: The Elements* (Cambridge: Cambridge University Press, 1920; reprinted as *Foundations of Science*, New York: Dover, 1957), chapter 5; F. P. Ramsey, "Theories" (1929), in Ramsey, *The Foundations of Mathematics* (London: Routledge and Kegan Paul, 1931); R. Carnap, *Foundations of Logic and Mathematics* (Chicago: University of Chicago Press, 1939), especially sections 21–25; R. B. Braithwaite, *Scientific Explanation* (Cambridge: Cambridge University Press, 1953), chapters I–III; R. Carnap, "The Methodological Character of Theoretical Concepts," in H. Feigl and M. Scriven, eds., *Minnesota Studies in the Philosophy of Science*, vol. I (Minneapolis: University of Minnesota Press, 1956), pp. 38–76; R. Carnap, *Philosophical Foundations of Physics*, ed. M. Gardner (New York, London: Basic Books, 1966), part V; Nagel, *The Structure of Science*, chapters 5 and 6.

[5] Campbell, *Foundations of Science*, p. 122; Ramsey, "Theories," p. 215.

where C is the set of formulas of the calculus and R the set of correspondence rules.

Whereas the bridge principles invoked in our initial characterization of a theory are conceived as a subset of the class of sentences asserted by the theory, the status of the correspondence rules in the standard construal is less clear. One plausible construal of them would be as terminological rules belonging to the metalanguage of the theory, which stipulate the truth, by definition or more general terminological convention, of certain sentences (in the language of the theory) that contain both theoretical and pre-theoretical terms. For this reason, no immediate analogue to (1b) is available as an alternative schematization of the standard view. The status of the correspondence rules will be examined further in section 6 below.

It would be a task of interest both for the history and for the philosophy of science to locate the origins of the standard conception and to trace its development in some detail. Such a study would surely have to take account of Reichenbach's characterization of physical geometry (i.e., the theory of the geometrical structure of physical space) as an abstract, uninterpreted system of "pure" or mathematical geometry, supplemented by a set of coordinative definitions for the primitives,[6] and it would have to consider Poincaré's and Einstein's views on the geometrical structure of physical space.

Campbell and some other proponents of the standard conception make provision for a third constituent of a theory—Campbell calls it an analogy, others (Nagel among them) call it a model—which is said to characterize the basic ideas of the theory by means of concepts with which we are antecedently acquainted, and which are governed by familiar empirical laws that have the same form as some of the basic principles of the theory. The role of models in this sense will be considered later; until then, the standard conception of theories will be understood in the sense of schema (2). I have myself relied on the standard construal in several earlier studies,[7] but I have now come to consider it misleading in certain philosophically significant respects, which I will try to indicate in the following sections.

[6] This idea is set forth very explicitly in chapter 8 of H. Reichenbach, *The Rise of Scientific Philosophy* (Berkeley and Los Angeles: University of California Press, 1951).

[7] For example, in my essay "The Theoretician's Dilemma," in H. Feigl, M. Scriven, and G. Maxwell, eds., *Minnesota Studies in the Philosophy of Science*, vol. II (Minneapolis: University of Minnesota Press, 1958), pp. 37–98.

Carl G. Hempel

3. The Role of an Axiomatized Calculus in the Formulation of a Theory

My misgivings do not concern the obvious fact that theories as actually stated and used by scientists are almost never formulated in accordance with the standard schema; nor do they stem from the thought that a standard formulation could at best represent a theory quick-frozen, as it were, at one momentary stage of what is in fact a continually developing system of ideas. These observations represent no telling criticisms, I think, for the standard construal was never claimed to provide a descriptive account of the actual formulation and use of theories by scientists in the ongoing process of scientific inquiry; it was intended, rather, as a schematic explication that would clearly exhibit certain logical and epistemological characteristics of scientific theories.

This defense of the standard conception, however, naturally suggests this question: What are the logical and epistemological characteristics of theories that schema (2) serves to exhibit and illuminate? Let us consider in turn the various features which the schema attributes to a theory, beginning with the axiomatized calculus.

What is to be said in support of assuming axiomatization? It might quite plausibly be argued that an axiomatic exposition is an indispensable device for an unambiguous statement of a theory. For a theory has to be conceived as asserting a set of sentences that is closed under the relation of logical consequence in the sense that it contains all logical consequences (expressible in the language of the theory) of any of its subsets. A theory will therefore amount to an infinite set of sentences. In order to specify unambiguously the infinite set of sentences that a proposed theory is intended to assert, it will be necessary to provide a general criterion determining, for any sentence S, whether S is asserted by the theory. Axiomatization yields such a criterion: S is asserted by the theory just in case S is deducible from the specified axioms, or postulates.

This criterion determines membership in the intended set of sentences unambiguously, but it does not provide us with a general method of actually finding out whether a given sentence belongs to the set; for in general there is no effective decision procedure which, for any given sentence S, determines in a finite number of steps whether S is deducible from the axioms. But in any event, the standard construal assumes axiomatization only for the formulas of the uninterpreted calculus C rather than for all

148

the sentences asserted by T,[8] so that the proposed supporting argument does not actually apply here.

One of the attractions the standard construal has had for philosophers lies no doubt in its apparent ability to offer neat solutions to philosophical problems concerning the meaning and the reference of theoretical expressions. If the characteristic vocabulary of a theory represents "new" concepts, not previously employed, and designed specifically to describe the theoretical scenario, then it seems reasonable, and indeed philosophically important, to inquire how their meanings are specified. For if they should have no clearly determined meanings, then, it seems, neither do the theoretical principles in which they are invoked; and in that case, it would make no sense to ask whether those principles are true or false, whether events of the sort called for by the theoretical scenario do actually occur, and so forth.[9] The answer that the standard construal is often taken to offer is, broadly speaking, that the meanings of theoretical terms are determined in part by the postulates of the calculus, which serve as "implicit definitions" for them; and in part by the correspondence rules, which provide them with empirical content. But this conception is open to various questions, some of which will be raised as we proceed.

As for the merits of axiomatization, its enormous significance for logic and mathematics and their metatheories needs no acknowledgment here. In some instances, axiomatic studies have served also to shed light on philosophical problems concerning theories in empirical science. One interesting example is Reichenbach's axiomatically oriented, though not strictly formalized, analysis of the basis and structure of the theory of relativity.[10] This analysis, which was undertaken some forty years ago, is technically distinctly inferior to more recent rigorous axiomatic formali-

[8] Note by contrast that in the investigations by Ramsey (in "Theories") and by Craig concerning the avoidability of theoretical terms in favor of pre-theoretical ones, axiomatization of the entire theory is presupposed. For a careful exposition and appraisal of those investigations, see I. Scheffler, The Anatomy of Inquiry (New York: Knopf, 1963), pp. 193–222. On Ramsey's method, see also Carnap, Philosophical Foundations of Physics, chapter 26; I. Scheffler, "Reflections on the Ramsey Method," Journal of Philosophy, 65 (1968), 269–274; and H. Bohnert, "In Defense of Ramsey's Elimination Method," Journal of Philosophy, 65 (1968), 275–281.

[9] Various facets of this problem are carefully presented and explored in Nagel, The Structure of Science, chapter 6, and in Scheffler, The Anatomy of Inquiry, part II.

[10] See H. Reichenbach, Axiomatik der relativistischen Raum-Zeit-Lehre (Braunschweig: Friedrich Vieweg und Sohn, 1924); and also Reichenbach's article, "Ueber die physikalischen Konsequenzen der relativistischen Axiomatik," Zeitschrift für Physik, 34 (1925), 32–48. In this article, to which Professor A. Grünbaum kindly called my attention, Reichenbach sets forth the main objectives of his axiomatic

zations; but it was nonetheless philosophically stimulating and illuminating, for it sought to clarify—much in the spirit of Einstein, I think—the roles of experience and convention in physical theorizing about space, time, and motion and the physical basis of the relativistic theory of spatial and temporal distances, of simultaneity, and so forth. More fundamentally, Reichenbach's investigations were intended as a critique, based on a specific case study, of the Kantian notion of a priori knowledge. Again, an axiomatic approach played an important role in von Neumann's argument[11] that it is impossible to supplement the formalism of quantum mechanics by the introduction by hidden parameters in a way that yields a deterministic theory.

Some contemporary logicians and philosophers of science consider the axiomatization of scientific theories so important for the purposes of both science and philosophy that they have expended much effort and shown remarkable ingenuity in actually constructing such axiomatic formulations. Some of these, such as those developed by Kyburg,[12] are small and relatively simple fragments of scientific theories in first-order logic; others, especially those constructed by Suppes and his associates, deal with richer, quantitative theories and formalize these with the more powerful apparatus of set theory and mathematical analysis.[13]

But some of the claims that have been made in support of axiomatizing scientific theories are, I think, open to question. For example, Suppes has argued that formalizing and axiomatizing scientific concepts and theories is "a primary method of philosophical analysis," and thus helps to "clarify conceptual problems and to make explicit the foundational assumptions of each scientific discipline," and that to "formalize a connected family of concepts is one way of bringing out their meaning in an explicit fashion."[14]

In what sense can an uninterpreted axiomatization be said to "bring out the meanings" of the primitive terms? The postulates of a formalized

---

efforts; on pp. 37–38, he rejects as irrelevant to his enterprise Hermann Weyl's objection that Reichenbach's axiomatization is too complicated and opaque from a purely mathematical point of view.

[11] J. von Neumann, *Mathematical Foundations of Quantum Mechanics* (Princeton, N.J.: Princeton University Press, 1955), chapter 4.

[12] H. E. Kyburg, *Philosophy of Science: A Formal Approach* (New York: Macmillan, 1968).

[13] A lucid and copiously illustrated introduction to this method of axiomatization by definition of a set-theoretical predicate is given in chapter 12 of P. Suppes, *Introduction to Logic* (New York: Van Nostrand, 1957).

[14] P. Suppes, "The Desirability of Formalization in Science," *Journal of Philosophy*, 65 (1968), 651–664; quotations from pp. 653 and 654.

theory are often said to constitute "implicit definitions" of the primitives, requiring the latter to stand for kinds of entities and relations which jointly satisfy the postulates. If axiomatization is to be viewed as somehow defining the primitives, then it is logically more satisfactory to construe axiomatization, with Suppes, as yielding an explicit definition of a higher order set-theoretical predicate. In either case, the formalized theory is then viewed in effect as dealing with just such kinds of entities and relations as make the postulates true.[15]

This construal may have some plausibility for axiomatized purely mathematical theories—Hilbert adopted it in regard to his axiomatization of Euclidean geometry—but it is not plausible at all to hold that the primitive terms of an axiomatized theory in empirical science must be understood to stand for entities and attributes of which the postulates, and hence also the theorems, are true; for on this construal, the truth of the axiomatized theory would be guaranteed a priori, without any need for empirical study.[16]

There are indeed cases in which axiomatization may be said to have contributed very significantly to the analytic clarification of a system of concepts. Suppes rightly mentions Kolmogorov's axiomatization of probability theory as an outstanding example.[17] But it should be noted that Kolmogorov's formal system admits of such diverse interpretations as Carnap's logical or inductive probability, Savage's personal probability, and the empirical construal of probability in terms of long-run relative frequencies. The latter, of central importance in empirical science, has presented vexing difficulties to philosophical efforts at a satisfactory explication. Von Mises, Reichenbach, Popper, Braithwaite, and others all have sought to explicate the concept of statistical probability, or to specify the

---

[15] For a careful and illuminating critical examination of the construal of postulates as implicit definitions for the primitives see chapter II of R. Grandy, "On Formalization and Formalistic Philosophies of Mathematics" (doctoral dissertation, Princeton University, 1967). Concerning the restrictions that the requirement of truth for the postulates imposes on the permissible interpretations of the primitives, Grandy notes that it "is not a restriction on the constants alone but on the set of constants plus the universe of discourse. A paraphrase of this is: The postulates implicitly define, if anything, the constants plus the quantifiers" (p. 41).

[16] Kyburg therefore divides the axioms of a theory into "material axioms" and meaning postulates (in the sense of Carnap and Kemeny) and stresses that "we cannot lump [these] together and regard them as an *implicit definition* of the terms that occur in them" (*Philosophy of Science*, p. 124). It is presumably the meaning postulates alone that provide implicit definitions; but the distinction of two kinds of axioms is beset by the same difficulties as the analytic-synthetic distinction.

[17] Suppes, "The Desirability of Formalization in Science," p. 654.

principles governing its scientific use. Some of these principles concern the pure calculus of probability, with which alone Kolmogorov's axiomatization is concerned; others—and indeed the philosophically most perplexing ones—concern its application. And Kolmogorov's analysis does not touch at all on this second part of the problem of "bringing out the meaning" of the term 'probability' "in an explicit fashion."[18]

Generally speaking, the formalization of the internal principles as a calculus sheds no light on what in the standard construal is viewed as its interpretation; it sheds light at best on part of the scientific theory in question. And as for the claim that formalization makes explicit the foundational assumptions of the scientific discipline concerned, it should be borne in mind that axiomatization is basically an expository device, determining a set of sentences and exhibiting their logical relationships, but not their epistemic grounds and connections. A scientific theory admits of many different axiomatizations, and the postulates chosen in a particular one need not, therefore, correspond to what in some more substantial sense might count as the basic assumptions of the theory; nor need the terms chosen as primitive in a given axiomatization represent what on epistemological or other grounds might qualify as the basic concepts of the theory; nor need the formal definitions of other theoretical terms by means of the chosen primitives correspond to statements which in science would be regarded as definitionally true and thus analytic. In an axiomatization of Newtonian mechanics, the second law of motion can be given the status of a definition, a postulate, or a theorem, as one pleases; but the role it is thus assigned within the axiomatized system does not indicate whether in its scientific use it functions as a definitional truth, as a basic theoretical law, or as a derivative one (if indeed it may be said to have just one of these functions).

Hence, whatever philosophical illumination may be obtainable by presenting a theory in axiomatized form will come only from axiomatization of some particular and appropriate kind rather than just any axiomatization or even a formally especially economic and elegant one.

[18] Suppes himself acknowledges that the "difficulty with the purely set-theoretical characterization of Kolmogorov is that the concept of probability is not sufficiently categorical" (ibid.), and he stresses that the interpretation of a formalized theory is logically much more complex than the talk of correspondence rules in "the standard sketch of scientific theories" would suggest (P. Suppes, "What Is a Scientific Theory?" in S. Morgenbesser, ed., *Philosophy of Science Today*, New York and London: Basic Books, 1967, pp. 55–67).

4. The Role of Pre-Theoretical Concepts in Internal Principles

The assumption, in the standard construal, of an axiomatized uninterpreted calculus as a constituent of a theory seems to me, moreover, to obscure certain important characteristics shared by many scientific theories. For that assumption suggests that the basic principles of a theory—those corresponding to the calculus—are formulated exclusively by means of a "new" theoretical vocabulary, whose terms would be replaced by variables or by dummy constants in the axiomatized calculus C. In this case, the conjunction of the postulates of C would be an expression of the type $\phi(t_1, t_2, \ldots, t_n)$, formed from the theoretical terms by means of logical symbols alone. Actually, however, the internal principles of most scientific theories employ not only "new" theoretical concepts but also "old," or pre-theoretical, ones that are characterized in terms of the antecedent vocabulary. For the theoretical scenario is normally described in part by means of terms that have a use, and are understood, prior to, and independently of, the introduction of the theory. For example, the basic assumptions of the classical kinetic theory of gases attribute to atoms and molecules such characteristics as masses, volumes, velocities, momenta, and kinetic energies, which have been dealt with already in the antecedent study of macroscopic objects; the wave theory of light uses such antecedently available concepts as those of wavelength and wave frequency; and so forth. Thus, the internal principles of a theory—and hence also the corresponding calculus C—have to be viewed, in general, as containing pre-theoretical terms in addition to those of the theoretical vocabulary. Accordingly, the conjoined postulates of C would form an expression of the type $\psi(t_1, t_2, \ldots, t_k, p_1, p_2, \ldots, p_m)$, where the t's again correspond to "new" theoretical terms, while the p's are pre-theoretical, previously understood ones. Consequently, the theoretical calculus that the standard conception associates with a theory is not, as a rule, a totally uninterpreted system containing, apart from logical and mathematical symbols, only new theoretical terms.

It might be objected, from the vantage point of a narrow operationism, that in this new context, the "old" terms $p_1, p_2, \ldots, p_m$ represent new concepts, quite different from those they signify in their pre-theoretical employment. For the use of such terms as 'mass,' 'velocity,' and 'energy' in reference to atoms or subatomic particles requires entirely new operational criteria of application, since at the atomic and subatomic levels the quantities in question cannot be measured by means of scales, electrometers, and the like, which afford operational criteria for their measurement at the pre-theoretical level of macroscopic objects. On the strict operationist maxim that different criteria of application determine different concepts, we would thus have to conclude that, when used in internal principles, the terms $p_1, p_2, \ldots, p_m$ stand for new concepts, and that it is therefore improper to use the old pre-theoretical terms in theoretical contexts: that they should be replaced here by appropriate new terms, which, along with $t_1, t_2, \ldots, t_k$, would then belong to the theoretical vocabulary.

But differences in operational criteria of application, as is well known, cannot generally be regarded as indicative of differences in the concepts concerned; otherwise, it would have to be held impossible to measure "one and the same quantity" in a particular instance—such as the temperature or the density of a given body of gas—by different methods, or even with different instruments of like construction; as a consequence, the diversity of methods of measuring a quantity, already at the macroscopic level, would call for a self-defeating endless proliferation and distinction of concepts of temperature, of concepts of density, and so forth.

Moreover, as long as we allow ourselves to use the notoriously vague and elusive notion of meaning, we will have to regard the meanings of scientific terms as reflected not only in their operational criteria of application, but also in some of the laws or theoretical principles in which they function. And in this context, it seems significant to note that some of the most basic principles that govern the pre-theoretical use (relative to the classical kinetic theory, let us say) of such terms as 'mass,' 'velocity,' and 'energy' are carried over into their theoretical use. Thus, in the classical kinetic theory, mass is taken to be additive in the sense that the mass of several particles taken jointly equals the sum of the masses of the constituents, exactly as for macroscopic bodies. Similarly, the conservation laws for mass, energy, and momentum and the laws of motion are—at least initially—carried over from the pre-theoretical to the theoretical level.

In fact, the principle of additivity of mass is here used not only as a pre-theoretical and as an internal theoretical principle, but also as a bridge principle. In the latter role, it implies, for example, that the mass of a body of gas equals the sum of the masses of its constituent molecules; it thus connects certain features of the theoretical scenario with corresponding features of macroscopic systems that can be described in pre-theoretical terms. Those different roles of the additivity principle are clearly presupposed in the explanation of the laws of constant and of multiple proportions, and in certain methods of determining Avogadro's number. These considerations suggest that the term 'mass' and others can hardly be taken to stand for quite different concepts, depending on whether they are applied to macroscopic objects or to atoms and molecules.

In support of the same point, it might be argued also that classical mechanics imposes no lower bounds on the size or the mass of the bodies to which the concepts of mass, velocity, kinetic energy, etc., can be significantly applied, and the laws governing these concepts are subject to no such restrictions either.[19] This suggests a further response to the operationist objection considered a moment ago: the application of classical mechanical principles indicates that macroscopic methods using mechanical precision scales, etc., are not sufficiently sensitive for weighing atoms, but that certain indirect procedures will provide operational means for determining their masses. Accordingly, the need for different methods of measurement indicates, not a conceptual difference in the meanings of the word 'mass' as used in the two contexts, but a large substantive difference in mass between the objects concerned.

Analogous arguments, however, are not applicable in every case where pre-theoretical terms are used in the formulation of theoretical principles. According to current theory, for example, the mass of an atomic nucleus is less than the sum of the masses of its constituent protons and neutrons; thus the principles of additivity—and of conservation—of mass are abandoned at the subatomic level. Are we to say that this "theoretical change" indicates a change in the meaning of the term 'mass,' or rather that there has been a change in certain previously well-entrenched general laws which, before the advent of the new theory, had been erroneously believed to hold true of that one quantity, mass, to which both the new theory and the earlier one refer?

This question has received much attention in recent years in the debate over the ideas of Feyerabend, Kuhn, and some others concerning theoretical change in science and the theory-dependence of the meanings of scientific terms.[20] As the debate has shown, however, a satisfactory reso-

---

[19] This point is made also by Achinstein, *Concepts of Science*, p. 114; indeed his discussion, on pp. 106–119, of ways in which theoretical terms are introduced in science presents many illuminating observations and illustrations which accord well with the view expressed in this section, and which lend further support to it.

[20] See, for example, T. S. Kuhn, *The Structure of Scientific Revolutions* (Chicago:

lution of the issue would require a more adequate theory of the notion of sameness of meaning than seems yet to be at hand.

5. The Role of a Model in the Specification of a Theory

As mentioned earlier, some adherents of the standard construal regard a theory as having a third component, in addition to the calculus and the rules of correspondence, namely, "a model for the abstract calculus, which supplies some flesh for the skeletal structure in terms of more or less familiar conceptual or visualizable materials."[21]

In Bohr's theory of the hydrogen atom, for example, the postulates of the calculus would be the basic mathematical equations of the theory, expressed in terms of uninterpreted symbols such as 'i,' 'r_i,' 'E_i.' The model specifies the conception, referred to earlier, of a hydrogen atom as consisting of a nucleus circled by an electron in one or another of the orbits available to it, etc. In this model, 'r_i' is interpreted as the radius of the ith orbit, 'E_i' as the energy of the electron when in the ith orbit, etc. The correspondence rules, finally, link the theoretical notion of energy emission associated with an orbital jump to the experimental concept of corresponding wavelengths or spectral lines, and they establish other linkages of this kind.

In discussing these three components of Bohr's theory, Nagel remarks that as a rule the theory is embedded in a model rather than being formulated simply as an abstract calculus and a set of correspondence rules because, among other reasons, the theory can then be understood with greater ease than the inevitably more complex formal exposition.[22] It seems,

University of Chicago Press, 1962); P. K. Feyerabend, "Explanation, Reduction, and Empiricism," in Feigl and Maxwell, eds., *Minnesota Studies in the Philosophy of Science*, vol. III, pp. 28–97; P. K. Feyerabend, "Reply to Criticism," in R. S. Cohen and M. W. Wartofsky, eds., *Boston Studies in the Philosophy of Science*, vol. II (New York: Humanities, 1965), pp. 223–261; N. R. Hanson, *Patterns of Discovery* (Cambridge: Cambridge University Press, 1958). For illuminating critical and constructive discussions of these ideas, and for further bibliographic references, see P. Achinstein, "On the Meaning of Scientific Terms," *Journal of Philosophy*, 61 (1964), 497–510; Achinstein, *Concepts of Science*, pp. 91–105; H. Putnam, "How Not to Talk about Meaning," in Cohen and Wartofsky, eds., *Boston Studies in the Philosophy of Science*, II, 205–222; D. Shapere, "Meaning and Scientific Change," in R. G. Colodny, ed., *Mind and Cosmos* (Pittsburgh: University of Pittsburgh Press, 1966), pp. 41–85; I. Scheffler, *Science and Subjectivity* (Indianapolis: Bobbs-Merrill, 1967), especially chapters 1, 3, 4.

[21] Nagel, *The Structure of Science*, p. 90.

[22] *Ibid.*, p. 95. Nagel's detailed discussion of the subject (chapter 5 and pp. 107–117) calls attention also to other functions of models in his sense, among them their heuristic role.

however, that in some cases the significance of models in Nagel's sense goes further than this, as I will try to indicate briefly.

The term 'model' has been used in several different senses in the philosophy of science. One of these pertains to what might be called analogical models, such as the mechanical or hydrodynamic representations of electric currents or of the luminiferous ether that played a considerable role in the physics of the late nineteenth and early twentieth centuries. Models of this kind clearly are not intended to represent the actual microstructure of the modeled phenomena. They carry an implicit 'as if' clause with them; thus, electric currents behave in certain respects as if they consisted in the flow of a liquid through pipes of various widths and under various pressures; the analogy lies in the fact that phenomena of the two different kinds are governed by certain laws that have the same mathematical form. Analogical models may be of considerable heuristic value; they may make it easier to grasp a new theory, and they may suggest possible implications and even promising extensions of it; but they add nothing to the content of the theory and are, thus, logically dispensable.

But this verdict does not seem to me to apply to what Nagel would call the models implicit in such theories as the kinetic theory of gases, the classical wave and particle theories of light, Bohr's theory of the hydrogen atom, the molecular-lattice theory of crystal structure, or recent theories of the molecular structure of genes and the basis of the genetic code. All these claim to offer, not analogies, but tentative descriptions of the actual microstructure of the objects and processes under study. Gases are claimed actually to consist of molecules moving about and colliding at various high speeds, atoms are claimed to have certain subatomic constituents, and so forth. To be sure, these claims, like those of any other scientific hypothesis, may subsequently be modified or discarded; but they form an integral part of the theory. For example, as I suggested earlier, if a model in Nagel's sense characterizes certain theoretical variables as masses, velocities, energies, and the like, this may be taken to indicate that certain laws which are characteristic of masses, velocities, and energies apply to those variables, and that, if some of those laws are suspended in the theory, the requisite modifications will be made explicit. This happened, for example, in Bohr's model, where—in contrast to classical electromagnetic theory— an orbiting electron is assumed to radiate no energy. Hence, the specification of the model determines in part what consequences may be derived from the theory and, hence, what the theory can explain or predict.

More specifically, it seems that when a scientific theory is axiomatized, the process is limited to the mathematical connections that the theory assumes between quantitative features of the scenario; other theoretically relevant aspects of the scenario are specified by means of a model. I therefore agree with Sellars who remarks in a very similar vein that "in actual practice . . . the conceptual texture of theoretical terms in scientific use is far richer and more finely grained than the texture generated by the explicitly listed postulates," and that, in particular, the "thingish or quasi-thingish character of theoretical objects, their conditions of identity . . . are some of the more familiar categorial features conveyed by the use of models and analogies."[23] Thus, a model in the sense here considered is not only of didactic and heuristic value: The statements specifying the model seem to me to form part of the internal principles of a theory and as such to play a systematic role in its formulation.

It must be acknowledged that this way of formulating part of the internal principles of a theory is not fully specific and precise, that it does not provide an unequivocal characterization of exactly what statements the theory is meant to assert. But axiomatization, in the form of a "calculus," of part of a theory does not satisfy this desideratum, either; for it does not cover the correspondence rules; and for these, too, it seems virtually impossible to provide a formulation that could be regarded as adequate and complete. Indeed, as Nagel remarks, "theories in the sciences . . . are generally formulated with painstaking care and . . . the relations of theoretical notions to each other . . . are stated with great precision. Such care and precision are essential if the deductive consequences of theoretical assumptions are to be rigorously explored. On the other hand, rules of correspondence for connecting theoretical with experimental ideas generally receive no explicit formulation; and in actual practice the coordinations are comparatively loose and imprecise."[24]

6. The Status of Correspondence Rules

In the standard construal, schematized by (2) above, R is conceived as a class of sentences that assign empirical content to the expressions of the calculus; and their designation as operational *definitions*, coordinative

---

[23] W. Sellars, "Scientific Realism or Irenic Instrumentalism," in Cohen and Wartofsky, eds., *Boston Studies in the Philosophy of Science*, II, 171–204; quotations from pp. 178–179.

[24] Nagel, *The Structure of Science*, p. 99.

definitions, or *rules* of correspondence conveys the suggestion that they have the status of metalinguistic principles which render certain sentences true by terminological convention or legislation. The sentences thus declared true—let us call them interpretative sentences—would belong to an object language containing both the calculus and the pre-theoretical terms employed in its interpretation. The theoretical terms in the calculus are then best thought of as "new" constants that are being introduced into the object language by means of the correspondence rules for the purpose of formulating the theory. The interpretative sentences might have the form of explicit definition sentences (biconditionals or identities) for theoretical terms, or they might be of a more general type, providing only a partial specification of meaning for theoretical sentences, perhaps in the manner of Carnap's reduction sentences or still more flexible devices.[25] But at any rate they would be sentences whose truth is guaranteed by the correspondence rules.

But such a conception of correspondence rules is untenable for several reasons, among them the following:

First, scientific statements that are initially introduced by "operational definitions" or more general rules of application for scientific terms—such as the statements characterizing length by reference to measurement with a standard rod, or temperature in terms of thermometer readings—usually change their status in response to new empirical findings and theoretical developments. They come to be regarded as statements which are simply false in their original generality, though perhaps very nearly true within a restricted range of application, and possibly only under additional precautionary conditions. Most sentences warranted by operational definitions or criteria of application are eventually qualified as, strictly speaking, false by the very theories in whose development they played a significant role. Much the same point is illustrated by the following example: To "define" in experimental terms equal intervals of time, some periodic process may be chosen to serve as a standard clock, such as the swinging of a pendulum or the axial rotation of the earth as reflected in the periodic, apparent daily motion of some fixed star. The time intervals marked off by the chosen process are then equal by convention or stipulation. Yet it may happen that certain laws or theoretical principles originally based on evidence that includes the readings of standard clocks give rise to the

---

[25] Such, perhaps, as interpretative systems of the kind I suggested in section 8 of "The Theoretician's Dilemma."

verdict that those clocks do not mark off strictly equal time intervals. One striking example is the use of ancient astronomical reports of an almost purely qualitative character—concerning the date and very roughly the time of day when a certain total solar eclipse was observed at a given place —to establish a very slow deceleration of the earth's axial rotation, with a consequent slow lengthening of the mean solar day (by no more than .003 seconds in a century).[26]

Thus, even though a sentence may originally be introduced as true by stipulation, it soon joins the club of all other member statements of the theory and becomes subject to revision in response to further empirical findings and theoretical developments. As Quine has said, "conventionality is a passing trait, significant at the moving front of science, but useless in classifying the sentences behind the lines."[27]

These considerations might invite the following reply: Of course a theory—including its correspondence rules—may well undergo changes in response to new empirical findings; the question at issue, however, does not concern the possible effects of scientific change on correspondence rules, but rather the epistemic status of the interpretative sentences of a given theory, "frozen," as it were, at a particular point of its development. If such a theory is systematically characterized by means of a calculus and a set of interpretative sentences, do not the latter have the character of terminological conventions?

Here it should be recalled, first of all, that a theory usually links a given theoretical concept to several distinct kinds of phenomena that are characterizable in terms of the antecedently available vocabulary. For example, contemporary physical theory provides for several different ways of determining Avogadro's number or the charge of an electron or the velocity of light. But not all the interpretative sentences thus provided for a given theoretical term can be true by convention; for they imply statements to the effect that if one of the specified methods yields a certain numerical value for the quantity in question, then the alternative methods will yield the same value, and whether this is in fact the case is surely an empirical matter and cannot be settled by terminological convention. This point has indeed been stressed by some proponents of the standard conception.

[26] See N. Feather, *Mass, Length and Time* (Baltimore: Penguin, 1961), pp. 54–55.
[27] W. V. O. Quine, "Carnap and Logical Truth," reprinted in Quine, *The Ways of Paradox and Other Essays* (New York: Random House, 1966), pp. 100–125; quotation from p. 112.

Thus, Carnap pointed out in his theory of reduction sentences that when a term is introduced (or interpreted, as we might say) by means of several reduction sentences, the latter taken jointly normally have empirical implications.[28]

Moreover, it is not clear just what claim is being made, in the context of a systematic exposition of a theory, by qualifying certain of its sentences as "true by convention." As was noted above, such designation may serve to make a historical point about the way in which those sentences came first to be admitted into the theory—but that is of no significance for a systematic characterization of the theory.

Nor, despite initial plausibility, can it be said that to qualify a sentence as true by rule or convention is to mark it as immune to revision in the eventuality that the theory should encounter adverse evidence. For, with the possible exception of the truths of logic and mathematics, no statement enjoys this kind of absolute immunity, as is illustrated by our preceding considerations, and as has been made very clear especially by Quine's critique of the analytic-synthetic distinction.[29]

The concept of bridge principle as invoked in our initial characterization of theories does not presuppose the analytic-synthetic distinction and treats the bridge principles as part of the theory, on a par with its internal principles. In fact, it should be explicitly acknowledged now that no precise criterion has been provided for distinguishing internal principles from bridge principles. In particular, the dividing line cannot be characterized syntactically, by reference to the constituent terms; for, as has been noted, both internal principles and bridge principles contain theoretical as well as antecedently available terms. Nor is the difference one of epistemic status, such as truth by convention versus empirical truth. The distinction is, thus, admittedly vague. But no sharp dividing line was needed for the use here made of the intuitive construal (1), namely, as a vantage point for a critical scrutiny of the standard conception.

[28] Cf. R. Carnap, "Testability and Meaning," *Philosophy of Science*, 3 (1936), 419–471, and 4 (1937), pp. 1–40, especially pp. 444 and 451. An analogous comment applies to interpretative systems, of course; see my "The Theoretician's Dilemma," p. 74.

[29] Perhaps Quine's earliest detailed attack on the distinction is mounted in his classical "Two Dogmas of Empiricism" (1951), reprinted in W. V. O. Quine, *From a Logical Point of View*, 2nd ed. (Cambridge, Mass.: Harvard University Press, 1961). Another early critique is given in M. G. White, "The Analytic and the Synthetic: An Untenable Dualism," in S. Hook, ed., *John Dewey: Philosopher of Science and of Freedom* (New York: Dial, 1950).

Carl G. Hempel

7. On "Specifying the Meanings" of Theoretical Terms

Our critical scrutiny, however, has suggested no solution to one central question which the standard construal sought to answer, namely the question of how the meanings of the "new" terms in a theory are specified. We found difficulties both with the conception that the postulates of the uninterpreted calculus provide implicit definitions for the theoretical terms and with the idea of correspondence rules as principles of empirical interpretation; but no alternative answer to the question has been offered. I believe now that the presumptive problem "does not exist," as Putnam has said and argued,[30] or, as I would rather say, that it is misconceived. In conclusion, I will briefly suggest some considerations in support of this view.

What reasons are there for thinking that the "new" concepts introduced by a theory are—or at least should be—specifiable by means of the antecedently available vocabulary? One consideration that influenced my earlier concern with the problem is, briefly, to this effect: A theory purports to describe certain facts, to make assertions that are either true or false. But a sentence will qualify for the status of being either true or false only if the meanings of its constituent terms are fully determined; and if we want to understand a theory, or to examine the truth of its claims, or to apply it to particular situations, we must understand the relevant terms, we must know their meanings. Thus, an adequate statement of a theory will require a specification of the meanings of its terms—and what other means is there for such a specification than the antecedently available vocabulary?

But even if, for the sake of argument, we waive questions about the concept of meaning here invoked, these considerations are not compelling. On the contrary; when at some stage in the development of a scientific discipline a new theory is proposed, offering a changed perspective on the subject matter under study, it seems highly plausible that new concepts will be needed for the purpose, concepts not fully characterizable by means of those antecedently available. This view seems to me to derive support from those studies of the language of science—especially in the logical empiricist tradition—which have led to a steady retrenchment of the initial belief in, or demand for, full definability of all scientific terms by means of some antecedent vocabulary consisting of observational predicates or the like. The reasons that led to countenancing the introduction

[30] Putnam, "What Theories Are Not," p. 241.

of new terms by means of reduction sentences, interpretative systems, or probabilistic criteria of application all support the idea that the concepts used in a new scientific theory cannot be expected always to be fully characterizable by antecedently available ones.

But the very relaxation of the requirements for the introduction of new scientific terms gave rise to such questions as whether we can claim to understand such partially interpreted terms; whether the sentences containing them can count as significant assertions or can be regarded at best as an effective, but inherently meaningless, machinery for inferring significant statements, couched in fully understood terms, from other such statements; and whether reliance on incompletely interpreted theoretical terms could be entirely avoided in science.

But this way of looking at the issue presupposes that we cannot come to understand new theoretical terms except by way of sentences specifying their meanings with the help of previously understood terms; and surely this notion is untenable. We come to understand new terms, we learn how to use them properly, in many ways besides definition: from instances of their use in particular contexts, from paraphrases that can make no claim to being definitions, and so forth. The internal principles and bridge principles of a theory, apart from systematically characterizing its content, no doubt offer the learner the most important access to an "understanding" of its expressions, including terms as well as sentences.

To be sure, all these devices still leave unanswered various questions concerning the proper use of the expressions in question; and this may seem to show that, after all, the meanings of those expressions have not been fully specified, and that the expressions therefore are not fully understood. But the notion of an expression that has a fully specified meaning or an expression that is fully understood is obscure; besides, even for terms that are generally regarded as quite well understood there are open questions concerning their proper use. For example, there are no sharp criteria that would determine, for any strange object an astronaut might encounter on another planet, or indeed for any object that might be produced in a test tube on earth, whether it counts as a living organism. Theoretical concepts, just like the concept of living organism, are "open-ended"; but that, evidently, is no bar to their being adequately understood for the purposes of science.

An Inductive Logic of Theories

1. Theoretical Inference

In their contributions to this volume, Professors Hempel and Feigl have both discussed the "layer-cake" model of scientific theories from the point of view of the meaning and interpretation of theoretical language. The problem of interpretation has also concerned Professor Maxwell, and he has in addition mentioned the problem of confirmation of theories, but only to assert that this is wholly independent of and irrelevant to the problem of meaning. It is this last assertion of independence that I wish to question in this paper. I believe that the problem of theoretical inference or confirmation is insoluble in terms of a deductive model of theories, and indeed in terms of any analysis which makes a radical epistemological and semantic distinction between theoretical and observation predicates. It follows that an analysis of the meaning of theoretical terms such as that given by Maxwell will be insufficient to support an adequate confirmation theory of theories, and that the questions of what a theory means and how it comes to be confirmed are not after all independent. The argument of this paper will be of reductio ad absurdum form: first developing expressions for a deductive theoretical system and its models in the usual way, and then showing how they fail to allow for inductive inferences of a kind generally regarded as justifiable.

The problems of induction and confirmation have been given a new lease of life in recent discussions, but little has yet been done toward a logic of inductive inference in relation to scientific theories. Nevertheless, among those who still hold that there is such a thing as inductive logic, I suppose it would be generally agreed that there ought, in principle, to be some way of explicating inductive inferences to theories, and to the further observational consequences of theories. Putnam[1] has given a striking

[1] H. Putnam, " 'Degree of Confirmation' and Inductive Logic," in P. A. Schilpp, ed., The Philosophy of Rudolph Carnap (LaSalle, Ill.: Open Court, 1963), pp. 761–793.

example of the kind of inference involved here, which would surely be regarded by most scientists as inductively justifiable. In an attempt to show that any confirmation theory of Carnap's type is bound to be inadequate for theoretical inference, he considers inferences of the kind that were made when the first atomic bomb explosion was predicted and subsequently tested with success. Here there was a body of evidence drawn from physics and chemistry which supported the nuclear theory, and from this theory were derived in great detail, and with very great confidence, predictions about what would occur in an experimental situation so far wholly unrealized in any previous experience, namely the slamming together of two subcritical masses of uranium to produce an explosion of given magnitude. Schematically the situation is this: Given total initial evidence e_1, we use this to support a theory t in which we have sufficient confidence to deduce future predictions e_2, namely the results of the atom bomb test. Putnam's own argument to the effect that no confirmation theory can explicate this inference seems to me insufficiently general, but there is a perfectly general argument which shows that no probabilistic confirmation theory of any type yet developed will allow us to infer with higher than prior confirmation from e_1 to e_2 *merely in virtue of the fact that both are deductive consequences of some theory,* where "theory" is understood in the deductive-model sense. This theorem has, in fact, already been mentioned by Carnap,[2] in his discussion of some points in Hempel's paper "Studies in the Logic of Confirmation."[3]

In this paper, Hempel expresses a principle which seems to underlie the kind of inference Putnam exemplifies. The principle is what Hempel calls the consequence condition for confirmation, namely "If some evidence e confirms a hypothesis t, then e confirms every L-consequence of t" (C_1). If this condition were satisfied it would cover Putnam's example, because e_2 would then be confirmed by e_1 which confirms t. Consider the case in which $t \rightarrow e_1 \cdot e_2$. This is the situation we have when we say that t explains the data e_1 and predicts e_2. If we now add a further condition C_2, "If t L-implies e_1, then e_1 confirms t," which Hempel calls the converse consequence condition, it is easy to show that C_1 and C_2 taken together are

[2] R. Carnap, *Logical Foundations of Probability* (Chicago: University of Chicago Press, 1950), especially pp. 471–472.

[3] C. G. Hempel, "Studies in the Logic of Confirmation," *Mind*, 54 (1945), 1–26, 97–121. Hempel has modified some of these points in his addendum to a reprint of the paper in C. G. Hempel, *Aspects of Scientific Explanation* (New York: Free Press, 1965), p. 50.

counterintuitive. Suppose $t \equiv e_1 \cdot e_2$. Then certainly $t \rightarrow e_1 \cdot e_2$, and using C_1 and C_2 we conclude that e_1 confirms e_2. But this is absurd, because e_2 may be any statement whatever. If t is produced just by arbitrarily conjoining any other statement e_2 to e_1, we should certainly not want a confirmation theory in general to allow e_1 to confirm e_2. Some further conditions must be imposed upon admissible e_1 and e_2, and either C_1 or C_2 or both must be modified.

This argument is very general, and concerns any confirmation theory which satisfies C_1 and C_2. In particular, any probabilistic c-theory satisfies C_2, but Carnap[4] has shown that in his probabilistic c-theory, confirmation does not satisfy C_1. Therefore he avoids the counterintuitive inference, but apparently at the cost of being unable to explicate inferences of Putnam's type. In general, in a probabilistic c-theory we cannot have

$$(1) \qquad c(e_2, e_1) > c_0(e_2)$$

unless there is some probabilistic dependence between e_1 and e_2, and such dependence is not guaranteed by the fact that e_1 and e_2 are both implied by some t, because in the absence of further conditions upon t, this is trivially the case for every pair of statements e_1, e_2. I have suggested elsewhere[5] a method of rescuing Putnam-type inferences within a probabilistic c-theory, which involves specifying a relation between e_1 and e_2 that ensures the satisfaction of (1) only in case e_1 is intuitively relevant to the confirmation of e_2. The relation which constitutes such relevance between e_1 and e_2, I suggest, is a relation of *analogy*. I cannot discuss in detail here how this would be defined, but I will describe the conception briefly by taking an example which I develop more fully in the other paper.

2. The Analogical Character of Theories

Consider the inference made by means of Newton's theory of gravitation (t) from some initial data (e_1) which we will suppose comprise Kepler's laws, to the prediction (e_2) that a falling body will fall in the neighborhood of the earth with a certain acceleration. We suppose that the historical situation is that we accept Kepler's laws, but do not yet know the acceleration relation for falling bodies. This is not too far from the actual case, because, as has been pointed out many times, Galileo's law is contra-

---

[4] Carnap, *Logical Foundations of Probability*, p. 471.
[5] M. B. Hesse, "Consilience of Inductions," in I. Lakatos, ed., *The Problem of Inductive Logic* (Amsterdam: North-Holland, 1968), pp. 232–257.

dicted by Newton's theory, and we do in fact have more confidence in the relation e_2 predicted by that theory than in Galileo's earlier empirical approximation to it. This confidence seems to come from the support given to Newton's theory by the other data entailed and explained by it, which support we take to be in some way passed on to the prediction. But in general, as I have shown, this is not a justified inference. Unless more is said about the relation between e_1 and e_2 we shall not have the desirable increase of $c(e_2, e_1)$ compared with $c_0(e_2)$. When the condition (1) is satisfied I shall say that the corresponding inductive inference is justifiable.

Let us express Kepler's laws and the predicted acceleration relation very schematically as

$$(2) \quad \begin{aligned} &e_1: \ (x)(F(x) \cdot G(x) : \supset P(x) \cdot Q(x)) \\ &e_2: \ (x)(G(x) \cdot H(x) : \supset Q(x) \cdot R(x)) \end{aligned}$$

Here a relation of "analogy" has been assumed between e_1 and e_2 in the following sense: The predicates F, G represent properties of the planets asserted by Kepler's laws to have certain motions which we denote by the conjunction of predicates P, Q. (It would of course be necessary to use metric predicates if this were a realistic reconstruction, but we simplify drastically for purposes of exposition by considering only monadic predicates.) Expressed in similar fashion is e_2, for the bodies referred to in e_2 share some properties with those referred to in e_1, but not all. All these bodies are solid, massive, opaque, and so on; but bodies near the earth differ from planets in size, shape, chemical composition, and so on. In the same way, Kepler's laws can be thought of as describing motions which are in some respects the same as and in some respects different from the motions described in e_2: All the orbits are ideally conic sections, but they are traversed at different speeds, and about different foci.

My suggestion is that the confidence we have in the prediction of e_2 is due to the relation of analogy between e_1 and e_2 which is constituted by the repetition of predicates G and Q in the expressions of e_1 and e_2. We regard e_2 as confirmed by e_1 because the bodies described by e_2 are sufficiently similar in some respects to those described by e_1 to justify the inference that their behavior will also be similar. Explication of this inference therefore requires a c-theory which will yield the inequality (1) in particular when e_1 and e_2 are as specified in (2), and in comparable cases. There is an example of such a c-theory (the η-theory) in Carnap and Steg-

Mary Hesse

müller,[6] where Carnap's earlier theory is developed in such a way that this condition is satisfied. Explication of analogical inference of this kind has therefore been shown to be possible within a probabilistic confirmation theory, although the η-theory itself is not wholly satisfactory in other respects.

It is not necessary here to go into the question of how this relation of analogy would be defined in general. It is sufficient to say that the best tactics do not seem to be to try antecedently to define what we would mean in all cases by the analogy between two objects or two systems in virtue of the inferences we regard as intuitively justified. Such a procedure is liable to strain intuition too far. Rather, we should take some simple cases, such as the one just discussed, in which it is clear that the inference would generally be regarded as justified, find what conditions these cases would impose upon a c-theory, and then use the weakest possible c-theory satisfying these conditions to define the analogy relation in cases which are too complex to give much help to intuition. That is to say, whenever the inequality $c(e_2, e_1) > c_0(e_2)$ is satisfied for observation statements in such a c-theory, we shall say that there is necessarily a relation of analogy in the sense intended between e_1 and e_2. This relation of course has to be consistent with an assertion of analogy or its absence in cases where we do intuitively recognize the existence of analogy and the justifiability of analogical argument, or their absence.

The question that immediately arises, however, concerns the place of the theory t in this explication. We have not needed to mention t either in the expressions of e_1 or e_2, or in the statement of inequality of c-functions. Is t then wholly redundant? Further inspection of (2) reveals that this is not the case, for it is implied in (2) that there is a theory, indeed more than one theory, which has the traditional relation to the data and prediction of entailing their conjunction. In particular, if t is

$$(3) \qquad (x)(F(x) \supset P(x)) \cdot (G(x) \supset Q(x)) \cdot (H(x) \supset R(x))$$

then $t \to e_1 \cdot e_2$. Furthermore, t has the desirable characteristic of "saying more than" $e_1 \cdot e_2$, since it is not the case that $e_1 \cdot e_2 \to t$. What t does in effect is to pick out from e_1 and e_2 the predicates G, Q which are in common between them, and to assert that the essential correlation in both cases is that bodies which are G are also Q, and that the properties of the two domains of phenomena which are different are due to two other laws,

[6] R. Carnap and W. Stegmüller, *Induktive Logik und Wahrscheinlichkeit* (Vienna: Springer, 1959), Appendix B.

one of which (relating F and P) applies only to the e_1-domain, and the other (relating H and R) only to the e_2-domain.

It might be noted at this point that when we speak of Newton's theory as "explaining" Kepler's laws and the law of falling bodies, we do not as a rule claim that Newton's theory includes a deductive explanation of *all* the differences between planets and falling bodies, that is, we do not include in the explanatory theory laws $(x)(F(x) \supset P(x))$ and $(x)(H(x) \supset R(x))$ which mention all the properties the bodies do *not* share. Newton's theory contains laws explaining why some features of the motions of planets are different from those of falling bodies, but not all such features are mentioned in the theory; for example, their different chemical compositions do not appear in the antecedent of any law of Newton's theory, nor do their different initial velocities appear in the consequent of any such law. Kepler's laws, on the other hand, if they are considered as data to be explained, do not imply any distinction between properties which are in the later light of Newton's theory "relevant" or "irrelevant" to the search for explanation. Kepler's own understanding of planets in the assertion "All planets move in ellipses" certainly included for example the assumption that planets have magnetic properties, which he considered specially relevant to explanation of their motions. This aspect of the explanandum is, however, not mentioned in Newton's "explanation" even in the initial conditions for Kepler's laws. It follows that the expression (3) for t which was used above in deference to the requirement that the explanandum be *deducible* from the explanans plus initial conditions is too strong to reproduce the real situation, in which an "explanation" is *not* required to entail the explanandum as *that was originally formulated*, but is already the result of assumptions of relevance and irrelevance which are rarely made explicit in deductivist accounts of theories. Before the deductive account can be made to work at all, irrelevant features must in fact be dropped from the explanandum as unexplainable by that theory (although they may of course be explained by another theory). The analogical account of theories which has just been suggested has the merit of making these assumptions of irrelevance explicit from the beginning. According to this account, we should regard the theory t, not as in expression (3), but rather as $(x)(G(x) \supset Q(x))$, together with the statements of initial condition which differentiate the e_1-domain as an application of t from the e_2-domain.

The pattern of theoretical inference we have been studying now takes

169

on a different aspect. We are no longer concerned with a dubious inductive inference from e_1 up to t and down to e_2, but with a direct analogical inference from e_1 to e_2. And t does not provide the upper layer of the cake, but rather, as it were, extracts the jam from e_1 and e_2, that is to say it reveals in these laws the relevant analogies in virtue of which we pass from one to the other inductively.

3. The Function of Models

In the light of these confirmation conditions for theoretical inference, let us investigate the adequacy of a typically deductivist construal of theories and their interpretive models. In particular, I want to consider how the use of models for theories provides examples of inference in which we need stronger logical relations between e_1 and e_2 than can be included in the usual deductivist scheme.

Consider an expression of a theoretical system for a domain of entities, in which all the constant theoretical terms ($T_1, T_2, \ldots$) whose "meanings" are problematic have been replaced by the variables $\tau_1, \tau_2, \ldots$:

$$(4) \qquad (x)(y) \ldots (x,y, \ldots \epsilon S)\phi(\tau_1, \tau_2, \ldots O_1, O_2, \ldots)$$

This is a representation of the theoretical calculus, together with a set of observation predicates $O_1, O_2, \ldots$, so that at this stage it is only a partially interpreted system, the τ's remaining uninterpreted. Now there will be a model of this system (let us call it the Q-model) which is represented by replacing the τ's again by the problematic theoretical terms $T_1, T_2,$ $\ldots$, although it is rather difficult to see that this is in the ordinary sense an *interpretation*, because the problem of the meaning of theoretical terms arises precisely from the fact that we do not know what constant predicates the T's are, and so do not know what domain of entities and predicates satisfies this model. However, *if* we knew this, then in the usual logician's sense the Q-model would be a model of the system (4). In what follows "model" will be used of *linguistic* entities (systems of laws, theories, etc.) *not* of the sets of entities and predicates which satisfy these systems.

It seems to me that what the physicist normally means by a model for a theory is not the Q-model, but rather a system of laws satisfied by a set of entities and predicates different from the set of entities and predicates which were to be explained when he set up his theory. When he refers to a set of Newtonian particles as a model for gas theory, this is a set of entities different from the gases whose behavior he is endeavoring to explain by the gas theory. When the crystallographer builds a structure of colored balls and steel rods on the laboratory bench, this is a set of entities different from the organic molecules he is attempting to construct a theory for. So it is necessary to talk in terms of two domains of entities: S will now be, not a universal domain, but what I shall call the *domain of entities of the Q-model*, and I shall denote by S* the *domain of entities of the P-model*, where the P-model is a model in the physicist's sense just indicated.

Let us elaborate expression (4) in order to take account of the relation of the Q-model to the P-model. We shall suppose that there are two sets of observation predicates, $O_1, O_2, \ldots$, and $O_1', O_2', \ldots$, where the first set enters into laws known to be true in S and S*, and the second into laws known to be true only in S*, and not yet examined in S. We shall suppose the Q- and P-models to have an *analogical relation* in the following sense: They share the O- and O'- predicates, and the Q-model involves also predicates $M_1, \ldots$ not applicable to S*, and the P-model involves predicates $N_1, \ldots$, not applicable to S (the "negative analogy" of the two models). We assume that the laws known to be true of S* and S are respectively

(5) e_1^*: $(x)(y) \ldots [(x,y, \ldots \epsilon S^*)\psi(N_1,O_1, \ldots)]$
(6) e_1: $(x)(y) \ldots [(x,y, \ldots \epsilon S)\psi(M_1,O_1, \ldots)]$

and the laws known to be true of S* and unexamined in S are respectively

(7) e_2^*: $(x)(y) \ldots [(x,y, \ldots \epsilon S^*)\psi'(N_1,O_1', \ldots)]$
(8) e_2: $(x)(y) \ldots [(x,y, \ldots \epsilon S)\psi'(M_1,O_1', \ldots)]$

The P- and Q-models can be expressed as

(9) t^*: $(x)(y) \ldots [(x,y, \ldots \epsilon S^*)\phi(P_1,N_1,O_1,O_1', \ldots)]$
(10) t: $(x)(y) \ldots [(x,y, \ldots \epsilon S)\phi(T_1,M_1,O_1,O_1', \ldots)]$

where the P-model is a *true interpretation* of the partially interpreted expression corresponding to (4), and we have

(11) $(x)(y) \ldots [\phi(P_1,N_1,O_1,O_1', \ldots) \rightarrow \psi(N_1,O_1, \ldots) \cdot$
        $\psi'(N_1,O_1', \ldots)]$

All predicates in these expressions are constants.

The problem is to show that there is a justified analogical inference, in the type of confirmation theory we have discussed, from $e_1^* \cdot e_1 \cdot e_2^*$ to e_2. But before considering this, there is an obscurity about the notion of observability which ought to be cleared up at this point. There is not only the very important distinction between observable *predicates* and observable entities but there is also a distinction concerning observability of predicates in *different domains*. It may very well be the case that in the P-model we have predicates (the P's), such as "mass," "velocity," "radius," which are observable in the domain of macroscopic physical objects, but not in that of microscopic objects. Where the P-model is used as a model for the theory about gases, the corresponding predicates in the Q-model (the T's) are not observable in any domain, and are only given as it were courtesy titles when we refer to them as "mass," "velocity," etc. If we are to use these adjectives at all in relation to the Q-model, we must at least recognize that they name properties unobservable in the domain S, though observable in S*. The domains of both Q-model and P-model, however, contain the O-predicates, and these are observable in both domains. (The O'-predicates may not be observable in S*, as we shall see presently.) In S* the O's will include the average pressure of a cloud of macroscopic particles hitting a surface, such as hailstones striking a wall horizontally. In S they will include the pressure of the gas measured by manometers. "Pressure" is the same predicate, observable in both domains.

The difference between the deductivist's construal and my own emerges when we consider the T-predicates. He generally wishes to say that these are entirely undetermined except by the Q-model, whose status as an interpretation is, as we have seen, highly problematic. Although there may be another model, the P-model, of the same calculus, there is for the deductivist no relation between the Q- and P-models other than that they are models of the same calculus and share the same O-predicates. Therefore, in this view, if we do use the words "mass," "velocity," etc., in relation to the T's, this is an equivocal use when compared with their use in relation to the P's. We cannot know whether for God the T's are the same predicates as the P's or not. Talk of the T's therefore seems to me at best to define a *class* of models of the partially interpreted system (4), and it is not clear that replacing the variable τ's by constant T's has added anything to the content of (4), because we do not know what these putative constants are and have in principle no means of finding out (unless of course the T's later become *observable*, but in the deductivist's view this could not be a *general* solution to the problem of the interpretation of theoretical terms, because not all such terms will become observable—if they did *his* problem would dissolve).

If we consider the deductivist construal in the light of the considerations about theoretical inference in the preceding sections, it is not clear that it has any resources for explicating this kind of inference. Even if we waive the difficulties about interpretation for the moment, and suppose that the Q-model is an interpreted theory in the usual sense, we have shown above that inferences from one subset of observable consequences of this model to another subset are not in general inductively justifiable. In particular there would be no justifiable prediction from a set of experimental laws about gases, say Boyle's and Charles's laws, by way of the kinetic theory to other laws about gases if the kinetic theory is understood as a Q-model, that is, if there is no more than an equivocal sense in which we can speak of its being "about" masses and velocities of molecules. Suppose, however, we now bring in the P-model. In the gas example this is a model of Newtonian particles whose laws of motion are known to be true, or at least accepted for purposes of exploitation in the theory of gases. We suppose the models expressed as in (9) and (10), and that ϕ entails the known experimental laws shared by both models as in (11). It should incidentally be noticed that since $\psi'(N_1, O_1', \ldots)$ is entailed by ϕ, it is known to be true in S* even if it has not been directly examined, or even if it is for all practical purposes unobservable in S*. For example, it is unlikely that the analogue of Boyle's law in Newtonian particle mechanics has ever been observed to be true; nevertheless it is believed because Newton's laws are believed in that domain.

In a confirmation theory of the type described in section 2 we may have a relation of analogy between e_1* and e_1 as expressed in (5) and (6). (Compare the e_1, e_2 of section 2 (2), where N_1, ..., stand for F, P; M_1, ..., stand for H, R; and O_1, ..., stand for G, Q.) It is very important to be clear at this point that the relation of analogy here spoken of is *not only* the formal analogy in virtue of the fact that both models are models of the same calculus, but includes what I have elsewhere[7] called a *material analogy* in virtue of the sharing of the O-predicates. Similarly, we may have material and formal analogies between e_2* and e_2 expressed in (7) and (8), in virtue of the sharing of O'-predicates and their relations expressed in ψ'. Since e_2* is true, and if the negative analogy between S and S* is not too strong, there will then be a justifiable analogical inference to e_2, which is strengthened by the truth of both e_1* and e_1. Moreover, the same argument yields a justifiable inference to

[7] M. B. Hesse, *Models and Analogies in Science* (London: Sheed, 1963; Notre Dame, Ind.: University of Notre Dame Press, 1966).

(12) $(x)(y) \ldots [(x,y, \ldots \epsilon S)\phi(P_1,M_1,O_1,O_1', \ldots)]$

where the P-predicates are observable in S* but unobservable in S.

Comparing (12) with (10), we see that the analogical inference leads to the suggestion that the T-predicates should be *identified* with the P-predicates, rather than being regarded as problematic constants whose reference is in principle unknown. Although their referents in S are unobservable, their "meaning" is derived from the observables of S*, that is, they mean the same as they do in the P-model, and satisfy the same laws. With any construal of the T's not involving some identification of this kind, it is not clear that there can be justifiable analogical inference to T-statements in S. (We shall return to this point in the next section.) With the identification, however, the inference to predictions can be made even stronger, for as was remarked above, it is not necessary that the O'-predicates should be observable in S*, only that the P's should be. In such a case it is clear that the inference to e_2 depends essentially upon knowing the truth of t* empirically, and making an analogical inference to the probable truth of t and hence e_2, and that this depends upon the identification of the P- and T-predicates in S.

Diagrams may help to elucidate the structure of these inferences and to relate them to what has been said in section 1 about the status of the theory in predictive inference. In the diagrams arrows on the lines indicate alleged justifiable inferences and dotted lines relations of analogy. In the case of the deductive construal, we have seen that these are illusory, and that there is no justifiable inference from e_1 to e_2 unless there is an analogical relation between these laws independent of t. If there is no such relation, however, we may be able to make the inference if we can find a P-model as represented in the right-hand diagram. Here broken lines indicate analogical relations, and the inductive inference depends on the analogy between the laws e_1, e_1^* in the two domains, and the assertion of the truth of t*. Then there is a justifiable inference to t, and hence to e_2, since $t \rightarrow e_2$. Truth is, as it were, fed into the *theory* of S from the P-model,

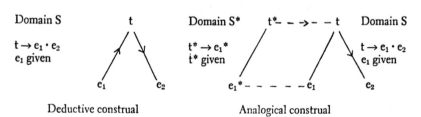

Domain S   t   Domain S* $t^*- - \rightarrow - - t$ Domain S

$t \rightarrow e_1 \cdot e_2$ $t^* \rightarrow e_1^*$ $t \rightarrow e_1 \cdot e_2$
e_1 given t^* given e_1 given

        e_1 e_2 $e_1^* - - - - - e_1$ e_2

     Deductive construal   Analogical construal

and so passed on to e_2, whereas in the deductive construal there is no justifiable inference through the theory of S because this theory acquires no probable truth from any source other than its own entailments. Theories cannot be pulled up by their own bootstraps, but only by support from external models.

It will immediately be objected to this account that there are many examples for which analogical inferences of the kind described not only would be unjustified in the light of further evidence, but would never be regarded as justifiable even before further evidence is collected. Analogical arguments are notoriously weak and liable to failure and must generally be treated with extreme caution. This of course is true, but it must be borne in mind that the examples above have presupposed the principle of total evidence. *If* all the evidence we have is summarized as in t* and e_1, then the inferences may be intuitively reasonable. But if we have other evidence to the effect, for example, that S and S* differ from each other in many further characteristics, or if we know of other domains in which the inference to ψ' breaks down, then such information may well weaken the inferences to the point of disconfirmation. I suspect that when apparent counterexamples to these inferences are produced, they will be found to involve one or other of these types of additional evidence. In principle an adequate confirmation theory must be capable of dealing with such complexities, and must explicate the weakness as well as the strength of analogical arguments.

4. Identification of Theoretical Predicates

In the absence of a detailed and adequate confirmation theory it is not possible to be precise in reply to objections of the kind just mentioned. But it is perhaps permissible to speculate a little further upon the characteristics which an adequate c-theory might exhibit. In particular, it may be possible to suggest some compromise between the position outlined here and the deductive account, though still within the framework of a confirmation theory. There are two ways in which my position has been opposed too sharply to deductivism and should now be modified.

First, it may be doubted whether we wish to *identify* a theoretical predicate such as "mass" of a molecule or electron with "mass" of a macroscopic particle. On the other hand I think it has been correct to say that "mass" cannot be simply *equivocal* without destroying the possibility of theoretical inference. What we need to reconstruct is a notion of analogical meaning of the word "mass" in the two domains, where "analogical" is used as a middle term between "univocal" and "equivocal," as in some Thomist philosophy. "Mass" is not always used "in the same sense" when predicated of different systems, but it is not on the other hand a pun when it is used of positrons, neutrons, quasars, and the like.[8] The problem of determining how far the meaning of "mass" can be extended analogically and the problem of deciding what analogical inferences are justifiable are closely related. Predicates can be stretched just as far as analogical argument remains justifiable, and conversely. How far this is would have to be decided by looking at the whole complex of evidence in all domains in which the predicate is applied. If, for example, the difference in domain is only one of scale, we shall probably be quite satisfied simply to identify the P- and T-predicates, but when the difference of scale is accompanied by other differences as radical as those between, say, the macroscopic and the nuclear domains, we may become increasingly unwilling to allow any analogical extension of meaning from one to the other.

The second respect in which my account might be modified in the direction of deductivism is in regard to the attributes of the S- and S*-domains which are allowed to weigh in analogical inference. It is convenient here to refer to a discussion by Sellars[9] in which he argues that to identify the theoretical predicates with the P-predicates from an antecedent observation language is to fall into the "myth of the given," and to misrepresent the *novelty* which may be introduced by using P-models in connection with theories. To the construal of T-predicates favored by Nagel[10] Sellars objects that it makes the T's new but not meaningful, and to my construal in my *Models and Analogies in Science* he objects that it makes them meaningful but not new.

If I have understood Sellars correctly, his proposal for the resolution of

[8] I have discussed analogy, context meaning, and related questions in my "Aristotle's Logic of Analogy," *Philosophical Quarterly*, 15 (1965), 328–340; "The Explanatory Function of Metaphor," in Y. Bar-Hillel, ed., *Logic, Methodology and Philosophy of Science* (Amsterdam: North-Holland, 1965), pp. 249–259; "A Self-Correcting Observation Language," in B. Van Rootselaar and J. F. Staal, eds., *Logic, Methodology and Philosophy of Science*, vol. III (Amsterdam: North-Holland, 1968), pp. 297–309; and "Theory and Observation," to appear in the forthcoming vol. 4 of the University of Pittsburgh series in the philosophy of science edited by R. G. Colodny.

[9] W. Sellars, "Scientific Realism or Irenic Instrumentalism," in R. S. Cohen and M. W. Wartofsky, eds., *Boston Studies in the Philosophy of Science*, vol. III (New York: Humanities, 1965), pp. 171–204.

[10] E. Nagel, *The Structure of Science* (New York: Harcourt, Brace and World, 1961).

this dilemma is as follows. Not all analogy is analogy of *particulars* in virtue of their sharing identical attributes, as has been assumed in the discussion of analogy in section 2 above. Attributes themselves may be similar or analogous, that is, first-order predicates may themselves be predicated by second-order predicates, and may be analogous in virtue of sharing such second- or even higher-order predicates. Sellars gives as examples the second-order predicate "perceptible," which applies to first-order predicates, and the second-order predicate "transitive," which applies to first-order relations such as "before," "to the left of." Either such second-order attributes can be *mentioned*, as in " 'transitivity' is true of before," or they can be *shown*, as by exhibiting the transitivity postulate satisfied by "before." The function of a P-model is to introduce second- and higher-order predicates which are shared with the Q-model, and thus to convey some interpretation to the T-predicates: "Thus as a first approximation, it can be said that models are used in theory construction to specify new attributes as *the attributes which* share certain higher order attributes with attributes belonging to the model, fail to share certain others (the negative analogy)—and which satisfy, in addition, the conditions laid down by the relevant correspondence rules."[11] Thus, Sellars claims, both meaning and novelty are allowed for in the relation of Q- and P-models, and the P-model remains heuristically useful at least so long as its higher-order attributes remain implicit. When they are themselves formalized, presumably the P-model can be abandoned, and postulates representing higher-order attributes can be explicitly added to the Q-model and corresponding calculus.

I confess that many features of this suggestion remain obscure to me. In the first place, if the higher-order attributes can be referred to by intensional expressions such as "transitive," "perceptible," it is not clear that we have escaped the "myth of the given." These expressions are already in the descriptive language, and a more sophisticated analysis of "analogy," involving a type logic, could presumably take account of analogies depending on the sharing of these predicates, as it can of first-order predicates. If novelty depends on introduction of new predicates, there is no novelty here. If, on the other hand, the introduction of novelty depends essentially on the P- and Q-models merely *exhibiting* analogy of higher-order attributes, it is not clear that anything other than *formal* analogy has been introduced. *Each* model of a calculus "exhibits" such

[11] Sellars, "Scientific Realism or Irenic Instrumentalism," p. 181.

analogy with every other model of the same calculus; indeed once the calculus has been fully expressed, mention of models other than the Q-model is wholly redundant to this kind of analogy, since the higher-order attributes of the Q-model are already shown by the calculus itself. And in this case the construal of the T's as "the attributes which share certain higher-order attributes with attributes belonging to the [P-]model" does not seem to differ in principle from Nagel's account, which Sellars rejects, or from the deductivist's which we have seen reason to reject above. For in both these accounts it follows from the status of the Q- and P-models that there are *some* second- or higher-order attributes that they share, namely the relations exhibited in the calculus of which they are models.

It is possible, however, that Sellars has in mind a situation which is somewhere between the two extremes just mentioned, namely a P-model which tacitly introduces higher-order attributes in virtue of which we vaguely accept an "analogy" between it and the explanandum, but which have so far been unanalyzed, and for which we may not have names in the language. For example, we may recognize an analogy between a loud noise and a bright flash, and may exploit it in a P-model for light drawn from the phenomena of sound, without necessarily having in our language a concept "intensity" which applies to both noise and flash. However, the question now arises whether the P-model is introduced here because of antecedent recognition of an analogy, even though this was inexpressible in the existing language, or whether the adoption of this P-model itself *introduces* a new higher-order attribute which sound and light phenomena share.[12] The answer to this question is, surely, "six of one and half a dozen of the other." But such liberality must not be taken to the point of admitting *any* model as a candidate for the P-model. Unless *some* analogy of predicates, whether first or higher order, is recognized, which is not merely the relation of isomorphism between two models of some same calculus, use of a P-model in theoretical inference becomes vacuous, as I shall now try to show.

Lying behind Sellars's attempt to reconcile the meaningfulness of theoretical concepts with the possibility of novelty is some obscurity about the function of models in *inference*. Sellars has not given sufficient weight to the fact that my plea in *Models and Analogies in Science* for recognition of the logical role of models depends essentially upon taking *predictive*

---

[12] A closely related question is discussed in M. Black, *Models and Metaphors* (Ithaca, N.Y.: Cornell University Press, 1962), pp. 37–38.

power as a necessary condition for theories. From this point of view, acceptance of his shared higher-order predicates as ingredients in the role of models in relation to theories will depend upon whether or not we regard such shared predicates as justifying analogical inference. Take the example of transitivity, and assume that the P-model contains the first-order relation "before," and that the Q-model contains a relation R which is either said or shown to be transitive. According to Sellars we need know nothing about R except that it satisfies the postulates of the theory and is transitive. Waiving now the question of what relation R *is*, we must nevertheless ask how the P-model helps us to make analogical inferences in the theory. The answer is, surely, not at all. For suppose we risk an analogical inference from P-statements involving "before" to Q-statements involving R. There is a large class of relations all members of which are consistent with what is known about R, but clearly the analogical inference will not be valid for all of them. Suppose R is, in God's private eye, "larger than," then an analogical inference involving R would be equally as justifiable or unjustifiable as a similar inference involving "smaller than," but the conclusion cannot be true in both cases. Any analogical inference may of course lead to false conclusions for *empirical* reasons, but in this case one or the other conclusion must be false for *logical* reasons. On the other hand, if R is *known to be* a particular transitive relation having other affinities with "before," such as "to the left of," there might very well be a justifiable inference from the P-model to the Q-model if the theory were concerned, for example, with a geometry of space-time. In other words, whether higher-order predicates can function to justify analogical inference is a question only to be decided by examining particular predicates, and the possibility of doing this presupposes either that the predicates are already in the language or that they can be coined as required to name particular attributes in virtue of which an analogy is suspected. It cannot be the case that every shared higher-order predicate is sufficient to generate a justifiable analogical inference, for this would certainly lead to inconsistency, as in the "transitivity" example above, and would even be vacuous, if it could be shown that any system shares *some* higher-order attributes with every other system, a proposition which should not be too difficult to prove if our ontology is generous enough.

Finally, a remark about novelty and the myth of the given. Sellars's objection to identifying the predicates of the Q- and P-models seems to follow from his rejection (page 184) of the assumption (which he ascribes to Nagel) that there is a one-to-one correspondence between predicates and the extralinguistic attributes to which they apply. If this assumption were true, then indeed it would be difficult to see how analogical inference from observable predicates of the P-model would leave room for any theoretical novelty. Sellars instead wants to allow for enrichment and revision of the observational vocabulary by its interaction with theory. But use of models as analogues neither presupposes such a one-to-one correspondence nor rules out the kind of interaction of observation and theory that Sellars requires. Indeed, if we understand predicates as *analogical* in their applications to different situations rather than as either univocal or equivocal (as suggested above and discussed in the references there given), the possibility of novelty is safeguarded by the indefinite variety of analogical extensions of existing predicates. It is, after all, rare for new descriptive predicates to be *coined*, and much more common for new situations to be described by complex combinations of old predicates. I do not wish to deny, however, in anything I have said here, that totally new concepts may sometimes be required in theoretical science. It only seems to me to follow from the foregoing arguments that *if* totally new concepts are introduced, there is no possibility of theoretical inference of the kind discussed here, and consequently I suspect that such occasions of total novelty are rare.

To summarize the conclusions of this paper with regard to inductive and deductive construals of theories: Though one may admit that deductivism represents our existing knowledge about the theory of a given set of observable laws, it does not explicate the kinds of inference to prediction that we normally find justifiable. And these inferences ought not to be dismissed to the context of discovery or heuristics or psychology or history, but ought to be seen as part of the content of theories as we state them. Therefore they should feature in an adequate construal of theories. My alternative to the deductive account has been an attempt to show how this may be done.

Structural Realism and the Meaning
of Theoretical Terms

A theory, in the sense used in this paper, is a set of statements some of which both refer to unobservables and are capable of functioning essentially in the derivation from the theory of statements that are observationally decidable. Theories will be considered to be *semantically autonomous*; this means that there will be no (metalinguistic) rules whose function is to relate unobservables to observables (e.g., by relating theoretical terms to observation terms). This requirement amounts to the same as Carnap's later practice of eliminating "correspondence rules" ("C-rules") in favor of C-postulates (and, we might add, C-theorems). Such a C-statement is any nonmolecular statement that contains both theoretical terms and observation terms.

It will be assumed that, in spite of notorious difficulties, a distinction between the unobservable and the observable can be made. The program proposed in this paper will be applicable no matter how, within reasonable bounds, such a distinction is drawn. My own view on this matter, which I shall not try to defend here, is that all items should be considered theoretical unless they occur in direct experience; since I reject any form of direct realism, this means that the observable is instantiated only in inner events of observers. One question that the paper attempts to answer is this: Granted that a theory contains two kinds of terms one of which is relatively semantically unproblematic (commonly called observation terms) and the other (theoretical terms) for which an account of meaning seems to be required, how is such an account to be given? Russell's *principle of acquaintance* implemented by a device proposed by Ramsey provides, I believe, a satisfactory answer. I shall state the former in semantical terms as follows: All the descriptive (nonlogical) terms in any meaningful sentence refer to items with which we are acquainted. (Russell's usual formulation is to the effect that every ingredient of any proposition that we can understand must be an item with which we are acquainted.) I shall take observation and acquaintance to be coextensive; therefore the principle requires that all descriptive terms be observation terms. This has frequently been interpreted as a reductio for Russell's principle, but it is here of course that Ramsey's method comes to its rescue by transforming the statement of any theory into a sentence expressing all the essential features of the original theory but containing no theoretical terms. As far as I have discovered, Russell was not acquainted with this work of Ramsey, and I should like now to examine the relation of the two approaches to each other.

If I assert (truly, let us suppose) that somebody stole my car last night (but I don't know who it was), then, according to Russell, the person that stole my car is "not an ingredient of the proposition" that I assert. Semantically, this may be put by saying that no sentence expressing the proposition will contain a name (or any descriptive constant) that directly refers to the person who stole my car. My knowledge of that person is *knowledge by description* as opposed to *knowledge by acquaintance*. On one point, I shall now take a position with which Russell might disagree. This does not matter since my primary interest is in developing a theory of theory meaning rather than in Russell scholarship. I shall contend that the proposition does *refer* to the person who stole my car (again assuming that someone did steal it), in some reasonable sense of "refer." We may express the proposition as follows:

(a) $(\exists x)(Px \cdot Sxm)$

where "Px" means x is a person and "Sxm" means x stole Maxwell's car. The "ingredients" of the proposition are my car and the properties (or relations) of being a person and stealing—all with which I am acquainted. Nevertheless, in my sense of "refer" the proposition refers to the person who stole my car although he is not an "ingredient" of the proposition. In my interpretation, this means that this reference is not made directly by any descriptive term (or group of descriptive terms) but is accomplished indirectly by means of the *logical form* of the proposition and by descriptive terms all of which *directly* refer to some other entity. Thus in my interpretation of Russell's principle of acquaintance, we are able to refer to items by means of descriptions (definite and indefinite), even though we are not acquainted with them. This does not violate the principle of acquaintance since such reference is accomplished entirely by means of (1) terms whose *direct* referents are items of acquaintance and (2) items of a purely logical nature such as variables, quantifiers, and connectives. (Although Russell has developed a detailed theory of the nature of logical notions, I shall ignore this here and simply proceed as if they were unproblematic.)

The principle of acquaintance, thus interpreted, is obviously purely epistemological in character and has no ontological implications at all. The ontological status of the guy that stole my car is in no manner impugned by the fact that he is not an "ingredient" of the proposition referring to him and that my knowledge of him is by description rather than by acquaintance. In fact this particular proposition, if true, ensures that he has exactly the same ontological status as does any person with whom I am acquainted.[1] Although I shall not so argue at any length, I believe that all of this pretty much coincides with Russell's own views, especially with those developed any time after about 1927 (the publication date of *The Analysis of Matter*). For, in major works such as *The Analysis of Matter*, *Human Knowledge: Its Scope and Limits*, and *My Philosophical Development*, he explicitly and vehemently rejects phenomenalism and related views and defends a strong realism regarding the physical world while, at the same time, he strongly maintains the principle of acquaintance.

There is a slightly different way of talking about the referents of expressions of knowledge by description. Consider the propositional function expressed when the quantifier of the sentence in question is omitted:

(b)   Px · Sxm

Suppose it turns out that Smith stole Maxwell's car. Substitution of Smith's name for the free variable "x," then, would transform (b) into a sentence expressing the truth that Smith stole Maxwell's car.

(c)   Ps · Ssm

Consider one other example. Consider the assertion that someone stole Maxwell's car and someone else delivered it to the police:

(d) $(\exists x)(\exists y)[Px \cdot Py \cdot x \neq y \cdot Sxm \cdot Rym]$

and the corresponding function:[2]

(e)   Px · Py · x≠y · Sxm · Rym

---

[1] For convenience only, in this example, I am using the common-sense framework which permits acquaintance with material objects, other persons, etc. Strictly speaking, no such acquaintance exists according to Russell; and I agree.

[2] Throughout this paper, "function" is used in the sense of "propositional function."

Suppose that, in fact, Smith stole it and Brown returned it. Substitution of Smith's name "s" for "x" and Brown's name "b" for "y" again transforms the function expressed by (e) into a true proposition. This may be expressed, alternatively, by saying that Smith and Brown, in that order, constitute a *model* of the function (e). In general a model of a function is an ordered n-tuple that, in the manner indicated above, *satisfies* the function, i.e., transforms the function into a true proposition. (Quite trivially, of course, Smith is a model for the function (b).) Now, obviously, to assert (a), i.e., to assert that someone stole Maxwell's car, is equivalent to asserting that there exists at least one value of the variable "x" in the function (b) that names an individual who stole the car, which, in turn, is to say that there exists at least one *model* of the function (b). Similarly, to assert (d) is equivalent to asserting that (e) has at least one model. Quite generally, Russell's treatment of knowledge by description by means of existential quantifiers and other logical apparatus is equivalent to the model-theoretic treatment just sketched. In the first example, (a) tells us that someone stole the car, but contains absolutely no information about who did it; or, in *model-theoretic* terms, we know that the function (b) has a model but have no knowledge at all about what the model is.[3] Questions about matters such as how we come to know propositions like (a) assuming that we do, or, in general, how claims of knowledge by description are validated or confirmed, assuming that they sometimes are, are, of course, irrelevant here. They are problems for confirmation theory but have nothing to do with problems of meaning and reference.

Since I am equating *observing* with *being acquainted*, all knowledge of theoretical entities must be knowledge by description. The principle of acquaintance thus precludes the use of terms that refer directly to unobservables. All such reference must be accomplished by means of observation terms and logical terms. Theoretical *terms*, thus, are not permitted. But we shall see that this no more precludes our having knowledge about unobservables and referring (by description) to them than does my not knowing who stole my car prevent me from knowing (at a reasonably high degree of confirmation) that somebody stole it or from referring to whoever did it by means of description. The principle of acquaintance is per-

---

[3] Expressions containing *definite* descriptions may be treated in Russell's manner, e.g., "The author of Waverly is male" becomes $(\exists x)[Mx \cdot (y)(Wyw \equiv x = y)]$. Or they may be given the equivalent model-theoretic treatment either by asserting that $Mx \cdot (y)(Wyw \equiv x = y)$ has a model or, more concisely, that $Mx \cdot Wxw$ has exactly one model.

fectly consistent with realism regarding the unobservable and in no way should it be taken as providing aid or comfort to phenomenalism, instrumentalism, or operationism of any variety.

To extend the treatments outlined above to theories, it is convenient to begin with the model-theoretic approach. Models, of course, are ordered sets of elements, which, in addition to individuals (as above), may be—indeed usually are—properties and relations (or sets, including sets of ordered n-tuples). Consider for example the function

(f) $(x)(y)[(Sx \cdot Sy) \supset (Rxy \supset \sim Ryx)]$

where "S" and "R" are free predicate variables. If "S" is replaced by a predicate, let it be "M" designating the property of being a physical object and "R" by a predicate designating the dyadic property of being heavier than, let it be "O," then a true proposition is obtained:

(g) $(x)(y)[(Mx \cdot My) \supset (Oxy \supset \sim Oyx)]$

Thus the ordered set, (M,O), satisfies (f) and, consequently, is a model of (f).

Suppose that instead of singling out a particular model of (f), we merely assert that it has at least one model. As an alternative to asserting this in the manner of the preceding sentence, we can again use an extension of Russell's approach and employ descriptions constructed with the aid of existential quantifiers to obtain

(h) $(\exists S)(\exists R)(x)(y)[(Sx \cdot Sy) \supset (Rxy \supset \sim Ryx)]$

As with previous cases, asserting (h) is obviously equivalent to asserting that (f) has at least one model.

It will be helpful, at this point, to adopt a small amount of terminological apparatus. A propositional function that contains no descriptive terms (i.e., contains only logical terms—variables, quantifiers, and connectives) will be called a "pure function." (An example is (f).) One that contains one or more descriptive terms (and, of course, at least one free variable) is a "mixed function," e.g., (b). Let us call a model of a mixed function in which the descriptive terms retain their original meaning a "common model." The sense of "model" so far used in this paper has been just that of "common model," and I shall continue this practice unless there is explicit notice to the contrary. It is demonstrable, of course, that any pure function, unless it is formally inconsistent, has a model. But it obviously does not hold in general that mixed functions necessarily have common models; in most cases, whether or not such models exist will be a contingent matter of fact.

$$\text{x is king of France in 1968}$$

has no common model because of the contingent fact that France is no longer a monarchy.

$$\text{x is president of the U.S.A. in 1969}$$

has as its common model an individual named Nixon because of the contingent fact that the Democrats saw fit to nominate Hubert Humphrey as his opponent.

Let us now consider mixed functions in which some or all of the free variables are predicate variables (either monadic or polyadic or some of both), e.g.,

(i) $(x)[\psi x \cdot \phi x \supset (\exists y)Cy]$

where "ψ" and "ϕ" are predicate variables and "C" is a predicate constant. As before, of course, to assert that (i) has at least one common model is equivalent to asserting the proposition

(j) $(\exists \psi)(\exists \phi)(x)[\psi x \cdot \phi x \supset (\exists y)Cy]$

Now let "C" designate the property of being a click in an appropriately located and properly constructed Geiger counter. Then if we somehow knew the truth of a proposition such as

(k) $(x)[Ax \cdot Dx \supset (\exists y)Cy]$

where "Ax" stands for "x is a radium atom" and "Dx" for "x radioactively decays," then we would also know that (i) has at least one model, viz. (A,D), and, equivalently, that the proposition (j) is true. Both are logical consequences of (k)'s being true, although the converse entailment, of course, does not hold; (k) is logically stronger than its two consequences just mentioned. Now according to the principle of acquaintance, not only can we never know propositions such as (k), we cannot even understand them, for "A" and "D" purport to refer directly to unobservable kinds of particles and events. In no way, however, does it preclude our understanding propositions such as (j) or, even, our knowing that they are true or highly confirmed. (How this may be known is, again, a problem for confirmation theory, and does not concern us here.) Exactly the same holds for the assertion that (i) has at least one model. Now just as we knew that somebody, we called him x, stole Maxwell's car, although we did not know who did it, we also know that there are two properties (or, alternatively, two sets), call them ψ and ϕ, such that if anything exemplifies both of them, an audible Geiger counter click will occur, although we have no knowledge of *what* properties ψ and ϕ actually are. Our knowledge of them is by *description* and, as in all such cases, we refer to them not by predicate constants, but indirectly by means of purely logical terms plus an observation term, in this case, "C." As is well known, (j) is the Ramsey *sentence* of the little "theory," (k). In general, the Ramsey sentence of a theory is formed as follows: Form the conjunction of a set of statements sufficient to express the content of the theory, and then replace each theoretical term by an existentially quantified variable of appropriate type, making the scope of each quantifier the entire conjunction. The considerations above make it apparent that asserting the Ramsey sentence of a theory is equivalent to doing the following: Form the conjunction as above and replace each theoretical term by a free variable of appropriate type. The result will be called *the mixed function of the theory*. Finally, assert that this mixed function has at least one common model. (A *common model* of a theory's mixed function will be called an *intended model* of the original theory.)

Thus either the Ramsey device or the equivalent model-theoretic procedure will enable us to apply some of Russell's insights about descriptions and to maintain his principle of acquaintance for any scientific theory. The resulting Ramsey sentence (or model-theoretic assertion), if otherwise satisfactory, solves the notorious problems about the meaning of theoretical terms in a very decisive manner: The theoretical terms are completely eliminated; the only nonlogical terms that remain are observation terms. It is well known and easily demonstrated that the Ramsey sentence has exactly the same observational consequences as does the original theory. This fact, crucially important though it is, has been used as a basis for some mistaken and very misleading inferences. From it, together with the fact that theoretical terms are eliminated, it has been inferred that this approach also eliminates reference to theoretical entities, or at the very least represents an attempt to do so. From the considerations above it should be obvious that this is entirely mistaken. This approach offers no support whatever for the view that observations are the main concern of science and that the purpose of theories is simply to order them and to predict future ones. The Ramsey sentence of any good theory does this, but it, again like any good theory, does much more besides. It gives us knowledge about the universe and how it operates. It differs from an un-Ramsified theory only in that it does not attribute to unobservable things properties that, as far as we know, belong only to observables. It is just as amenable to a realist interpretation as is any un-Ramsified theory.

The Ramsey sentence refers to theoretical entities in exactly the way in which *any* description refers—by means of variables, quantifiers, logical connectives, and descriptive terms whose *direct* referents are other than the referents of the description. This, incidentally, may be taken as an explication of the claim of Russell and others that our knowledge of the theoretical is limited to its purely structural characteristics and that we are ignorant concerning its intrinsic nature. *Structural characteristics* may be taken to be just those that are not intrinsic and can be described by means of logical terms and observation terms. Intrinsic properties are those that are, or could be, direct referents of predicates. For example, *red* is an intrinsic property and *transitivity* is a structural property. However, such examples may mislead. For one thing, not all structural properties are also purely formal as is transitivity; in fact those referred to by scientific theories rarely are. Like transitivity, however, structural properties are always of a higher logical type; they are properties of properties, properties of properties of properties, etc. (or sets of sets, etc.), while intrinsic properties are almost always first-order properties.[4] In order to prevent another misunderstanding, it should be noted that although the Ramsey sentence gives us knowledge only of the structural properties of the theoretical and leaves us ignorant of its intrinsic (first-order) properties, it does refer by means of description to unobservable intrinsic properties.[5] That is, the ignorance concerning intrinsic nature is ignorance concerning what the intrinsic properties *are*; if we know the Ramsey sentence, we know that there are such properties and we know something about them even—we know something about their structural properties. Our knowledge of them is knowledge by description. Exactly analogous again: Our ignorance about the car thief was ignorance about who it was, not ignorance about whether there was such a person. We knew this and more perhaps; for example, judging from his footprints, we might know that he was very

[4] I say "almost" to allow for the possibility that such things as *being a color* are second-order properties and also are intrinsic, nonstructural properties. Were it not for such difficulties, it might be advisable to drop the word "intrinsic" here entirely in favor of "first order," for "intrinsic" is sometimes used to refer to monadic properties as opposed to relational ones; and this would be entirely out of order here, since there exist many polyadic intrinsic (in our sense) properties.

[5] We might say "to *sets* of unobservable individuals" for the benefit of the nominalistically inclined.

large. (The analogy holds for the car thief but it is one logical type lower. Here our ignorance concerns logical type *zero* and we have some knowledge at logical type *one*, e.g., the individual has the first-order properties of being a thief and being large.)

In principle, talk of structural properties, intrinsic properties, etc., is superfluous; for a grasp of the meaning of the Ramsey sentence provides the whole story. In practice, however, this manner of speaking is very convenient for discussing some of the implications of the Russell-Ramsey approach for a number of traditional philosophical problems. For example, since we do not know the intrinsic properties of the theoretical entities of brain physiology, the possibility is open that they are thoughts, feelings, percepts, etc. Such a mind-body monism has none of the air of paradox and absurdity of traditional materialistic monisms. For once it is seen that physical science cannot give us all the facts about the physical since it leaves us ignorant about its intrinsic properties, no prima facie obstacles prevent us from speculating that some of the brain's intrinsic properties may be mental ones. Exemplifications of such properties would be events that are just as much in the "physical" causal network and just as much in time *and* space as any other events. So, if one can understand (not believe, necessarily) the claim that what we are directly aware of are events in our minds and the claim that mental events *just are* exemplifications of certain intrinsic properties of the brain, then it should be easy to understand (again, not necessarily to believe) Russell's unjustly hooted contention that when a physiologist examines another person's brain, what he really sees is a portion of his own brain. The fact that an event ontology seems the best for contemporary physics makes Russell's claim more plausible and its meaning even more clear; according to such an ontology the brain, like any physical entity, would consist entirely of events, and there is nothing that forbids that some of its parts, i.e., some of its constituent events, be mental ones. "Physical" should be defined as *being in the causal, space-time network*; Russell's monism could then be put by saying that all mental events are physical events, and (redundantly) that some physical events (i.e., those just mentioned) are mental events, and that whether all physical events are mental or proto-mental is a matter about which we are ignorant.

After this digression, let me set up a few straw men and have a little fun with them. It might be objected that, according to this approach, before Ramsey all theories were meaningless; and if this is true it might be asked

189

what accounts for their considerable success and for the fact that intelligent women and men have thought them to contain not only meaning but truth as well? There are two replies, each of which is sometimes applicable depending on the case. One could grant that they were, strictly speaking, meaningless but would point out that they were formally well formed and consistent and formally yielded the same observational consequences as their Ramsey sentences. The other reply would hold that they were meaningful but false—false because they covertly attributed observational properties to unobservables; users of the theory incorrectly imagined theoretical entities as tiny replicas of and as having the same properties as common-sense observables. But although false, they were close to the truth (had high verisimilitude to use Popper's terminology), for although they attributed wrong intrinsic properties to the unobservables, they ascribed the true structural ones, so that, again, the observational consequences were correct.

Another apparent difficulty is monolithicity of the Ramsey sentence. For example, even to make one singular statement about an unobservable, e.g., to say something about the electron that just entered this counter (speaking pre-Ramseyese), we would have to expand the long conjunction under the scope of the quantified predicate variables to include the correlate of the singular statement and then assert the whole Ramsey-sentence-expanded-to-include-correlate-of-singular-statement mess. But the difficulty is only apparent. The simplest reply is that the Ramsey version of a theory, although the true version or, at any rate, the closer-to-the-truth version, is needed only to illuminate or solve certain kinds of problems. For most purposes, such as the practical ones of workaday science, we can just use the un-Ramsified theory, even making "graven images" of theoretical entities, to use a metaphor of Feigl's. The results will be satisfactory for the reasons given in the preceding paragraph. Moreover, there are at least two methods of removing the monolithicity. One is due to Carnap, who devised a method of *explicitly defining* each theoretical term (you heard me) in terms of Hilbert's "ϵ" operator and a modified portion of the Ramsey sentence.[6] When the definitions are substituted into the original theory for the defined terms, the Ramsey sentence is obtained. Thus we may just use the theory in its original form, recognizing that each theo-

---

[6] R. Carnap, "On the Use of Hilbert's ϵ-Operator in Scientific Theories," in Y. Bar-Hillel et al., eds., *Essays on the Foundations of Mathematics* (Jerusalem: Magnes Press, 1961).

retical term is merely an abbreviation for a (very long) expression that contains only logical terms and observation terms. The considerations above should make it abundantly clear, however, that this entirely satisfactory result of Carnap's no more eliminates reference to *theoretical entities* than does any use of the Ramsey sentence. For the other method of eliminating the monolithicity, see an excellent article by Herbert Bohnert in *Philosophy of Science*.[7]

Another objection to this approach holds that while the use of existentially quantified variables is, of course, legitimate, nevertheless it must not be *in principle* impossible that actual instances should be discoverable so that instantiations may be made. There is more than one reply. First of all, such a requirement seems totally unwarranted; to use one of Russell's examples: "There are finite numbers higher than anyone will ever think of" is perfectly meaningful and, incidentally, true, although it is impossible in principle that the existentially quantified variable can ever be replaced by the name of a number. Also, it is, in the case of the Ramsey sentence, questionable whether instantiation is, *in principle*, impossible. God or Omniscient Jones, who acquire knowledge by direct, noncausal cognizance, *would* be able to effect such instantiations.

Finally, it might be held that the approach runs afoul of the notorious difficulties in dividing terms into an observational class and a theoretical one. First of all, we should remind ourselves that the approach is applicable no matter how the division into two classes is made; as long as the meaning of the terms in one class is relatively unproblematic and as long as the truth values of at least some sentences containing these terms are determinable in a relatively unproblematic manner, this approach seems quite helpful for solving problems about the other class of (more problematic) terms. Whether the division is made on the basis of observability or whether, following the suggestion of Hempel in this volume, the classes are old terms (and thus relatively unproblematic) and new ones (and, so, more problematic), or whether still some other basis is used, the Ramsey approach can still be profitably applied. Moreover, if the common-sense observation framework is abandoned in favor of one in which the referents of observation terms are always ingredients of mental events, most of the difficulties that were present in distinguishing common-sense observability from unobservability disappear. Russell holds, of course, that in the only

---

[7] Herbert Bohnert, "Communication by Ramsey-Sentence Clause," *Philosophy of Science*, 34 (1967), 341–347.

viable sense of "observe directly," observables do have this nature and that the common-sense framework is incompatible with current scientific knowledge, and must, therefore, be abandoned when our goal is to get as close to the truth as possible. That he is entirely correct is. I believe, as certain as any of our other current knowledge, but this view will not be defended here.

In this paper an attempt has been made to apply Russell's principle of acquaintance and some of his work on descriptions, augmented either by use of the Ramsey sentence or by model theory, to some of the main problems of the theory of scientific theories. The problem of the meaning of theoretical terms has been eliminated while, at the same time, a means of explicating a realist interpretation of theories is provided. The approach also yields an explication of the view of Russell and others that our knowledge of the physical world is limited to its purely structural aspects. A framework for a view that might be called "structural realism" is thereby provided.

Notes on Feyerabend and Hanson

I should like to make some very schematic points, largely about Feyerabend's views, in part about Hanson's, with a view to raising certain fundamental questions.[1]

Feyerabend asks us to consider a case in which there are two incompatible theories T and T', from which, respectively, C and C' are derivable and are, in the language of T, observationally indistinguishable. In addition, another observational proposition M is derivable from T' but not from T. We are to suppose that M is true, from which we are to conclude, in this context, that T is refuted. Let us concede the case and consider the implications. One must suppose that if M refutes T and is an observational statement, then the meaning of M remains constant for T and T'. But if it remains constant for T and T', T and T' being incompatible, M cannot be theory-laden in the sense that the meaning of the observational terms of M cannot depend on the incompatible theoretical statements that are being tested. In general, if the competing theories are construed as entire systems that are tested by M, the observation statement M is theory-free in that the meaning of its terms does not depend on the meaning of the theoretical terms of those systems. Alternatively, the testing scientist must possess a language L in which the incompatibility of T and T' may be identified as well as the refuting force of M. This must be true even if M is incommensurable with the concepts of T, so that no expectations whatsoever regarding M may be formulated from the vantage of T. But to say this is to say that observation statements may be theory-free, in the relevant sense.

Now, it is altogether reasonable to say, as Feyerabend has said, that one has no satisfactory theory of meaning that squares at once with the change in theory from T to T', the indistinguishable nature of C and C' from the

[1] This note was prepared on the basis of my recollections of the remarks made by Professors Feyerabend and Hanson at the Minnesota Conference of May 1966.

point of view of T, and the fact that M refutes T, is incommensurable with the concepts of T, and is derivable from T'. This may be fair; but if it is, it is not in keeping with this very concession to insist that observation sentences are theory-laden or that theories somehow unilaterally generate their own observation vocabularies. These are simply reverse sides of the same coin: not to have an account of the one is not to be able to assert the other. Nevertheless, there may be a sense, utterly different from that which apparently is to obtain in Feyerabend's instance, in which observation sentences may be said to be theory-laden. The difficulty lies in making this sense precise, though it is important to note that its clarification cannot possibly depend on the sort of instance Feyerabend has asked us to consider. Furthermore, the question cannot possibly be a question for science since, in the respects in which science is at stake, as we have seen, observation sentences must be theory-free.

As to Hanson's account, I take it that he would not subscribe to the view I am attributing to Feyerabend. In any case, it strikes me that his own illustrative example, the development of fluid mechanics, argues that observational terms may be theory-free in the sense alleged, precisely because, on his own admission, there were no relevant theories available that might have borne a relationship to observation sentences at all like that obtaining in Feyerabend's example.

Please notice that I do not say "relatively theory-free" but rather "theory-free" simpliciter. The notion that an observation language is relatively theory-free or theory-laden must relate to some sort of philosophical analysis of the observation language of science that is, for such issues as Feyerabend and Hanson raise, theory-free simpliciter. The argument does not require, however, that one must be able to enumerate individual observation sentences that are or remain theory-free forever, nor does the argument deny that particular observation terms may be shown to depend on changes in the meaning of theoretical terms. Alternatively, to demonstrate that in fact certain observational terms are altered in meaning presupposes once again that, in the relevant sense, our observation language is theory-free.

The beauty of the argument, if it is a genuinely good one, is precisely that it avoids a hasty solution of the vexed question of the distinction between changes in meaning and changes in belief. To put the point in the form of a dilemma: Either T cannot be refuted on the strength of independent observations (however conveniently provided by attending to the implications of some competing or incompatible theory) or else the view, relevant in our context, that observation statements are theory-laden is itself incoherent.

On the philosophical or second-order question of whether observational terms are theory-laden, it must be conceded that if observation sentences are used to talk about the world, changes in the meaning of observation terms cannot be said, as such, to effect changes in the observables themselves. Moreover, observational terms whose meanings may admittedly change (and, in some cases, ascribably to the effect of changes in the meaning of theoretical terms) must be acknowledged to be theory-free in the sense that they are used in sentences in order to speak about observables that remain identifiably the same despite such changes in meaning. Feyerabend concedes that observables themselves, e.g., hands, operations, pointer readings, machinery, will not change in spite of the fact that our description of these observables, infected by the alternative theories to which we subscribe, may change as well as our theoretical concepts. He fails to see that if he concedes, as he must, that these observables will remain the same, there must be some residual descriptive language that is theory-free (in the relevant sense) in virtue of which the identity of such operations, machinery, and the like may be detailed.

As a first approximation to a satisfactory account, I suggest that when we speak of terms as theory-laden in the context of science we should consider the history and analysis of the use of these terms distributively; and that when, in the same context, we speak of terms as theory-free, we should consider the function of the entire set of observation sentences that compose our language as in testing the tenability of competing theories. "Theory-laden" and "theory-free," therefore, are not, in this context, coordinate expressions. But to say this is to take the bite out of the charge that our observation language is theory-laden.

The Crisis in Philosophical Semantics

While concern for the nature of theoretical concepts, so prominent in the recent history of philosophical inquiry, is currently at ebb, this is much less the quiescence of achieved consensus than it is an exhaustion of nerve. To be sure, the positivistic thesis that theoretical terms are cognitively meaningful only where they are equivalent to observational constructs is now dead and past mourning. But its execution and burial has exacted so severe a toll in analytic energies that little heart remains to acknowledge that this essentially negative achievement has left as obscure as ever what the cognitive properties of theoretical expressions in fact are. Which is a pity, for by the late 1950's the empiricist analysis of scientific theory had pushed to the brink of what could have been—and might still become—a revolutionary breakthrough in the philosophy of cognition. However, the dominant style of philosophical argument, persuasive and holistically critical rather than discovery oriented, has severely impeded realization of this prospect. By "persuasive and holistically critical," I mean dialectic which seeks primarily to recruit allegiance to some favored doctrine while treating any flaw or discomfiture in prima facie competing doctrines as sufficient grounds for their total dismissal. This is, however, a singularly inept way to seek insight into complex problems, for by failing to discriminate central issues from secondary details it remains blind to the rational structure of the problem. In contrast, the natural sciences have long since learned to appreciate the power of idealized approximations ("models") which highlight the essentials of phenomena too intricate to be grasped at the outset with errorless accuracy, and the importance of "robust" conclusions which are largely indifferent to the particular details of simplifying assumptions.

No matter how one approaches the analysis of theories, those questions which are central to the nature of theoretical concepts per se are so extensively laced through with more general problems of epistemology, ontology, and semantics that progress on the former can be achieved only by way of simplified working assumptions about the latter, resulting in an inevitable untenability of any such account in its entirety regardless of how sound its conclusions specifically about the theoretical aspects of language may be. Deeper penetration into the nature of theoretical expressions thus requires not the issuance and rebuttal of position papers but careful, nondoctrinaire problem analyses which attempt first of all to search out just what the various specific issues actually are, and then to see whether any significant conclusions regarding some of these can be found which are essentially independent of the stand one might take on the others. And if occasional simplistic presuppositions or not wholly realistic idealizations are needed to cut through otherwise impenetrable tangles of complexity, the edifying reaction to them is not holistically to spurn the inferences which follow with their aid but to ask just what difference, if any, they actually make for the latter, and how the idealized model can most insightfully be emended to approximate more closely the intricacies of literal reality. In what follows, I shall be concerned not so much with specific conclusions about particular theoretical expressions as with the structure of effective inquiry on this matter, and the semantical crisis which thereby emerges as a robust consequence of rejecting positivism.

What philosophical problems do theories present? Most can be regarded, in one sense or another, as an instance of the question *What do theoretical expressions mean?* Regardless of what might specifically be understood by "mean," here, this formulation contains a critical presupposition and implies a major directive. The presupposition is that there exist linguistic expressions of a "theoretical" sort which are effectively distinguishable from the remaining "nontheoretical" portion of our language, and whose semantical properties are in some ways more problematic than those of the latter. And the implied directive is that our primary objective is to clarify the status of theoretical expressions in these special respects, i.e., those wherein they are more problematic than their nontheoretical counterparts. Consequently, any allegation about the semantics of theoretical expressions of a given logical type which was assumed at the outset to be true in complete generality about all expressions of that type is not here a cogent conclusion; while conversely and more importantly, any flaw or implausibility to be found in a particular account of theoretical meaning need cast no aspersion on that argument's claims distinctively about *theoretical* concepts if this blemish lies in its universal background assumptions. Quite apart from the theory problem, philosophical analysis of cognition still has rocky going on even the simplest aspects of language, and it is essential that we remain sympathetically tolerant of idealized presuppositions about the semantics of nontheoretical expressions if these provide first-approximational leverage on distinctively theory-meaning questions.

We have noted that the "theory problem" presupposes a significant contrast between theoretical and nontheoretical concepts. But does such a difference actually exist? Or put operationally, how do we distinguish terms which are "theoretical" from those which are not. In the heyday of positivism the definition was straightforward enough: A descriptive, i.e., extralogical, term was "theoretical" if it purported to designate something which its user had not directly observed, i.e., if its referent was not in some sense an experiential "given." But now that the doctrine of sense data is in ill repute while observability is increasingly conceded to be a matter of degree rather than a categorical absolute, the old notion of "theoretical" will no longer suffice, especially in the face of current contentions that all so-called observational reports are in reality theory-laden. Just the same, de facto scientific practice does authorize an observational-theoretical distinction even if only a relative one. Empirical research in the natural sciences consists of (a) noninferential acquisition of certain beliefs which, though not incorrigible, command a high degree of conviction, and (b) derivation from these of an appropriate degree of confidence in various other propositions for or against which the former serve as evidence. These *inferentially basic* propositions of a science are, by definition, its "datum" beliefs, while the language generated from the terms and syntax of its datum beliefs, including all concepts explicitly definable from these, is its "observation language." However, the propositions upon which the more advanced sciences attempt to pass inferential judgment are generally not in the science's de facto observation language insomuch as they often contain terms which do not occur, nor can they be definitionally reduced to those which do, in any of the science's datum beliefs. Such terms are by definition "theoretical," and any grammatically well-formed phrase or sentence which includes one or more theoretical terms, i.e., any expression which, though not in the observation language, is in the language formed by adding the science's theoretical terms to its observational vocabulary, is a de facto theoretical expression for that science.

To be sure, this delimitation of "theoretical" terms by reference to extant inferential practice has been highly simplified. For one, I have spoken glibly of datum beliefs as though these are clearly and unambiguously recognizable as such, whereas in fact the propositions accepted as a basis for inference in the indicative mood vary considerably from person to person and even from moment to moment within the same person (and not merely in that new observations continue to be made while older ones fade from memory, either), so that the observational-theoretical distinction is at best relative to a particular person at a particular time. Again, we can easily imagine that some terms which count as "theoretical" by the structure-of-inference criterion might be perfectly capable of occurring in a given person's datum beliefs at a given moment even though he chances not to hold any which utilize it just then. Further, there is the nuisance of common-sense observational terms, i.e., those in which everyday perceptual judgments are couched, whose lack of clarity usually disqualifies them for use in technical datum reports even though they should be explicitly definable in the technical observation language. For example, we would not wish to count "tall" as a theoretical term when determining whether John is tall from the datum that John is 74 inches in height, for whether or not a person is "tall" should be analytically decidable from his quantitative height even though this term's imprecision thwarts our making this inference with complete deductive confidence. Even so, despite such complications, the empirical differences in technical vocabulary between a science's noninferentially believed datum assertions on the one hand and its nondemonstratively inferred conclusions on the other strongly urge that de facto language practices at the highest levels of cognitive sophistication do indeed sustain an epistemically significant observational-theoretical distinction, one which can as a first approximation be put as follows:

> *Heuristic Simplification 1* [HS-1]: If the total vocabulary of a language L is partitioned into two subsets V_T and V_0, the terms in subvocabulary V_T are all "theoretical" with respect to subvocabulary V_0 (which is then "observational" with respect to V_T) if (1) none of the terms in V_T are analytically equivalent to any expression constructed wholly from terms in V_0, and (2) the credibility of any sentence[1] S_T in L containing one or more terms in V_T derives entirely from the credibilities of certain sentences containing only terms in V_0 by way

---

[1] Throughout this paper I shall understand 'sentence' in the sense of sign-design-cum-meaning, so that it is legitimate to speak of "sentences" as having cognitive properties such as credibility.

of S_T's logical structure and logical relations to these sentences. *Heuristic Simplification 1a:* Some extant languages do, in fact, contain theoretical terms in this sense.

For simplicity, *HS-1* construes "logical structure" and "logical relations" to include not merely formal properties which are syntactically explicit but also those which appear under meaning analysis. For example, if "a" is synonymous with "b," "a = b" has the logical form $x = x$ while "P(a)" logically entails "P(b)." Presumably, all logical terms in L—"and," "or," "all," etc.—belong on the observational side of any observational-theoretical partition of L, though we should not ignore altogether the possibility of theoretical logical terms. Inasmuch as we shall make no technical use of *HS-1*, its lack of explicit detail is not here a crippling defect.

It is important to note that *HS-1* views the observational-theoretical distinction as relative, not absolute. Specifically, it leaves open the possibility that an observational subvocabulary V_o may be further partitioned into subsets V'_o and V''_o such that V''_o is theoretical with respect to V'_o. Thus *HS-1* is entirely compatible with the prospect that theoreticity is a matter of degree rather than of kind, and in particular that a language may well be hierarchically stratified in such fashion that the concepts in each stratum are theoretical with respect to the layers below it but observational with respect to those above.

Now consider the basic empiricist intuition about the structure of cognition, which logical positivism carried to a well-intentioned but untenable extreme. Purely for the sake of convenience, I shall speak as though

> *Heuristic Simplification 2 [HS-2]:* A person believes, disbelieves, thinks, hopes, understands, etc., that p only if there is a sentence in his language which asserts that p.

(This is almost certainly not literally true. People, and probably the higher animals, most assuredly have occasional beliefs which cannot be expressed by the linguistic machinery then available to them. But, insomuch as propositions and concepts, i.e., meanings, bereft of words to convey them scandalize many contemporary philosophers, while in principle a person can always enrich his language to express any concept or proposition which he can think, *HS-2* is essentially harmless.) Stripped to fundamentals, the classical empiricist thesis is that all knowledge derives from experience. However, since "knowledge" in the stringently ideal sense does not literally exist,[2] this is best put with the help of HS-2 as a claim that (1) the vocabulary of any person s contains a subset V_0 such that all s's beliefs depend for their various credibilities entirely upon s's belief in propositions expressible wholly in V_0 terms if they are not themselves so expressible, and (2) all descriptive terms in V_0 refer to entities with which s has experiential acquaintance. Exactly what is to be understood by 'experiential acquaintance' is hard to say, but here this is no matter; the important point is that empiricism envisions the existence of a *credibility base*, the concepts in which have a "given" status, even if justified belief is not necessarily restricted to what can be expressed by these givens alone. Moreover, the empiricist credibility thesis has a semantical counterpart: *All expressions in a person's language either depend wholly for their meanings upon their usage in relation to expressions whose descriptive terms refer only to entities with which that person is experientially acquainted, or are themselves such expressions.* Again the interpretation of 'experientially acquainted' is much less important than the notion that every language has a *meaning base* from which all its expressions draw their cognitive significance, and that this meaning base consists essentially of concepts found in the language's credibility base. The connection between concept development and knowledge acquisition is so intimate that the empiricist view of meaning is intuitively a corollary of the empiricist credibility thesis. But if this intuition is unconvincing, how do symbols acquire cognitive meaning? Even allowing for a possibility of innate concepts, any term whose meaning is neither innate nor given by an "ostensive" relating of words to experience must be cognitively dependent upon intralinguistic processes wherein an initially arbitrary sign design somehow acquires meaning from its role in linguistic contexts with terms already meaningful —or at least so we must assume until we conceive of some plausible fourth alternative for concept formation. Add to this the further premise that innate concepts either do not exist (this being at best a factual claim since it can be shown not to be logically or nomically necessary) or are in some sense ingredients of nonsensory experience, e.g., given by introspective awareness, and the empiricist semantical thesis results.

Although empiricist epistemology has traditionally emphasized "experience" as the source of knowledge, one could quarrel with this and still accept a liberalized empiricism in which "is an experiential acquaintance

[2] W. W. Rozeboom, "Why I Know So Much More Than You Do," *American Philosophical Quarterly*, 4 (1967), 281–290.

of" is weakened to "is cognitively primitive for." That is, the structure of credibility and meaning acquisitions envisioned by empiricism is indifferent to what may be the nature of this structure's base. With the cognitive role of "experience" thus split off as a separate issue, the structural core of the empiricist meaning thesis becomes

> Technical Idealization 1 [TI-1]: If V_O and V_T are subvocabularies of a language L such that all terms in V_T are theoretical with respect to V_O, then the meanings of terms in V_T depend wholly upon their joint usage with terms in V_O.

While TI-1 intends "theoretical" to be understood in the sense of HS-1, this is not actually essential. The terms in a vocabulary V_T might alternatively be defined as "theoretical" with respect to a vocabulary V_O if (1) all terms in V_T derive their meanings entirely from their usage with terms in V_O even though (2) no V_T-term is analytically equivalent to any expression constructable from terms in V_O. Then TI-1 is analytically true while the empiricist claim now becomes that the new terms which show up in the conclusions of certain scientific arguments are theoretical in this latter sense. Moreover, if there be several epistemically distinct kinds of theoretical concepts (though HS-1 describes the only one I know), we need not suppose that TI-1 applies to all; we can profitably explore the implications of TI-1 for this kind of theoretical concept while remaining perfectly free to recognize additional senses of "theoretical" which TI-1 may not fit.

It should be observed that like HS-1 before it, TI-1 speaks of theoretical terms only in a relative sense. That is, it is compatible with the prospect that terms which have acquired meaning from their couplings with antecedently established terms may in turn themselves provide a meaning base for a higher level of theoretical conceptualization.

I submit that any fruitful dialogue on the nature of theoretical concepts must first of all reach some degree of consensus on the acceptability of TI-1 as a point of departure. To reject it amounts to denying that there is anything distinctive about the meanings of theoretical terms, or at least that any terms exist which are "theoretical" in a sense under which TI-1 is true. Such denial was characteristic of logical positivism, and it also seems to lurk within the recently burgeoning doctrine that all terms, "observational" or otherwise, are really theory-dependent[3] and hence presum-

---

[3] P. K. Feyerabend, "How to Be a Good Empiricist—A Plea for Tolerance in Matters Epistemological," in B. Baumrin, Philosophy of Science: The Delaware Seminar, vol. 2

ably all of a kind. But *TI-1* is entirely compatible with the hypothesis that virtually all terms (or even all without exception, if *HS-2* is dropped) are theoretical with respect to *some* meaning elements; and to make much of the uncontrovertible point that theoretical presuppositions can always be found within the de facto datum beliefs of everyday life and technical science neither discredits the existence of still other theoretical terms whose meanings are in some fashion induced by couplings other than explicit definition with this de facto observation language nor in any way illuminates what is involved in such meaning derivations. In any event, until *TI-1* or something like it is accepted, inquiry into the cognitive status of theoretical terms remains largely a search for what the problem is. In contrast, acceptance of *TI-1* immediately confronts us with a question ordinarily so hedged about with distracting side issues that scarcely anyone has heretofore managed to grasp hold of it, namely, what are the meanings which theoretical expressions acquire from their usage in observation-language contexts?[4]

Undoubtedly one important reason why so few attempts have been made to say *what* theoretical concepts mean is the high degree of sophisticated foresight required to prepare a metalinguistic framework within which a significant answer to this question can even be expressed, let alone defended. For without careful advance planning, the analysis soon begs the question or succumbs to a horde of controversial irrelevancies. In the next few paragraphs I shall list the major strategies available for such an analysis and briefly note their most serious limitations. These strategies differ first of all in what semantical properties of theoretical expressions they attempt to illuminate, notably, theoretical meanings per se versus the referential consequences thereof, and secondly, in the type of metalinguistic assertion by which they seek to convey this information.

In principle, the most direct way to say what is meant by a given theoretical term *t* would be to describe its meaning *as such*, i.e., to say what

(New York: Wiley, 1963); N. R. Hanson, *Patterns of Discovery* (Cambridge: Cambridge University Press, 1958); T. Kuhn, *The Structure of Scientific Revolutions* (Chicago: University of Chicago Press, 1962), especially chapter 10.
⁴ While the philosophic and scientific literature is gorged with discussions of *whether* theoretical terms are meaningful when not analytically reducible to observational expressions, to my knowledge only Carnap and I have actually tried to say *what* the semantical properties of such terms may be. R. Carnap, "The Methodological Character of Theoretical Concepts," in H. Feigl and M. Scriven, eds., *Minnesota Studies in the Philosophy of Science*, vol. I (Minneapolis: University of Minnesota Press, 1956); R. Carnap, "On the Use of Hilbert's ε-Operator in Scientific Theories," in Y. Bar-Hillel et al., eds., *Essays on the Foundations of Mathematics* (Jerusalem: Magnes Press,

happens in a person's head when he understands t. Such an approach would attempt to identify what psychological features of a person's cognitive functioning are the grounds on which expressions containing t are *about* aspects of extralinguistic reality.[5] However, we still know so little about the psychological nature of meanings in even the most primitive psycholinguistic processes that to offer *this* kind of explanation is far beyond our present scientific capacity.[6]

A second way to characterize the meaning of an expression t is by synonymy claims of the form 't means e' or 't is analytically equivalent to E' where E is an expression whose meaning, e, is relatively unproblematic.[7] This is the most traditional mode of meaning clarification in analytic philosophy; however, in the present instance such assertions can only be trivial or question-begging. For if term t in language L is theoretical with respect to subvocabulary V_0 of L while being synonymous with some other expression E, then by definition E cannot be constructed from terms in V_0. Hence either E is itself theoretical in L with respect to V_0 (which may be instructive if E is a composite of theoretical terms other than t, but only if we can then say what these in turn mean), or E must be an expression in some language other than L. Now in the latter case, the statement 't means the same as E' clarifies t's meaning only if we understand E, i.e., only if 't means e' is a metalinguistic assertion at our command in which 'e' is equivalent in meaning to E and hence to t. But if 'e' has the same meaning as L-term t, then 'e' will inevitably be theoretical with respect to the metalinguistic translation of V_0 and we have merely replaced the question of t's meaning in L with the equivalent question of what

---

1961); W. W. Rozeboom, "Studies in the Empiricist Theory of Scientific Meaning," *Philosophy of Science*, 27 (1960), 359–373; W. W. Rozeboom, "The Factual Content of Theoretical Concepts," in H. Feigl and G. Maxwell, eds., *Minnesota Studies in the Philosophy of Science*, vol. III (Minneapolis: University of Minnesota Press, 1962); W. W. Rozeboom, "Of Selection Operators and Semanticists," *Philosophy of Science*, 31 (1964), 282–285.

[5] W. W. Rozeboom, "Intentionality and Existence," *Mind*, 71 (1962), 15–32, especially pp. 22ff, and "The Factual Content of Theoretical Concepts," pp. 339ff.

[6] Although technical psycholinguistics continues to progress, psychology has yet to learn even what differentiates cognitive from noncognitive states of the organism, much less to produce a viable theory of aboutness. Worse, many philosophers seem curiously resistant to recognition that meanings are aspects of what goes on inside people's heads and would hence not appreciate a psychological account of meaning even were a good one available.

[7] For clarification of the 'x means y' schema, see my "Intentionality and Existence," especially pp. 21ff. Essentially, the difference between the two forms 't means e' and 't is analytically equivalent to E' is that the expression 'e' used in the first case is *designated* by the expression 'E' used in the second.

theoretical term 'e' means in our metalanguage. To put the point more briefly, if we seek to resolve uncertainties about an expression's meaning by proffering a metalinguistic *translation* of it, we are just playing the quotes-dropping game, 'e' means e, which is like arguing "Oh, get off it, fellows, we all know what 'e' means."

A third, more devious but ultimately more prosperous route to the meanings of theoretical expressions is through their *referents* or, for theoretical sentences, their truth conditions. For while two expressions may differ in meaning even while sustaining the same extralinguistic connections, the latter are the cognitive payoff of meanings. There are at least three ways to implement this strategy for an expression E_T containing theoretical terms. The first is to seek conclusions of the form 'E_T designates e' or, more weakly, when E_T is a sentence, of the form 'e is a necessary and sufficient condition for the truth of E_T.' My experience with this approach[8] shows it to require rather specific working assumptions on controversial points of ontology which are best avoided wherever possible. Also essential are some semantical principles relating language to reality in virtue of which it becomes possible to arrive at conclusions of this sort. In particular, we need to know the referents or truth conditions of the observational expressions from which E_T draws its meaning. The Carnap-Tarski quotes-dropping gambit—i.e., 'e' designates e, and s is a necessary and sufficient condition for the truth of sentence 's'—furnishes an appropriate if perhaps idealized semantics for the observation language, but of course we cannot treat theoretical expressions this way without begging the question. A further complication is that there may not be any entity designated by expression E_T even though its *meaning* is in no way inferior to that of observational expressions of the same logical type (cf. 'Pegasus'; 'Adam and Eve'; 'Phlogiston'; and the like). Nonexistent referents do not disastrously cripple the usefulness of this approach, however, for we can still derive subjective conclusions of the form 'If e were to exist it *would* be designated by E_T.' While prima facie similar to the futile E_T-has-the-same-meaning-as-e approach, subjunctive assertions of reference differ from synonymy claims in being able to exploit the possibility that two expressions may necessarily, not just contingently, have common designata even though their meanings differ.

Alternatively, we can look for conclusions of the form 'E_T designates the referent of X' (or, when E_T is a sentence, 'E_T and X have the same

---

[8] See my "The Factual Content of Theoretical Concepts."

truth conditions'), where X, which is not necessarily in the same language as E_T, is an expression whose semantical properties are better understood by us than are those of E_T. This strategy is somewhat less beset by onto-logical problems than is the E_T-designates-e approach, but it requires a theory of semantical relations between the language L containing E_T and the language L^* wherein we hope to find E_T's referential counterpart X. (The possibility that L^* is a different language from L must be allowed because we cannot assume that an expression with the same referent as E_T is always constructable from the vocabulary relative to which E_T is theo-retical without positivistically presupposing at the outset that theoretical concepts never expand a language's referential scope.) In particular, while it is probably safe as a first approximation to stipulate that observational expressions in L are synonymously translatable into L^*, we cannot assume this about L's theoretical terms without trivializing the argument—whence it importantly follows that this approach must provide for refer-ential and truth-conditional equivalences between expressions which are not identical in meaning. Side issues about translatability and other com-plications of interlinguistic semantics can here be minimized by letting L^* be an observational enrichment of L, i.e., L^* is generated from L by adding more observational terms to the latter, but then we must guard against the possibility that theoretical terms may undergo shifts in mean-ing as the language containing them expands. And regardless of what language L^* we choose to scan for a referential counterpart to E_T, we must be prepared to find that it may contain no such expression unless L^* is strongly idealized.

Finally, the referents of theoretical expressions can be specified by citing the conditions which such a referent must satisfy, i.e., by arriving at con-clusions of form 'E_T designates entity e iff e satisfies conditions C' or, inter-linguistically, 'Expressions E_T and X have a common referent iff X has property P.' In view of IC-1, below, this approach has perhaps the bright-est prospect of all, but how far it can avoid the technical problems inher-ent in the others is uncertain.

Let us now return to TI-1 and sharpen its cutting edge. What is it for theoretical terms in V_T to have a "joint usage with terms in V_0"? In gen-eral, there seem to be four major ways in which linguistic expressions, theoretical or otherwise, can be "used": (1) They can be grammatically concatenated into more complex expressions. (2) Some can be used to designate extralinguistic entities. (3) Some can be used for description and predication. And (4) some can be used to convey the objects of "propositional attitudes" or "mental acts" such as believing, doubting, desiring, asserting. Most if not all other uses of language, including such interpersonal functions as informing, questioning, commanding, are derivative from these. And since use (1) is but ancillary to uses (2)–(4) while (3) can be subsumed under (4), we may say that the "usage" of theoretical terms in V_T with terms in V_O consists essentially of employing grammatically appropriate expressions constructed jointly from vocabularies V_O and V_T for making reference and expressing the objects of mental acts. Even if this inventory of language uses seems absurdly oversimplified, precisely *how* linguistic expressions are used is not critical just now; the important point is that the usage of V_T-terms through which the latter acquire meaning is by way of compound expressions in which they are conjoined with terms of V_O. And if so, what distinguishes a particular usage for V_T-terms in virtue of which these acquire one array of meanings rather than some other which an alternative usage might have given them? This must reside in what particular expressions are used in these ways, so that TI-1 may be clarified as

> *Technical Idealization 2 [TI-2]*: If V_O and V_T are subvocabularies of a language L such that all terms in V_T are theoretical with respect to V_O, then L contains an ordered set E of criterion expressions, i.e., those through whose "usage" the V_T-terms draw their meanings, such that for any semantic relationship R_s and L-expression E_T containing one or more terms in V_T, there exists a criterion relationship R_c such that whether or not E_T stands in relation R_s to an entity e is determined by whether or not e stands in relation R_c to E. This holds for interlinguistic semantical relations as well as relations of language to extralinguistic reality.

The set E is here described as "ordered" to allow for the possibility that the various expressions in E may differ in their "usage" and hence not participate symmetrically in the determination of a given theoretical expression's R_s-relata. For example, if E consists of three sentences (S_1, S_2, S_3), the theoretical meanings which are generated by jointly believing S_1, doubting S_2, and hoping S_3 are not necessarily the same as those generated by jointly believing S_3, doubting S_1, and hoping S_2. It should be noted that whereas all previous logical-empiricist writings, including my own, have assumed the meanings of theoretical terms to accrue only from *accepted*, i.e., believed, theoretical postulates, TI-2 and its consequences below are entirely open with respect to what propositional attitudes are involved in

207

the concept-definitive uses of theoretical sentences, nor do they require that the set E of criterion expressions comprise only complete sentences. How much need there may be for this increased generality is problematic, but it helps allay suspicion that idealizing the de facto treatment of scientific theories as an acceptance-nonacceptance dichotomy may be far too simplistic to have useful issue.

Now consider the import of saying that theoretical terms acquire their meanings from their context of usage. One implication of this is surely that with other psycholinguistic factors held constant, whether or not an entity e satisfies the R_c-relatedness-to-E criterion for being an R_s-relatum of E_T remains unaffected if the theoretical terms which are given meaning by this context are distinctively replaced, i.e., a one-one interchange, by any other heretofore meaningless sign designs.[9] That is, the V_T-terms in E bring no meaning to this context other than their purely syntactic properties of sameness and difference, so that in respect to whether or not e is R_c-related to E, the V_T-terms in E are dummies which can just as well be distinctively replaced by logical variables or other semantically empty syntactic place-holders without affecting this relationship. Hence,

> Idealized Conclusion 1 [IC-1]: If V_O and V_T are subvocabularies of a language L such that all terms in V_T are theoretical with respect to V_O, then L contains an ordered set E_O of expression schemata containing only terms in V_O such that for any semantic relationship R_s and L-expression E_T containing one or more terms in V_T, there exists a criterion relationship R_c such that whether or not E_T stands in relation R_s to an entity e is determined by whether or not e is R_c-related to E_O.

IC-1 has two extremely important implications. One precipitates the semantical crisis alluded to previously, and will be discussed later. The other concerns the metalinguistic resources required to identify the extralinguistic entities which theoretical expressions are about. Having no grounds on which to suspect otherwise, we may assume that the meaning imparted to a theoretical term by its usage in observation-language contexts depends only upon the *semantical* character of the latter. Consequently, the criterion relation R_c, whose coupling of an entity e to set E_O

---

[9] This assumes, of course, that any grammatical structure which in the vernacular would be combined with theoretical root terms to form lexicographic words would remain unaltered by this exchange. Thus 'Blishes gorp darbishly' may be transformed by replacement of theoretical terms into, for example, 'Cleeves brum tarkishly' but not into 'Cleevish brums tark.'

of observational expression schemata determines whether e is R_s-related to theoretical expression E_T, must itself be a semantical relation of some sort. If so, IC-1 implies that a metalanguage sufficiently rich in semantical concepts further needs to cite only expression schemata constructable from the observational vocabulary V_0 of language L (or to have the use of meta-linguistic translations thereof) in order to describe those conditions which are necessary and sufficient for an entity's standing in a given semantical relation to any particular L-expression which is theoretical with respect to V_0. Or, to make the same point at the object-language level, *a theoretical expression E_T designates or is otherwise semantically related to an entity e if and only if e is semantically related in some to-be-determined way to certain expressions containing only terms in the vocabulary relative to which E_T is theoretical.* Thus an observational vocabulary suffices to characterize the referents of terms which are theoretical with respect to it even when it is unable to designate them.

Further unfolding of IC-1's implications requires considerable assistance from supplementary hypotheses, especially about the specific observational contexts by which particular theoretical expressions are introduced. Even so, we can push the argument a little farther through a few mild assumptions about semantical criterion relations. The class of "semantic" relationships can roughly be delimited as follows:

> *Definition:* A relation R holding between a linguistic expression E and another entity e is *semantical* iff E's standing in R to e depends only upon E's meaning and logical structure,[10] i.e., iff every other expression E', whether in the same language as E or not, which has the same meaning and logical structure as E is also R-related to e.

However, this definition is woefully insufficient to itemize what, specifically, are the various instances which fall under it. While it is altogether possible that there exist basic semantical relations of which we are still ignorant, those meaning-mediated couplings of language to extralinguistic reality which are reasonably well recognized today are the following:

(1)   An expression E may designate (refer to, name, signify, symbolize, represent) an entity e. In particular, this possibility arises when E is a noun or noun phrase while e is a (possibly abstract) particular; when E is a predicate while e is a property, class, or re-

---

[10] The conjunction "meaning and logical structure" is probably redundant, inasmuch as it can plausibly be argued that the logical structure of a complex expression is an aspect of its meaning.

lation;[11] or—though this is more controversial—when E is a sentence or gerundized sentential phrase (e.g., 'John's winning of the contest') while e is a fact, event, or state of affairs.

(2)   When E is a predicate for which a name of e is a grammatically acceptable argument, E may *denote* (describe, apply to, be satisfied by) e.

(3)   When e is a property or class, E may *refer* to an entity which possesses or belongs to e. This is essentially the inverse of denotation.

(4)   When E is a sentence while e is a fact, event, or state of affairs, e may be a *truth condition* of E, i.e., be necessary for E to be true; e may *verify* E, i.e., be sufficient for its truth; or e may *refute* E, i.e., be sufficient for its falsity.

Pending new metalinguistic wisdom, let us assume that these are all the non-interlinguistic types of semantical relations there are, or, more precisely, that any semantical relation between a set E of linguistic expressions and an extralinguistic entity e is a logical construction out of relations of types (1)–(4). Now, an expression E designates an entity e iff e satisfies the predicate '__ $= E$.' (More exactly, '__ $= E$' is the predicate in L formed by concatenation of '$=$' and E, and similarly for subsequent equivocations in this paragraph between mention and use of E.) Likewise, when e is a property or class, E's referent possesses or belongs to e iff e satisfies the higher-level predicate '__(E)' or 'E is a(n) __' or 'E ε __.' Semantical relations of types (1) and (3) may thus for technical convenience be subsumed under denotation, and the question then arises whether any semantical relation between a linguistic expression E and an extralinguistic entity e which is not a state of affairs can be constructed out of basic relations of types (1)–(4) in a way not equivalent to some instance of type (2). (States of affairs are here excluded because it is not at all obvious how type (4) relations might be reduced to denotation.) While rigorous treatment of this question would be excessively tedious here, inspection of representative examples strongly urges that its answer

---

[11] While it is perhaps moot whether 'designation' is the best word for the aboutness relation which holds between a predicate 'P' and the property P which it ascribes to its arguments, recognition of this relation is indispensable to the semantical theory of any reasonably complex language, especially if the latter permits quantification over predicate terms, and I see no reason why it should not be classed under "designation" even if it is conceivably a different variety of reference from that holding between a proper name and a zero-level particular. The reader who wishes to protest that properties are the meanings of predicates rather than what predicates refer to by means of their meanings should first consider my argument to the contrary on pp. 27–30 of my "Intentionality and Existence."

is negative. Clearly this is so for any construction based only on relations of types (1)–(3), such as

(i) E either designates or denotes e; i.e., $\mathrm{Des}(E, e) \vee \mathrm{Den}(E, e)$
(ii)   Predicate E designates the conjunction of e and another property; i.e., $(\exists\phi)\mathrm{Des}(E, \phi \cdot e)$
(iii) E denotes a property of e; i.e., $(\exists\phi)[\mathrm{Den}(E, \phi) \cdot \phi(e)]$
(iv)   Sentence E designates a state of affairs of which e is a component; i.e., $(\exists s,\phi)[\mathrm{Des}(E, s) \cdot s = \phi(e)]$, or $(\exists\phi)\mathrm{Des}(E, \phi(e))$

for in every such case there is a straightforward way to replace the semantical verb and reference to expression E by use of E in a corresponding syntactic context so as to yield a predicate E^*, containing only E and logical terms, such that an entity e satisfies (i.e., is denoted by) E^* iff e stands in the original semantic relationship to E. Thus (i)–(iv) are respectively equivalent to

(i*) e either is or has E; i.e., $(E = e) \vee E(e)$
(ii*) E is the conjunction of e and another property; i.e., $(\exists\phi)(E = \phi \cdot e)$
(iii*) e has a property which possesses E; i.e., $(\exists\phi)[E(\phi) \cdot \phi(e)]$
(iv*) E is a state of affairs of which e is a component; i.e., $(\exists\phi)[E = \phi(e)]$,

where symbol 'E' in (i*)–(iv*) indicates a usage-occurrence of the same linguistic expression designated by 'E' in (i)–(iv). Not so evident, unfortunately, is what can be said of constructions such as

(v) e is a component of a state of affairs which verifies sentence E; i.e., $(\exists s,\phi)[s = \phi(e) \cdot \mathrm{Ver}(s, E)]$, or $(\exists\phi)\mathrm{Ver}(\phi(e), E)$

which incorporate a relation of type (4) by quantifying over the fact variable. So far as I can tell, however, the type (4) ingredients in any such construction always occur trivially. Thus if sentence E has any verifier, s, at all, then for any entity e there exists a verifier of E in which e is a component, namely, the conjunctive state of affairs $s \cdot (e = e)$; hence e satisfies (v) iff it satisfies

(v*) E is the case and $e = e$; i.e., $E \cdot (e = e)$,

wherein 'E' indicates an assertion-occurrence of the sentence designated by 'E' in (v). For now, therefore, it is reasonable to suppose that semantical relations between linguistic expressions and extralinguistic entities other than states of affairs which can be constructed from basic types (1)–(4) can be equivalently constructed from types (1)–(3) alone. If so, we may then provisionally state

William W. Rozeboom

Technical Idealization 3a [TI-3a]: If language L has normal syntactical resources, including the standard logical connectives and operators, then for any L-expression E and semantical relation R_c there exists a predicate E^* in L, containing at most terms in E and additional logical terms, such that for any extralinguistic entity e which is not a state of affairs, e stands in relation R_c to E iff E^* denotes, i.e., is satisfied by, e.

Now consider how an extralinguistic entity e might be semantically related to an ordered set, E, of linguistic expressions. The only possibilities I can conceive of for this are logical constructions out of e's semantical relations to particular expressions in E, such as

(vi) e stands in relation R_s to the first element of E, and either stands in relation R_s' to the second element of E or does not stand in relation R_s'' to the last.

(vii) e stands in relation R_s to at least one of the elements in E.

According to *TI-3a*, we may assume that E is an ordered set of predicates and that whether or not an entity e stands in the given relation to E is a logical construction out of whether or not e is denoted by the various predicates in E—or more precisely, that if this is not true at first it can be made so by certain modifications of the expressions in E without introducing any new extralogical terms. But so long as E contains only a finite number of predicates, any logical construction out of an entity's satisfying or not satisfying the various predicates in E is equivalent to its satisfying a single predicate correspondingly constructed out of the E-predicates. Thus if E is the ordered set P_1, P_2, P_3, P_4 while all the semantical relations in (vi) and (vii) are "is denoted by," (vi) and (vii) are respectively equivalent (with the same shift from mention to use as in previous examples) to

(vi*) $P_1(e) \cdot [P_2(e) \text{ v} \sim P_4(e)]$
(vii*) $P_1(e) \text{ v} P_2(e) \text{ v} P_3(e) \text{ v} P_4(e).$

Once again, then, having no good reason to suspect otherwise, we may extend *TI-3a* to

Technical Idealization 3b [TI-3b]: If language L has normal syntactic resources, then for any finite ordered set E of expressions in L and any semantical relation R_s, there exists a predicate E^* in L containing at most logical terms in addition to terms in E, such that for any extralinguistic entity e which is not a state of affairs, e stands in relation R_s to E iff E^* denotes e.

How likely it is that future semantical theory will discover need for significant emendations to *TI-3* I am unable to say. But no one can profitably spurn its adoption as a working assumption until he can offer a convincing alternative at a comparable level of technical power. Meanwhile, *TI-3* provides a workable reduction for the open-ended intuitive notion of "semantic relation," has strong arguments in its support, and in no way depends for its plausibility on any special presuppositions about the nature of theoretical terms.

Given *TI-3*, *IC-1* leads directly to a profound technical conclusion regarding the semantical properties of theoretical expressions. Since the "usage" which imparts meaning to language *L*'s theoretical terms presumably comprises only a finite number of psycholinguistic events, each of which can involve only a finite number of words, the criterion set E_0 of observational expressions cited in *IC-1* must be finite. Consequently, assuming that any expression schema formed by replacing theoretical terms in an *L*-expression with logical variables also counts as an "expression" in *L*, we have from *IC-1* and *TI-3b* that

> *Idealized Conclusion 2* [*IC-2*]: For a language *L* with normal syntactical resources, if E_T is an *L*-expression containing terms which are theoretical with respect to *L*'s subvocabulary V_0, then for any semantical relation R_s there exists an *L*-predicate E^*_0, containing nonlogical terms only from V_0, such that for any extralinguistic entity *e* which is not a state of affairs, E_T stands in relation R_s to *e* iff *e* satisfies E^*_0.

In particular, it is a corollary of *IC-2* that for any theoretical term *t*, there exists an observational criterion predicate P_0 such that for any extralinguistic entity *e* other than a state of affairs, *t* designates *e* iff *e* is denoted by P_0.

What is perhaps most basic of all a scientific theory's semantical challenges, namely, explicating the truth conditions of theoretical sentences,[12] is badly confounded by our present lack of any clear consensus over the extralinguistic sources of truth value and how the truth-determinative relationship is constituted out of a sentence's logical structure and the referential properties of its component terms. Without qualms or apologies, I shall here assume that what make sentences true or false are *facts*, *events* (a special kind of fact), or *states of affairs*, and that for metalinguistic purposes, states of affairs (facts, events) are represented (designated, signified, referred to) by linguistic expressions with the logical structure of a proposition. However, this still leaves obscure the semantical principles

---

[12] See pp. 302ff of my "The Factual Content of Theoretical Concepts."

by which states of affairs determine truth values. In particular, any reasoned thesis about the factual commitments of theoretical sentences must start with some assumptions about what it is for the state of affairs represented by a sentence S to be a *sufficient condition* for the truth of another sentence T. This notion of "sufficient condition" is considerably more problematic than most philosophers of language appear to realize. Certainly we mean by it much more than just the truth-functional implication $S \supset T$, since otherwise the fact signified by any true sentence would be a sufficient condition for the truth of any other which happens to be true. Yet the other alternative which comes most readily to mind, that S must formally or at least analytically entail T if the truth of S is to suffice for the truth of T, is far too strong, for S may be about a state of affairs which verifies T even when this cannot be deduced solely from the formal structure and/ or meaning of S and T. For example, if a and b are different concepts with the same referent, so that $a = b$ is true factually though not analytically, then for any predicate P, sentence $P(a)$ represents the very same state of affairs as does $P(b)$ and the truth of $P(a)$ thus necessitates the truth of $P(b)$ even though the latter is not a logical consequence of the former. More generally, if sentences S and S' signify the same state of affairs in nonsynonymous terms—and the possibility that this can occur must be admitted by any semantical theory which does not confuse meaning with reference—the truth of S is a sufficient condition for the truth of any analytic consequence T of S' even though T does not follow analytically from S. Unflinching recognition of this point is essential to the semantics of theoretical expressions, for not until we are able to identify sufficient conditions for the truth of a sentence *without* resorting to other sentences which analytically entail it can we specify the truth conditions of theoretical sentences without falling back upon metalinguistic concepts which are essentially synonymous with the theoretical expressions whose referents are in doubt.

On a loose, intuitive level, the truth of sentence S is a "sufficient condition" for the truth of sentence T iff, given adequate information about the semantical relations among the terms in S and T, we can see that the truth of S necessitates that T be true as well. Technical explication of this notion, however, must reckon with heretofore unsuspected complications regarding the nature of reference and will not be attempted here inasmuch as present purposes require only the intuitively evident principle that if the truth of a sentence S entails that there exists some state of affairs which

214

is a sufficient condition for the truth of sentence T, then the truth of S is itself a sufficient condition for the truth of T. That is,

> Technical Idealization 4a [TI-4a]: State of affairs s verifies sentence T, i.e., s is a sufficient condition for the truth of T, iff the truth of sentence 's' is a sufficient condition for the truth of T.
>
> Technical Idealization 4b [TI-4b]: If state of affairs s is a sufficient condition for there to exist a state of affairs which verifies sentence T, then s also verifies T.

Suppose, now, that S_T is a language-L sentence containing one or more terms which are theoretical with respect to L's observational vocabulary V_O, while s is some state of affairs which verifies S_T. In a sufficiently adequate metalanguage, s can be signified by a sentence formalized as '$F(e_1, \ldots, e_n)$' in which 'F' abbreviates a sentence schema containing only logical terms while each 'e_i' is a nonpropositional descriptive term or phrase which designates a component e_i of s which is not itself a state of affairs.[13] Assuming that these expressions are actually at our metalinguistic command, our premise is then that state of affairs $F(e_1, \ldots, e_n)$ verifies theoretical L-sentence S_T. Now consider any other state of affairs $F(e_1', \ldots, e_n')$ with the same logical structure F as $F(e_1, \ldots, e_n)$. Is or is not $F(e_1', \ldots, e_n')$ also a verifier of S_T? If it is not, then by IC-1 there must be some semantical relation in which $F(e_1, \ldots, e_n)$ but not $F(e_1', \ldots, e_n')$ stands to the criterion set E_O of observational expressions; whence, inasmuch as these two states of affairs differ only in that $e_i' \neq e_i$ for at least one value of index i, there must be some semantic relation R_s to E_O which is satisfied by component n-tuple $(e_1, \ldots, e_n)$ but not by $e_1', \ldots, e_n'$. According to IC-2, this relation R_s is equivalent to denotation by some predicate which is either in E_O to begin with or can be defined as a logical construction out of terms in E_O. Thus one state of affairs with logical structure F can be a verifier of theoretical sentence S_T, while others with this same structure are not, only if there exists some observational predicate P in L such that for any n-tuple $(e_1, \ldots, e_n)$, state of affairs $F(e_1, \ldots, e_n)$ verifies S_T iff $(e_1, \ldots, e_n)$ satisfies P. But then the existence of an n-tuple which satisfies both P and 'F' suffices for the existence of a state of affairs which verifies S_T and is hence, by TI-4b, itself a verifier of S_T. So long as L includes the needed logical grammar, more-

---

[13] While it is unnecessary to assume that entities $e_1, \ldots, e_n$ are "logical atoms" in the Russell-Wittgensteinian sense, I do here stipulate that if sentence '$F(e_1, \ldots, e_n)$' is molecular, 'F' pulls out enough of the sentence's logical structure to ensure that none of the e_i are themselves states of affairs.

William W. Rozeboom

over, this existential state of affairs can be signified by an L-sentence containing only logical terms in addition to the V_0-constructable predicate P, namely, the language-L equivalent of '$(\exists x_1, \ldots, x_n)[p(x_1, \ldots, x_n) \cdot F(x_1, \ldots, x_n)]$,' where '$p$' is the metalinguistic translation of P. Hence,

> *Idealized Conclusion 3 [IC-3]:* For a language L with normal syntactical resources, if S_T is an L-sentence containing terms which are theoretical with respect to L's subvocabulary V_0, there exists in L a sentence S_0, containing extralogical terms only from V_0, such that the truth of S_0 is a sufficient condition for the truth of S_T.

It requires only the most modest extension of this argument to conclude that *the truth conditions of a theoretical sentence S_T can be expressed wholly in terms of the observational vocabulary relative to which S_T is theoretical.* Contrary to first impression, this is *not* a return to positivism, for we need not—must not—unthinkingly presume that two propositions which have the same truth conditions necessarily *signify* (refer to, represent, are about) the same state of affairs. What is signified by a sentence $Q(t)$ in which Q is an observational predicate and t a theoretical term is surely some state of affairs $q(e)$ whose components q and e are designated by Q and t, respectively. But by *TI-2*, $Q(t)$ can discriminate $q(e)$ from another state of affairs $q(e')$ only if e but not e' possesses those observational properties which the usage of t requires of an entity designated by it (cf. *IC-2*). If e and e' are alike in all relevant observational respects, then $Q(t)$ must stand in the very same semantical relation to $q(e)$ as it does to $q(e')$. That is,

> *Idealized Conclusion 4 [IC-4]:* For a language L with normal syntactical resources, if $Q(t_1, \ldots, t_n)$ is an L-sentence in which $t_1, \ldots, t_n$ are terms which are theoretical with respect to L's observational vocabulary V_0, while Q is a predicate containing only terms in V_0, then there exists a V_0-constructable predicate P such that $Q(t_1, \ldots, t_n)$ signifies (represents, refers to, is about) a state of affairs $q(e_1, \ldots, e_n)$ iff Q designates q while n-tuple $(e_1, \ldots, e_n)$ satisfies observational predicate P.

It can then be argued that a necessary and sufficient condition for the truth of $Q(t_1, \ldots, t_n)$ is expressed by observation sentence $(\exists x_1, \ldots, x_n)[P(x_1, \ldots, x_n) \cdot Q(x_1, \ldots, x_n)]$. However, the existence of joint satisfiers of P and Q is not what theoretical sentence $Q(t_1, \ldots, t_n)$ is *about*; rather, the latter signifies certain states of affairs wherein the property designated by Q holds for an n-tuple of entities $(e_1, \ldots, e_n)$ respectively designated by theoretical terms $t_1, \ldots, t_n$. And since these e_i are

216

not, in general, designatable by observational expressions in L, neither can the states of affairs signified by $Q(t_1, \ldots, t_n)$ generally be signified by observation sentences. Although the present argument for *Idealized Conclusions 1–4* is admittedly informal, its basic structure should be sufficiently visible to make clear that a more rigorously deductive version requires only some support from technical postulates such as *TI-3, 4* on semantical and ontological matters essentially independent of the distinctive problems of theory meaning. Once *TI-2* is accepted, these conclusions or something very like them are extraordinarily robust under alternative choices for the supplementary assumptions.

It will be observed that *IC-2* and *IC-4* do not inherently prevent a given theoretical expression from standing in the designation relationship to more than one extralinguistic entity. That is, *theoretical expressions do not, in general, have unique referents*. It is, to be sure, always possible that only one object happens to satisfy the criterion for being referred to by a given theoretical expression E_T, but this is in no way intrinsic in the meaning E_T acquires from its usage.[14] THE IMPORTANCE OF THIS CONCLUSION CANNOT BE OVERESTIMATED. Note that the basic argument for nonuniqueness of theoretical reference resides directly in *TI-2* rather than in the supplementary assumptions invoked to reach *IC-2* and *IC-4*. For irrespective of what particular semantical ties to the observation language determine the semantical relata of theoretical expressions, the only such relation which past semantical theory has believed to be inherently many-one is designation. But if the referents of theoretical terms were always *designated* by the latter's observational criterion expressions, theoretical concept formation could not enlarge a language's referential scope. We must now face up to the realization that unless positivism is correct after all, designation too is in principle a many-many relationship, at least for theoretical expressions. And if it is true that most de facto concepts are theoretical to some degree, we must accordingly expect that multiple designation is not the exception but the rule.

That a descriptive term on any type level may simultaneously refer to more than one entity of that same type, so that, for example, a given predicate may *designate* a variety of properties over and above its denoting of the various particulars which exemplify these properties, is a breathtakingly new prospect which, to my knowledge, the philosophy of language has

---

[14] That is, unless E_T's usage gives it the properties of a definite description whose meaning forbids it to designate at all if it fails to designate uniquely.

never before considered. The only suggestion of a traditional counterpart lies in the doctrine of "common names," which construes certain predicate-forming nouns (namely, those which can replace 'x' in 'is a(n) x,' such as 'dog,' 'man,' 'hammer') as names shared by a plurality of particulars. But while it could conceivably turn out that "common names" are indeed genuine exemplars of multiple designation, the ease with which their usage can be assimilated to ordinary predication—for common names never occur as grammatical subjects,[15] and 'is a(n) x' seems entirely equivalent to 'belongs to x-kind' wherein 'x-kind' names a class of objects—makes it more reasonable to regard the prima facie multiple reference of common names as no more than a variant of the multiple *denotation* of things by the predicates they satisfy.[16] What must now be done is to rethink the entirety of technical semantics, not just the semantics of theoretical terms, to devise an account of language in which for *no* term or phrase, regardless of logical type, is it legitimate to speak metalinguistically of *the* entity to which it refers. The amount of classical semantics which will remain entirely untouched by this reworking of foundations is probably smaller than first impression would expect, for not until one seriously begins to probe the ramifications of multiple designation does the havoc this wreaks upon orthodox truth theory and traditional logic become apparent. But distasteful or frightening as this prospect may be, the sanctity of classical semantics cannot be preserved merely by ignoring the case for multiple designation or dismissing the notion as one man's personal perversity, any more than it would have salvaged classical mathematics to treat the logical paradoxes as amusing but idle curiosities.

History shows that most intellectual disciplines undergo periodic crises in the coherence or applicability of their tenets, often resolved only at the price of revolutionary alterations in their foundations. The philosophy of

---

[15] Except for rare cases in which the common name strongly appears to designate a *class*, as in 'Man is a rational animal,' which presumably intends to make a statement about mankind. That this is an anomalous usage may be appreciated by considering the oddity of, for example, 'Dog is a four-legged mammal' and 'Hammer is a tool for pounding.'

[16] Recent technical developments of the common-name concept (V. Sinisi, " 'ϵ' and Common Names," *Philosophy of Science*, 32 (1965), 281–286) are easily translated into more conventional semantics whereby the "common names" are seen to be the isomorphic counterpart of orthodox predicates while the common-naming relationship stands revealed as a form of denotation. The logic of genuine multiple designation—in which, for example, a sentence P(a) is, presumably, true if there exists a property designated by P which is possessed by some designatum of a, so that P(a) and Q(a) can each be true separately even while their conjunction is false—is altogether different from the logic of common names (cf. a review of the latter in Sinisi).

language has now entered upon such a phase, and the sole intent of this paper is to make evident the reality of this crisis, for which cause all pretense at technical precision and deductive rigor has been forsaken. It is no desire of mine that we should be obliged to create a new semantics and a reworked logic from which all presumptions of unique reference have been expunged, but neither can I justify a docile retreat to our only other viable alternative, namely, concession that cognitive access to the external world is positivistically bounded by the referential capacity of observational concepts. I shall gratefully accept chastisement by any critic who can point out where my reasoning on this matter has gone astray and whose proffers a convincing argument that the process by which theoretical terms acquire meaning does indeed single out unique referents for them. But until such reassurances have actually been accomplished, refusal to take the non-uniqueness of theoretical reference seriously is the most disastrous sort of if-I-don't-look-it-can't-hurt-me foolishness.

Discussion at the Conference on Correspondence Rules

HANSON: It seems to me that Professor Hempel was twisting a little bit when he made that reference to every use of a ruler giving rise to a different conception from every other use of a ruler. Surely what is at stake is not the fact that one chap might use a Hooke-type spring scale and another an Archimedes beam balance, but rather that the techniques are so different as to actually make us justify the entire undertaking in terms of quite different conceptual frameworks. For example, what happens when you weigh yourself in the morning in the bathroom on the local scale surely is quite different from what was undertaken by astronomers in the seventeenth, eighteenth, and nineteenth centuries when they undertook to weigh the earth. This was a standard locution, and there were serious problems involving both undertakings. Nonetheless, it seems to me that they are different in kind, and it isn't just a question of the fact that the techniques are different. Therefore, it seems to me that the conceptions ought, in all seriousness, to be thought somewhat affected.

HEMPEL: The remarks I made were in part directed against Bridgman,

EDITORS' NOTE: Although a large number of people attended the Conference on Correspondence Rules and made numerous contributions in private discussions, we shall list only those whose remarks have been included here. We again express our thanks to the Center's staff and visitors for their kind cooperation in surmounting the many difficulties involved in the holding of this conference and the preparation of this account. We especially thank for their lack of umbrage those whose contributions have, for one reason or another, been omitted. And we also reaffirm our gratitude to those participants whose edited remarks we present here: Peter Achinstein, Johns Hopkins University; Ian G. Barbour, Carleton College; May Brodbeck, University of Minnesota; Roger Buck, Indiana University; James W. Cornman, University of Pennsylvania; William Craig, University of California at Berkeley; Herbert Feigl, University of Minnesota; Paul Feyerabend, University of California at Berkeley, Yale University, and University of London; Adolf Grünbaum, University of Pittsburgh; Norwood Russell Hanson, Yale University; Carl G. Hempel, Princeton University; Mary Hesse, University of Cambridge; Edward L. Hill, University of Minnesota; Ernan McMullin, University of Notre Dame; Grover Maxwell, University of Minnesota; William W. Rozeboom, University of Alberta, Canada; Wesley Salmon, Indiana University; Carel M. Van Vliet, University of Minnesota; and Frederick M. Williams, University of Minnesota.

but Professor Hanson made me realize that one might raise the same doubts about his views. The problem as I see it is this: What criterion can one use for deciding when the meaning of a term changes? Hanson certainly gave impressive examples, and there are even stronger ones than the weighing case for showing that what he called the conceptual framework, in terms of which one uses one and the same word, is quite different in the one case than in the other. There is, therefore, some quite plausible motivation for saying: "Really, the meaning of the term has changed." But is there any satisfactory criterion for saying: "Up to here it is a change in empirical belief about one and the same concept?" We just think that this man has lost some weight, or that this determination of the weight of the earth was not correct. We have now changed the meaning of the word, but not the content of our empirical assertion. It seems to me that there is no clear way of making these distinctions, and it seems to me also that here Nagel is right: There is some vocabulary in terms of which one ultimately tests one's scientific assertions. Such a vocabulary has a fairly fixed use, and a fairly fixed meaning, notwithstanding changing theoretical connotations and their use in changing theoretical networks. So, once more, I ask the specific question: "How would one even tell whether there is a change of meaning or whether there is a change of assumptions about whatever it is one is talking about?"

HANSON: Clearly, what one does to determine the weight of a celestial object is to calculate, in terms of the theory of perturbations, what its deflection would be in some fairly well understood gravitational field. I should have thought that undertaking this technique just in order to see whether you gained or lost a few pounds (of course it is logically possible that one could do this) is farfetched.

Braithwaite mentions that the meaning of the term 'electron,' on the experimental level, might be quite different in an electrolytic context, on the one hand, as against a straight alternating current context, and both of these different again from what you would find in cloud chamber work. I think that this argument is not altogether implausible. I think that there is a meaning difference here.

HEMPEL: Wouldn't one say that in all of these cases this is just a wonderful consequence of our having thought about electrons? Before we had this enlightened notion, we thought we had three totally different kinds of phenomena. Now we see suddenly that, basically, they are the same kind of thing. This was a great deepening and broadening of our insight, and it came about because we saw that these basic entities were involved in all of the phenomena. Therefore, 'electron' has the same meaning in the three instances.

HANSON: In a word: No. I certainly wouldn't agree with that. If there are difficulties concerning what kind of entity an electron is, they would focus right on a point of this kind. After all, many of us know the wretched semantical history through which the notion of the electron has passed. To the extent that this is so, it isn't that we can simply make a reference to the little things that wiggle. One must note the way in which the 'e' figures inside the electromagnetic theory or the elementary particle theory, and notice how it is coordinated, or aligned, or how one steps toward the roots of the experience. It is exactly this which is the determinant of the meaning. To the extent that one can say that these techniques and indeed that the theories are involved (although they may be compatible and parallel, but with differences), it seems to me one ought not to settle in advance the question of whether or not they all have the same meaning, or pertain to the same thing. This is the open issue about which I think there should be more conversation or at least less confidence.

FEIGL: Don't you agree that we have situations in science which roughly correspond to my central image here, namely, that we get different lines, different fixes, on the same thing? The degree of approximation may cause some headache. Nevertheless, the scientists have a hunch that various indications, possibly probabilistic ones, point to the very same kind of entity, and therefore furnish confirmation for statements that occur in the theory. Theoretical advance usually is based on convergences of that sort. Whether we should speak of strict identity of meaning is another question, but that these convergences occur, I think, can hardly be denied in the face of the history of science.

HANSON: No one in his right mind would deny it. But also, nobody in his right mind would suggest that the history of science shows that all cases are the same case. But in point of fact many theories have shown a divergence. When you have this divergence you have this issue opened.

HEMPEL: One of the great prides of physics is that the determination of charge and similar things can be obtained from experiments of very different kinds. This, precisely, is one reason for saying that there really are such things. Here you have a great variety of supporting evidence that all points to there being one underlying factor. The very fact that there are different phenomena and that you can determine certain magnitudes by measuring one kind of thing and then something totally different is considered to be support for the underlying factor. Therefore, to say here "In these contexts that term means something different" would seem to give away the show and to give away just what science would pride itself most on: having cut nature at the joints.

GRÜNBAUM: If there is a single unitary thing underneath these diverse manifestations, then as you add new theoretical information you are enlarging your beliefs about the thing. This supports the conception that it is useful to talk about these terms denoting certain things, and your beliefs about them changing.

HEMPEL: Here I would say that I don't quite know where to draw the line sharply. It is sometimes assumed that a sharp line can be drawn. For example, it is assumed by Carnap when he speaks of frame principles for a theory which have to be fixed by fiat. We cannot change our previous frame principles, let us say the three-dimensionality of space or the availability of points for each real number triad, and so on, and then ask what is the evidence for or against them. One cannot profitably describe this as a result of empirical investigation. It is just a matter of replacing one possible and tenable system by another which affords a more simple, efficacious, economical way of giving a theoretical account. On the other hand, once the theoretical frame has been fixed, one decides other questions in terms of experimentation. This is an empirical matter.

My point here is that I do not think that this line exists. One can also give arguments for the choice of frame principles. So one thing shades over into the other, and all of this basically hangs together with the question of the analytic-synthetic distinction. One cannot be sharply separated from the other. I think, however, that in certain contexts it is useful to put emphasis on one, and in some others, emphasis on the other. I am not saying Hanson is wrong; but I am baffled about why one should describe the situation in his way. He claims that one has changed the meaning, when it seems to me so clear that in this example the scientist would say the term has the same meaning in these different contexts.

HANSON: Historically, the reference to the designation 'electron' referred to no particle at all. This was a way of referring to the fundamental unit of electrical charge. No dynamical sort of tiny billiard ball notion was in it at all. This is something that later came to it through cathode work, J. J. Thomson, and so forth. If one considers the electron, as it was referred to in the latter part of the nineteenth century, and its relevant electrodynamics, and the electron as it was referred to in the Bohr atom and the new quantum mechanics of the early 1920's, and then also collects this with the tremendous current difficulties in elementary particle theory, this combination is just too obscure. Certainly there are advantages to seeing to it that they are all roughly in the same ball park, but it doesn't follow from that that they are all the same players.

ACHINSTEIN: Bohr introduced two essentially new postulates into his theory. They contained terms such as 'angular momentum,' 'orbit,' 'energy,' 'frequency.' These were all available terms; he wasn't introducing any new terms, nor did he have to explain in his original 1913 paper the meanings of these terms. In fact, there was no new term, even in the full development of his theory, with the possible exception of 'resonance potential,' which was not antecedently available. It appears that this theory simply contains terms from the antecedent vocabulary. I want to ask Professor Hempel whether he had this in mind, or if this is a possibility.

Second, is to ask whether electrons exist tantamount to asking whether a theory about electrons is true? It seems to me that that couldn't be the case. We might imagine Bohr putting forth his theory about the atom, and then saying: "I wonder whether atoms or electrons really exist?" He could have asked this and not meant, "I wonder whether my theory is true, or even approximately true?" That is, he could have imagined that his theory was quite false and still have wondered whether there were electrons or whether there were atoms.

Finally, on the question of meaning, it seems to me that Hanson and Feyerabend owe us a theory of meaning, in terms of which we could decide whether it is true that terms have changed their meaning whenever a theory has been modified or replaced by another. All that one gets consists of references to examples from physics. They say: "Look at the notion of velocity in quantum mechanics, or mass, and so on, and look at it in classical mechanics; just see that they are different."

HEMPEL: I haven't really thought about whether there might be a theory which has no characteristic terms. I would, offhand, have thought not, for if there were none then one wouldn't speak of a theory. But I want to think about this, and I would have to look at Bohr's formulation to see what new terms there might be. There is no syntactical way of distinguishing between theoretical principles and bridge principles, precisely because the theoretical principles would contain antecedently available terms. The analysis which I proposed would always have to be formulated in such a way that when one asked "Do x's exist?" one would have to say, "X's in the sense of some particular theory; or else referred to or characterized or assumed by some particular theory." Bohr could have asked whether electrons exist in the sense of the then current theories. He might have said: "Now, I want to add something to this theory. Of course, my further assumptions might be false, but there might still be electrons." He would, then, refer to electrons as conceived by the already available theory. Surely, one would always have to indicate, in principle, what assumptions one makes about these things. But then to say they exist is, I think, tantamount to saying these assumptions are correct.

SALMON: There are two basic points which seem to me to be of fundamental importance. One is that when we approach a scientific theory by way of an uninterpreted calculus, then it very often turns out to be the case that we see the possibility of formulating precisely the same physical theory in terms of vastly different formal calculi. We see, moreover, that if we pay careful attention to the way in which the primitives in the systems are interpreted, that we can get equivalent physical theories as a result. Historically, this had the value that people have supposed that there are absolutely contradictory statements being made about nature, for example, about physical space, when, as a matter of fact, the statements are not contradictory at all. At the other extreme, people have supposed that all of these statements are equivalent to one another, and that anything one says about physical space is perfectly compatible with the facts. Then, one runs into a complete conventionalism. The important fact that has emerged from this situation is that when we look at the formal system or systems, and the correspondence rules or the coordinating definitions, we get a tremendous amount of illumination about the nature of the physical system that we are considering as well as the various ways in which we can correctly describe the physical facts in the situation. Moreover, the same uninterpreted system may very well have many quite diverse applications if one introduces different sets of correspondence rules for it. The example that comes immediately to my mind is group theory, which has very many applications in physical science as well as in mathematics. We learn something extremely important about the physical theory by realizing that the same formalism is usable in these very diverse ways. This leads me to suggest that perhaps one can think of theories in terms of an uninterpreted calculus, C, and some rules, R, which do not have to be confined to rules which utilize only observational terms. I am perfectly happy to go along with the idea of an antecedently available vocabulary. Nevertheless, when we write ϕ ($\tau_1, \ldots, \tau_n; P_1, \ldots, P_n$) there isn't really any distinction here between the τ's and the P's except for the fact that one can, if one wishes, give a very direct interpretation to some of the τ's simply in terms of the antecedently available vocabulary. That is, there can be a virtual identification of an uninterpreted term in the system with one of the terms of the available vocabulary. But this does not make, at least on this very primitive level, any great difference between those τ's which are thus directly interpreted and those τ's which require a somewhat more complicated kind of interpretation. All of the terms have to be interpreted. We may decide that 'm' means mass, or that 'p' means pressure, but this is as much an interpretation of a primitive and undefined term in a formal calculus as if we had something that was much fancier and much more difficult to interpret. One of the points on which Quine has placed very great emphasis is one's inability to state in any nonarbitrary way what constitutes a definition or a semantic rule. It seems to me that there is no reason to expect that there should be a nonarbitrary way of doing so. That is, if we want to give some definitions or rules of correspondence, there seems to me to be no reason why we shouldn't simply say: "The following are going to be the definitions and the rules of correspondence." If that is arbitrary, I can't see anything wrong with that type of arbitrariness. Moreover, when this attitude is introduced into the interpretation of physical theory, then one can perfectly well say: "If that's what you mean by the term, then a certain statement in the theory is true, but if something else is what you mean by the term, then a certain statement in the theory will be false." Of course, there is the possibility of having the same formal system interpreted in various ways, some interpretations being true, some false. For that matter, with alternative formalizations there is no reason why a formula which is a definition in the one case shouldn't be a physical postulate in another. While I am strongly sympathetic with the ideas of the antecedent vocabulary and the utilization of it for the interpretation of theories, I still think that the distinction between the uninterpreted calculus and the rules of correspondence or coordinating definitions may be very useful in the rational reconstruction of scientific theory.

HEMPEL: Precisely because one and the same body of theory, as you find it, let us say, published in an article or in a textbook, can be axiomatized in so many different ways, one has to be very careful about drawing any conclusions from a particular axiomatization that has been proposed. I argued that nothing about the epistemic status can be inferred from a given axiomatization. I agree with Professor Salmon that one can lay down certain correspondence rules. But this would suggest the idea that the scientist takes these to be correspondence rules and takes these to be definitions. If something is a definition, then it is analytically true. I would then say that there isn't anything in physics that corresponds to it. A physicist will not stick with a particular thing as being a definition; he might change it. Physicists started with an operational definition of one meter as given by the scratches on the standard international meter. Later on they revised this because theory led to further assumptions. The old definition is wrong; not only is it not definitionally true, it is empirically false. I would, therefore, say that it doesn't help very much to give any particular axiomatization if one wants to make these further claims. I would agree that one sees alternative ways of formulating things. Reichenbach provides an example to illustrate the fruitfulness that axiomatization can have, namely, in reference to geometry. There, however, it was very intelligently and illuminatingly used to argue against the Kantian position and to take issue with Poincaré in certain ways. This was very helpful, but I think that it required not just some axiomatization, but a peculiar one, and then too, a combination of this with certain general philosophical ideas which were independent of just the formal logical procedures employed.

SALMON: To the extent that we are concerned with anything that vaguely can be characterized as an ontological question, I think that the point is absolutely essential that there are equivalent descriptions. There are alternative axiomatizations, and in many cases one can say that what appears to be "really real" under one formalization turns out to be quite derivative, or a matter of definition, in another formalization. The very possibility of such alternative formalizations should give us a good deal of pause when it comes to drawing any conclusions about what a given scientific theory is committing us to in the way of existence of theoretical entities or the status of particular theoretical terms. The alternatives which emerge from this process of alternative axiomatization seem to me to have a very important philosophical moral point for us here.

BUCK: I am puzzled what it means to say: "Yes, it is the same concept of mass, but it was not specified in all respects before it occurred in this new theory." How would this differ from saying: "No, it is a different concept of mass now that it enters into the new theory, because, although it is still not completely specified, it now has some more specification"?

Discussion

HEMPEL: First of all, one difficulty really has to be put at the doorstep of Hanson and others. Namely, they have to give us a criterion in terms of which we can tell when there is a change of meaning, and when the concept is still the same and one is now making empirically different assertions about it. Here I just don't know what to say. I do have the feeling that the concepts which we use in order to describe the evidence have a certain stability, but I have to say this just intuitively. They have stability in their meaning; otherwise it wouldn't be really possible, even theoretically, to say that we introduce new and totally different theories which are, however, supported by the evidence.

Also, I felt some sympathy for Hilary Putnam's view that the notion of temperature has the same meaning now as it had in Galileo's time. Since I have not been provided with a concept of sameness of meaning, and since I am not able to offer one, I can only say that I find this suggestive and interesting. I find that it is helpful to look at it this way without being able to provide any clear theoretical elaboration.

HANSON: Professors Hempel and Achinstein have both called for a criterion for difference in meaning. I think consistency is a fairly reasonable and classical criterion. I will just give two examples. In the classical theory a moving charge radiates. In the Bohr theory, the moving charge, the orbital electron in the hydrogen atom, does not radiate. Now, here is a candidate for difference in concept. Both are called 'charge,' yet are different. Another example is the notion of a particle in seventeenth- and eighteenth-century mechanics, and the notion of a particle in contemporary microphysics. It is an essential feature of a quantum mechanical particle that it spreads, that it is not uniformly locatable as a geometrical mass point, contrary to a classical particle. Because these two characterizations are incompatible, therefore, they provide different contexts: 'Particle' is different in these two contexts; 'charge' is different in these two contexts.

GRÜNBAUM: I just want to make a general remark about the significance of this quest for general criteria for sameness of meaning and general criteria for change in belief about the facts versus a change in meaning. I have very definite axes to grind on this. I find that there is a lot of work being done in the quest for general criteria. This work often ends up with negative results, and then produces a sort of a general skepticism. All that this yields, then, is the knowledge that many clever proposals do not work. Hence, our efforts lose a great deal of the potential relevance which they may have to the actual logical concerns which scientists have with specific scientific theories. I do not see now any hope for a program that is going to produce general criteria of sameness of meaning. I believe that in very specific cases and in specific theories such as elementary particle theory and geometry, one can, contrary to the general Quinian conception, say: "In this case, in a very interesting sense, there has been a change in the meaning of the term," or "In this case, in a very interesting sense, there has been a change in the belief about what the facts are."

After all, where does this notion come from? It comes from daily life situations. Suppose a physician were to tell the mother of a child that the child has acute appendicitis, and the child were then examined and found to be feeling just fine. The physician would surely not say: "What I really meant by this is that the child is smiling." The mother would say: "You have just played with words; you have not done anything but waste my time." Precisely this kind of thing comes up in discussions of the foundations of geometry. There has been confusion among people between saying that relative to a certain congruence standard, the geometry is such and such, and saying that if you play with what the congruence criterion is, then you can have any geometry you want. Here I would say that whatever the merits of the results of the work of Quine and others, I find it profoundly interesting to say to somebody: "You have played with what you understand the congruence criterion to be, and, therefore, you have been able to come up with the statement that the geometry can be anything you want," or to say on the other hand: "Relative to this congruence criterion the geometry is such and such." I would plead for this: that we enter specific contexts, and find that in these specific contexts we can articulate, very usefully, some of these distinctions, and carry out piecemeal work on these specific scientific theories. We will, then, helpfully clarify issues that scientists are arguing about and are interested in, without a general program of criteria.

FEIGL: Let us consider the axiomatization that Reichenbach provided for the theory of relativity. I think that this shows that the position of Duhem and Quine about the testing of theories as totalities is logically obvious. That is, when we have a conjunction of propositions and, *modus tollendo tollens*, refute that conjunction, then, of course, we do not know which of the propositions are false. But I maintain that experimental techniques serve precisely to pinpoint the dubious postulates. That requires at least relatively independent testable postulates. Now, I am not sure whether we should drag in the question of whether there is a fundamental distinction of analytic and synthetic propositions. In any case, I think that if we were to defend the work of the logicians of science to the extent that would be useful for the working scientists themselves, then at least a measure of independent testability must be available for the postulates. That shows that the mathematicians' idea of an elegant, highly compressed axiomatization, such as Caratheodory's of relativity, is extremely elegant, but absolutely not perspicuous as far as the empirical consequences are concerned. If we can show which experimental results are relevant for which postulates, then such an analysis will clarify and make fully explicit what goes on in the thinking of the scientists when they are forced to modify their theories. Moreover, in regard to change of meaning, I believe that one task of the logician of science is to take a snapshot of a current theory at a given stage, and try as best as he can to formulate explicitly what this theory asserts in terms of postulates and correspondence rules.

VAN VLIET: It always occurs to me that we overlook the fact that Bohr's theory has been replaced by quantum mechanics. In quantum mechanics there is no problem, because you can only have radiation if the electron will go in an orbit. However, if you take the proper theory, we have to take the distribution of an electron over a sphere. In an S-state there is, therefore, no angular momentum and thus there is no radiation. It seems to me that the strange attribute of an electron has been removed again. It is similar with respect to the parity of the particles. One has to ask what the reality of a definition in physics is. The point that has not been brought up here is that the physicist, as a rule, will try to couple his definition to a mathematical concept. For instance, an electron is defined as being the smallest quantity of charge measurable. Now you might add other properties to it. But the definition has always been this, and will remain this, in my opinion. Suppose that someone had defined an electron as having a Landé factor of two, that is, that the wave magnetic moment is about twice what you'd expect classically. That would have been wrong. The definition would have to be revised, because we now know that it's not two; it's 2.000 . . . whatever it is. But it was not defined that way; it was defined as the smallest quantity of charge. It is not a matter of whether definitions were different in previous times. The question is whether they are the same right now.

A physicist will try to satisfy what is called Hankel's principle. In other words, when we extend our theories, we want to find somewhere a core of knowledge common with what has been said before. Hankel's principle is not given as a truth for all times, but is an ideology that we try to satisfy. If we cannot satisfy it, if the concepts are divergent, then we need a new concept. This is also done in quantum mechanics. After all, we do have Bohr's correspondence principle.

ACHINSTEIN: If to learn the meanings of the terms in the theory one has to learn the theory, and if every time the theory changes the meanings of the terms change, then this seems to produce various paradoxes. First, how does one learn the theory if one has to learn the theoretical concepts first, and they're an integral part of the theory? Second, how does one theory contradict another? Bohr thought he was contradicting classical electrodynamics. If the terms in his theory mean something different from those in classical electrodynamics, he cannot really have been said to have denied classical electrodynamics. Third, all propositions in a theory would be analytic. That is, if an atom is just the sort of thing that satisfies Bohr's postulates, then the propositions of the theory that describe atoms would be necessarily true. Certainly, I am not denying that meanings change; I think sometimes they do. Sometimes they do more radically than at other times. But sometimes we want to describe what has gone on simply as a change in our beliefs about the world.

WILLIAMS: In the past, the philosophers have done a great service by pointing out to the scientists that they should take care to distinguish between their observables and their theoretical concepts. Here we turn around and, as I understand Professor Hempel, put into a theoretical construct antecedent terms, which in part contain what we would call observables. It seems to me that by including these observables we are violating the main lesson that the scientist learned before, namely, that he has to keep his theory separate from the observables or the operational language. It seems to me that these antecedent concepts or antecedent terms really must be part of the theoretical language. What we are doing is turning around, and instead of trying to keep this separation for the edification of the scientist, we are trying to look at what the scientist is doing and incorporate all his lack of distinction between theory and principle into the philosophy.

ROZEBOOM: If an electron is whatever is true of it, then this makes electron theory analytic. But a statement about electrons is not analytic in the sense of being a logical truth. This does not make electron theory logically true, because there may not be anything about which the things which are said of electrons are true. It does not follow that to say electrons are such and such is to make a logically true statement. Nevertheless, this is perfectly compatible with saying that by the term 'electron' we mean whatever it is that has these properties. This is just one facet of the extensive revisions of basic semantical theory that the analysis of theoretical concepts requires. A statement can be analytically true, if it is true, and not be true at all. Namely, if the statement about electrons is true, it is true analytically, because that's what you mean by 'electrons.' It need not be true at all, because there need not be any such things.

BUCK: Professor Feyerabend has, at various times, insisted on the incommensurability of the terms in competing theories. At the same time he has urged that different theories will very often yield incompatible predictions within some domain. Some account of incompatibility different from that which one usually learns in logic is going to be required in order to make good on the claim that a pair of incommensurable theories can yield incompatible predictions. The standard situation for incompatibility is that one gets a prediction, P, out of one theory, and a prediction, not-P, out of the other. In the second theory, either the statement P or the statement not-P will not be formulatable because of the incommensurability of the concepts of the two theories.

I want to put the point again in this way: How is one to know that the latter, and more successful, of two theories is dealing with the same domain which was being dealt with by the earlier theory? When one speaks of the domain, and takes it for granted that each of two theories do indeed apply to it, in terms of the concepts of what theory does one identify the domain? Perhaps oxidation theory is just dealing with a different domain from phlogiston theory. Instead of being in the position to assert that oxidation theory is doing a better job of explaining certain sorts of chemical reactions than is phlogiston theory, perhaps what we ought to say is that the followers of Lavoisier have identified a whole new set of properties of chemical elements and compounds, and have been able to develop a thesis about these. Inasmuch as these concepts and those of the phlogiston theory are incommensurable with each other, there is absolutely no reason to suppose that phlogiston and dephlogisticated gases do not exist. The two theories just talk about different things or features. To prove that they do not would be to reintroduce some strong sense of incompatibility.

FEYERABEND: I have been advancing certain concrete arguments which I believe can be made very clear and quite compelling. If these concrete arguments lead to a situation in which one does not know what to do, one has two choices: Either drop the arguments or try to adapt one's way of thinking to the new situation. I would say: Let us wait and see what will come about. When I talk about theories, I mean general points of view which are concerned with everything there is. For example, 'all ravens are black' is not a theory in that sense, because it is only about ravens. The theory of relativity is a theory in that sense because it considers extension, and extension is a property, at least according to some theorists, of everything there is. As soon as one has such theories, one faces the problem of the interpretation of one's observation language. If the theory is a universal theory, then it should be applied universally. This means that its conceptual apparatus should also be applied to the description of observational processes. Hence, one has to give up the idea of an observation language which remains unchanged, whereas only the superstructure changes. This conception has to be given up not only because of general considerations, but also on the basis of this very concrete consideration: Why should one not interpret the theories directly when one teaches them? For example, why should not the materialist who is convinced of the correctness of his point of view teach his child the terminology of brain processes, pain, and so forth, rather than teach him the old observation language which contains an old theory? I think that these are very simple questions and that the answer to them is that we ought to interpret theories in that way.

Consider two radically different theories, one a theory which describes the fixed stars as divine beings and describes their paths as the paths of divine beings, and the other, an astronomical theory. One can very easily translate from the one to the other. One translates the observation statements of the people who hold the theory of stars as divine beings into the terminology of modern astronomy. It seems to be quite possible to jump from one of two radically different conceptual systems into the other. One does not refute certain statements in astronomy by reference to an observation language which is common to both. Conceptual systems are obtained, which, from the point of view of meaning, have nothing to do with each other, although they can easily be translated from one into another. Somebody who goes into a primitive tribe and learns the language learns to do this. He can learn a new language from scratch. That one can translate from one into another seems to indicate that what is needed is a new kind of theory of meaning, or perhaps one not too much concerned with meaning.

Discussion

BUCK: Isn't this astronomical case very easily handled by someone like Nagel? The Babylonian astronomers observed gods, and it takes only a little reference to what they said to know that their observations coincide with the observations of Tycho, who was observing something quite different. This is a simple problem, if one is willing to admit that there is some overlap between the observational vocabularies.

FEYERABEND: You are describing what these people observed, tacitly using the point of view which we believe in today. They described what they observed using terms which had no extension, at least according to what I believe in, because I think that there are no gods—not even one. If someone who was inside the conceptual system of these people pointed to something, then it seemed that he was talking about something. At the same time he described it by terms which have no extension, so he seemed to be talking about nothing. The transition from the one terminology to the other is really quite a tremendous one. But you can make it, despite the fact that there is no conceptual similarity between the two systems.

CRAIG: Is this your argument: that an actual phenomenon and the value of the phenomenon as predicted by a certain hypothesis may be indistinguishable observationally, whereas when additional theories are applied, they are no longer observationally indistinguishable?

FEYERABEND: The theories which are needed are theories which are incompatible with the theories from which the original hypothesis was obtained.

CRAIG: Do you then conclude that sometimes one has to construct new theories even when the old one is not already contradicted by experience?

FEYERABEND: Yes. Sometimes refuting instances for a certain theory can be found only with the help of an alternative to that theory. Therefore, the construction of alternatives is not as foolish as some people would say.

CRAIG: I do not think many people would hold that the only situations in which we start constructing new theories are those in which experience disconfirms present theories. I think almost everyone would maintain that we start looking for new theories if the present theory does not account for sufficiently many of the phenomena.

FEYERABEND: Yes, but that is not an easy affair, because no theory ever accounts for all the phenomena. In every theory we have, there are some open places. This being the case, some quantum physicists, having a successful theory, object to the introduction of new theories. As a matter of fact, they claim to have shown that their theory explains all the things which can be found by experiment. But another theory might point out that there are certain difficulties of the current theory which have been overlooked. There are difficulties concerning any theory at any time. In the history of scientific theories, it is never the case that a theory is completely in accordance with all the experimental findings existing at a certain time. Usually these experimental findings are put aside. They are not yet regarded as refuting instances; but it is said that at some time they will be explained.

One example was the great inequality of Jupiter and Saturn, which, for about a hundred years after Newton's *Principia*, could not be explained on the basis of the principles of Newtonian mechanics. People did not give up Newtonian mechanics on that account, but said that at some time the inequality would be explained, and Laplace explained it. Today, Newtonian mechanics is, in domains which are outside relativistic effects, inconsistent with many observational results for the following reason: Physicists use, in different domains, different approximations of Newtonian mechanics. They try to make these approximations mutually consistent and gravitationally consistent. This means that one tries to adapt them to each other in such a way that an approximation times the law of gravitation does not lead to an inconsistency with another approximation. The result is that the approximations do not cover the facts as well as they would if one did not make them gravitationally consistent. So we have rather large deviations from Newtonian mechanics in the domain in which everyone would say there are no quantum effects and there are no relativistic effects. This being the case, the idea that a simple fact which we discover and which deviates from the theory will refute it becomes completely unrealistic.

Usually people say that Newtonian mechanics was given up because of the anomalous advance of the perihelion of Mercury. That is not at all correct. The whole movement is about 5600 seconds, and all sorts of disturbances are subtracted. In the subtraction of these effects all approximations are made gravitationally consistent. Hence it has to be expected that, somewhere, a discrepancy will arise. Why, finally, has it been said that there are difficulties with classical mechanics? Because an altogether different theory, namely the general theory of relativity, from quite different principles provided a value which filled in the missing 43 seconds, which is a very, very curious affair. Therefore, physicists said there must be something wrong with classical mechanics, because, if so many different principles explain this effect, it cannot be possible that it is explained on the basis of Newton's theory. Relativity says it is caused in a different fashion.

HANSON: Although I am, in general, in great agreement with Professor Feyerabend, I think his examples are just letting him down, and to that extent they are letting the rest of us down too. The aberrations in the perihelion of the planet Mercury were taken very seriously in the middle of the nineteenth century. In 1859 Leverrier went out of his way in order to establish that there really was an effect here. There were at least five different attempts to correct the discrepancy between the theory and the observations by serious overhauls inside classical celestial mechanics and the theory of perturbations. There were at least three attempts which consisted in suggestions that Newtonian mechanics was false and should be abandoned.

A second point is that the perihelion of Mercury problem is a relatively incidental item in Einstein's papers of 1916–17. The major objective of the work of Einstein was not to treat this particular thing. Despite the seriousness with which this imperfection in classical mechanics was regarded, I do not think that it was this which triggered the new theory at all. This was, in a sense, an additional "confirmation" of general relativity. So, in the absence of an alternative theory, the discrepancy was regarded very, very seriously. Individuals did try to find a way of correcting this by adjusting the structure of the theory, or by finding an alternative theory. The theory which finally did the job was not designed to do that job. It was designed to do a different job.

FEYERABEND: I completely agree. However, there were other difficulties, for example, the motion of the moon and the reciprocal masses of Jupiter and Saturn. That one paid special attention to the perihelion of Mercury may, perhaps, have been triggered by the fact that one had some success with the motion of Uranus and the discovery of Neptune. At any rate, when you have a new theory, whatever the reason for which you have introduced it, if this new theory explains a particular deviation of observations from the old theory that gives you an extra strong reason for regarding the old theory with suspicion. It is a stronger reason than if you just have the difference between observations and the old theory.

ROZEBOOM: In your view, how is it possible for two theories to be inconsistent, and how is it possible for a theory to have a refuting instance?

HEMPEL: If one follows your advice, then one would describe the world in terms of the theoretical apparatus. Provided your observations would check with this, there would be nothing left to worry about. The questions which arise and which eventually might lead one to scuttle the theory are often formulated in terms of concepts which are not construed in terms of the theory. These are construed in terms of observations. But your view seems to imply that my observations would then come out as the theory requires.

FEYERABEND: I have argued that we should interpret the terms of observation languages by means of the most recent theories we possess. The arguments I give for this are, first, simplicity and, second, revision of the principles of the observation language, which should not be exempt from revision. These are two quite straightforward reasons. In the first place, it is simpler not to have two different levels of observation. When the theory is detailed enough to give an account of macroscopic objects and of their behavior, the syntactical apparatus of the theory is detailed enough to be taught as an observation language. Secondly, we should have means of correcting the principles of any observation language. From the fact that I interpret concepts in a relativistic fashion, it does not follow that all the predictions I make with the help of the theory that has these concepts will be as the theory says they are. It does not follow that if I interpret the observation language so that it fits the postulates of the theory, then I will not find any observational result incompatible with the theory. Of course, I put in initial conditions. The initial conditions will also be in terms of the theory. If I had put in two different sets of initial conditions, one of these sets of initial conditions will give me different predictions from the other. One of them will agree with what I actually do observe, also interpreted in terms of the theoretical concepts.

BUCK: If you are going to describe observations and initial conditions solely in terms of the theoretical vocabulary, then the only sorts of events that you are going to be able to describe which do not conform to the theory are going to be straightforward anomalies with respect to the theory. There is something wrong with an account of our capacity to describe phenomena if that account guaranteed that every phenomenon described had to be exhibitable either as a predictable consequence of the theory or as incompatible with it. Then there could not be any third class, namely, phenomena with which the theory simply did not deal. Or, if there were, as there must be, in what terminology are they described?

FEYERABEND: I would describe them in terms of a theory with which I try to explain them. Many people assume that macroscopic phenomena should be described by some variation of general relativity, and that microscopic phenomena should be described by the quantum theory. Then there is a question of what will be the observation language in both cases. In one case it will be a relativistic observation language; in the other case it will be a quantum-theoretical observation language. There are no elements common to the different theories. Even the observation languages have no common elements. Usually that will not be the case, because most theories which are put forth are not so general and do not have such a great scope.

Assume that one theory is being taught to a little child, e.g., general relativity, and another theory is being taught to another little child. How many problems will one have, and how many problems will the other have? The one with fewer problems is better off than the other. There does not need to be any overlap of theories.

ACHINSTEIN: Furthermore, in your review of his book you seem to have attributed to Nagel the idea that one ought never to introduce, for any purpose whatsoever, a theory which is incompatible with well-confirmed theories. It seems to me that no one who holds the layer-cake view is committed to this position at all. Nagel, by reference to the third element of theories, namely, models or analogies, certainly was conceiving of the possibility and indeed endorsing the idea that theories which are inconsistent or incompatible with theories which we currently believe should be introduced for certain purposes.

The position I take Nagel to be denying would be this: One ought not to be fully committed to *believing* or *accepting* theories which are inconsistent with theories which have a high degree of confirmation. But you did not attribute this view to Nagel. You attributed to him the view that one ought never to think about introducing in any way a hypothesis or theory which is inconsistent with current theories. This is certainly not what Nagel says and certainly not what any of the layer-cake people need be committed to.

FEYERABEND: The reason I attributed this view to Nagel is that he says, and repeats again and again, that science proceeds by explanation; the principle which advances and amplifies science is explanation. Then he says that explanation consists of inventing primary sciences for the reduction of secondary sciences. Reduction simply consists of logical derivation. From that it follows that the only method which he talks about in the sciences, and which seems to him to be the only method worth discussing, is the method of deduction of secondary sciences from primary sciences. This means that the primary and secondary sciences must be consistent. I infer that this is a correct description of his view because, when it was pointed out to him by Smart that in actual cases they are not consistent, he said he must revise his model and assume that there are probability connections between the two. Nagel said that the laws connecting the various theories, not the laws connecting the theories with the observational bases, are probability statements.

The probabilistic weakening of the connection between the various layers of the model is not so desirable. It takes away just that method of criticism for which we sometimes introduce what is called the primary science. Furthermore, there should not be any layers at all, because the last theory on top should be strong enough to give an account of everything the previous theories have given an account of. If it is not quite sufficient for such an account, then this will be simply explained by saying that the account given by the lower level theories should not be preserved. Let me give you an example.

Let us consider an explanation of why the sun rises at a certain time on the eastern horizon. First, it doesn't rise at that time. What appeared at the horizon at that time was an optical image; the sun is below the horizon. Second, we explain why the optical image appears there. Very often such an explanation consists of throwing out the old explanandum and replacing it by a new explanandum. We throw out the old explanandum on the basis of a theory which we hold, and which we have tested or believe to be correct in different domains.

I think this principle should be applied more generally. This is not to say that we must retain the lower levels, which always seems to be Nagel's move. Sometimes we have to throw out the lower levels, because our assumptions contained in the lower levels are not correct. Therefore, these are not the correct explananda. You seem to imply that if you explain, there must be something to be explained. Certainly, but what decides what is the thing to be explained, the theory introduced for the explaining, or the observational statements existing before? It may sometimes be one, sometimes the other. Very often it is the former, especially when we are dealing with very general theories. Then the theory tells us the kinds of things which happen and the kinds of observations which we have to expect, although the particular observations which we have when we rise at five in the morning are not fixed by the theory. But what the rising of the sun is, whether it is the rising of an image or of the physical object, is determined by the theory.

ROZEBOOM: What is the meaning that a theory gives to its observation base? Can you argue from premises to the possible correctness or incorrectness of a theoretical conclusion if the premises depend upon the conclusion for their meaning? This is a question of the possibility of having an observation basis, or something formally akin to an observation basis, which we use in order to pass judgment on the theory. If the sources of our inference to a theory depend upon our acceptance of the theory to give them meaning, then it doesn't seem to be possible to doubt the theory, because in so doubting we destroy the premises on which we base our argument.

FEYERABEND: It is possible to doubt celestial mechanics once I have decided to interpret all observations of fixed stars as statements about the behavior of gaseous balls, and all the observations of planets as statements about the behavior of masses of stone with gaseous atmospheres around them reflecting sunlight. If the planets don't turn up where they are expected to, according to the theory, then it is possible to doubt the theory.

BUCK: Can you explicate this idea of their coordinates independently of the theory?

FEYERABEND: Their coordinates would be given by a pair of numbers, alpha, delta, ascension, declination, respectively, which at a certain time have two values, A and B. The theory indicates how I am to interpret these numbers. From the theory I derive A, B; observational results give me A', B'. Now it is quite possible that the theory gives me A, B, when I interpret alpha, delta as required by the theory and observation gives me A', B' when I interpret alpha, delta as required by the theory. If these two pairs of numbers are different, then I shall have reason to doubt the validity of the theory. What you seemed to assume before was that once I interpret the observation language in terms of a theory, observation will never give me a statement inconsistent with the prediction made from the theory. That is not at all the case.

BUCK: I think the layer theorists would say, among other things, that the description of the position of the planet, when it turns up where it's not wanted, is generated in a language whose adequacy and clarity is in no way dependent on the celestial mechanics that you are using to predict the position. It appears that it follows from your view that the worse the predictions were, the less reason one would have to take them as overthrow-

Discussion

ing the theory. It looks as though it would be difficult to confront a theory with a really bad prediction because you wouldn't know what it meant.

FEYERABEND: One can turn the whole thing around the other way. Take a normal observation language, which we have not formulated in accordance with some recent scientific theory, e.g., the physical thing language of Carnap, mentioning tables, chairs, macroscopic objects, and so on. This language has a certain ontology, it has a certain structure, and it is very clear to us. However, we do not know under what circumstances this language will turn out to be incorrect. Thus, we are in a very bad position with respect to that language. Now one who works in an observation language formulated in terms of a recent scientific theory is better off because he knows under what circumstances he will be forced to doubt his conceptual system.

BARBOUR: On your view, what sense would it make to speak of the Newtonian case as a limiting case of relativistic physics?

FEYERABEND: I think that the answer to this question of limiting cases is not difficult. From general relativity, we derive a formula which is identical with a certain formula used in classical mechanics. But this formula cannot be interpreted as it is interpreted when it expresses a law of classical mechanics. For example, positions and temporal intervals will not be absolute, but will be relativistic ones. Something is in common: the formula. Something is different: the interpretation.

BARBOUR: Not only the formula part, but many of the terms are in common, too. Velocity is measured in the same way in both systems, even though the ways in which it is defined or derived are not the same.

FEYERABEND: There is no absolute velocity anymore; there is no absolute distance anymore. There is something different. From the fact that the same measuring instrument and the same measuring procedures are there, it doesn't follow that considerable conceptual changes have not taken place.

BARBOUR: Could you indicate in what sense the observational language of the Einsteinian and the Newtonian would differ in reporting on the path of Mercury?

CRAIG: Do they have different notions of pointers?

FEYERABEND: You have a pointer and you have a certain distance from zero. There would be no difference as regards the number which is read. There will be no difference as regards the procedure for ascertaining this number; it is a macroscopic thing.

Discussion

GRÜNBAUM: There is a presystematic sense of theoretical versus non-theoretical which has crept into this whole discussion. What we are appealing to in these disavowals is the supposition of some fundamental dichotomy of the old protocol versus nonprotocol type in trying to convey what sorts of things we are including in the antecedent or theoretical vocabulary. When Professor Hanson says that something is theory-laden, he is saying that when we use physicalistic language of the original Carnap type, we are using terms which, in the framework of a sense-data versus something else dichotomy, involve more than talk about sense data.

ROZEBOOM: What I have been trying to do is to set up a notion of the observation language, in a way which minimizes this kind of commitment. I did not want to say anything about the nature of the meaning of the observation terms. In other words, I want to leave this open, without starting with the statement that these are theory-laden. I want to allow the possibility of later agreeing with somebody who says that the observation terms also contain theoretical meanings.

GRÜNBAUM: If somebody says retroactively about the antecedent vocabulary that this vocabulary also has theory-involving terms, is he using the term 'theory' in your new sense?

ROZEBOOM: Let me just suggest that in another scientific context the propositions that in this context are taken as a basis for argumentation might be taken as an object of argumentation.

FEYERABEND: If you take what one roughly calls observational terms as a basis for argumentation, there may be many people who agree to certain observational states of affairs. Equally, the starting point of some particular physicists may be at a 'theoretical' level.

ROZEBOOM: It seemed to me that when you spoke of the more abstract functions, you were doing something that was not at all relevant to the development of higher levels of theory. What you were doing was to say that given a set of functions, you can find a more general function form such that specification of different values for parameters in this function form will yield these more specific equations. I fail to see that this has any relevance whatsoever to ascending from observations to theory.

HANSON: In general it will be the case that there will be more than one higher-level generalization from which the lower-order algebraic expressions can be generated. If this is so, then there is a choice between these, and this choice has then to be made in terms of other considerations, about which I said nothing.

242

Discussion

SALMON: Did you mean to say that the Wright brothers did not understand what the word 'flight' means?

HANSON: There is a sense in which if you simply are sitting there, fat, dumb, and happy, and a bird takes off from under your nose, you can say of it that it is in flight now. Nonetheless, there are other things that might be said about it. If you mean that the gap between the ground and where the object is now is increasing, then this is going to apply without distinction to a large number of things which you would not want to call 'flight.' Some understanding of what it is that constitutes the physical interaction between the wing of the animal and the medium through which it is moving is involved in making reference to it as flight. There are many things that a bird can do with wings outstretched which will not constitute flight; it could be in a stall, it could be falling like a Steinway piano. These are possibilities and they are perfectly compatible with what one just said about the bird. To say that it is in flight is to commit oneself to further, and I think theoretical, claims concerning the nature of the interaction between the elements in the fluid medium and the elements in the solid medium.

FEYERABEND: One can apply the standard account to the old hydrodynamics. One has coordinating definitions between certain magnitudes in the theoretical calculus and certain measured magnitudes, and one finds out that the theory is false when one does that. As a matter of fact, that is a way of finding out that the theory is false. Perhaps nowadays one has not arrived at a general theory of flight. But this does not show that once one has arrived at it, describing it by correspondence rules would not be correct.

HANSON: My conjecture is that it might well be the case that the theory one ends up with may have the structure that it has in virtue of the commitments one had already made much earlier at the level of the coordinating definitions.

HESSE: I think that the following example indicates how one can have an observation statement that is theory-laden and yet use it to refute one of two competing theories. Paul Feyerabend gives this at the end of his *Minnesota Studies* article when he wrote about the meanings of the words 'up' and 'down.' Consider Anaximander, who thought of the earth as drum-shaped. He conceived of space as nonisotropic. That is, there is preferred direction everywhere, which is the direction along which bodies fall. If we suppose that he retained this theory of space, but was forced to

243

admit that the earth is a sphere, then, sitting in Athens, he would observe bodies falling along his preferred direction in space. His meaning of the word 'down' refers to this preferred direction. Now, consider a Newtonian, whose meaning of 'down' is radial, in agreement with his theory of force fields. If Anaximander were asked to predict what happens to falling bodies in New Zealand, he would say that bodies fall "down." Newton would say that in New Zealand bodies fall "down" in the sense of Newton. These two observation statements are theory-laden, each with one of the two theories of space. Yet there is surely no question that if the experiment were performed, there would be a decisive result. Anaximander would admit when he arrived in New Zealand that his theory had been falsified. His statement would be couched in the language of his own theory, and would look verbally the same as the prediction Newton would make. At this point they would have to get together and decide about the sense of each other's language. If they were both honest men, we and they should be able to determine which one is mistaken.

ROZEBOOM: At any one time it may turn out that where people were agreeing, suddenly they can no longer agree upon their beginning statements. When this situation arises, it has always been possible to arrive at a new basis for agreement, in terms of which the previous propositions can be debated as conclusions rather than as premises.

HANSON: At the level where the claim is being adjudicated there will be terms which function in exactly this way, and so on all the way down.

ROZEBOOM: We do not need to assume that there is a rock bottom.

HESSE: It is a mistake to think that there are levels, because you can never tell antecedently which of the terms are going to be the ones which you will have to withdraw from. Those terms retained are temporarily the observation language, any terms of which might subsequently have to be withdrawn from.

FEYERABEND: The conclusion that there is a more primitive observation language is a non sequitur. Assume that we start with two classes of people, the Newtonians and the Anaximandrians. The latter class will admit that they are wrong, not using the language of the Newtonians, and perhaps not even understanding it. The former class will claim that they are right. They, being convinced that they are right, will teach everyone else their language. The consequence that there is a more primitive language in terms of which they must converse with each other is completely unwarranted. One can learn any language from scratch. A little child can do that without any more primitive language.

MAXWELL: I believe that in a good reconstruction there will be a sharp dichotomy between the observation language and the theoretical language. The observation language will be theory-laden only in the sense that there will be lawlike connections with other entities in the observation language, not in the sense that there will be meaning connections with unobservables. Furthermore, this observation language will be an observation language of sense contents.

ROZEBOOM: Many of the examples about theory-ladenness applied to common-sense terms are spurious. These are not the kinds of observation sentences a scientist is trying to come up with.

FEYERABEND: Certainly. But the term 'down' as it was used in Anaximander's theory is a very abstract term. It is almost a cosmological term. It says something about the direction in the universe. It is a paradigm of what one would call a theoretical term, although one is accustomed to it. Furthermore, anyone who uses the terms in that way, will, when he realizes that he has gone around the earth, and that he is not on a flat earth, admit that he is wrong. That is sufficient for refuting the theory that everything falls down in that sense.

GRÜNBAUM: What is the concept of theory that is used when one says that observation terms are theory-laden? What is the contrast that is implied between being theoretically infested and being theoretically aseptic? The answers to these questions have to be provided in order to give an interesting sense to the theory-laden character of observation terms. Claims formulated in Carnap's physicalistic thing language can be regarded as theory-laden, in the context of the sense-data versus non-sense-data dichotomy. Is that what is intended here?

MC MULLIN: I would like to suggest that there are two different types of theory-ladenness. It is sometimes the case that the terms in which the "observation statement" itself is expressed are altered by a new theory, or differ between two theories. This raises a specific kind of difficulty about the notion of observation, or about the reporting of observations. On the other hand, there are frequently cases where there is a disagreement between two theories, and where the observation statements upon which both of these theories rest are in no sense affected by the differences between the two theories. Take, for example, the contemporary discussion of two theories of planetary origin, the catastrophic theory and the condensation theory. So far as I know, the observation statements in both cases are not at all affected by which of the two theories you take. There is, in fact, some hypothesis always involved in observation, but not necessarily the hypothesis that is at stake between two theories.

FEYERABEND: I completely agree that with some theories the problem of the change of observation language does not arise. Therefore, I restrict my discussion to theories which I have called universal theories. These are theories which enable one to say something about anything there is in the world. The theory of planetary origin is not a theory in this sense. Examples of such universal theories are the general theory of relativity, the special theory of relativity, classical thermodynamics, and statistical physics. These theories have an advantage over theories like that of the origin of the universe, for the reason that they give us an opportunity to correct our observation language.

GRÜNBAUM: Let us take a case that does not involve cosmological considerations. We have a spaceship that goes a velocity of 0.9c to the left with respect to a ground system, and a spaceship that goes to the right with a velocity of 0.9c with respect to the ground system. A Newtonian will say that if a man in one of the spaceships measures the relative velocity of the other ship with respect to him, he will get 1.8c by Newtonian velocity addition. An Einsteinian will say that what he will get is something like $(180/181)c$, which is less than c. Surely, when they are about to compare their findings, the comparison will involve not just language noises, but things that are done with clocks and meter sticks. In fact, there will be very similar things that are done with clocks and meter sticks. The similarity in the two cases cannot be characterized as syntactical. It cannot be characterized only in terms of numbers. In fact, numerically, there is no similarity. The similarity lies in the fact that they are both testing a velocity addition law, in a very interesting common sense of 'velocity,' namely, distance with respect to transit time. To be sure, the full-blown concept of time is different in the two cases, but it involves a difference in synchronized clocks. Clocks are not syntactical entities.

Moreover, if we consider the relevant formula for an angle in Riemannian geometry, the CONCEPT of angle is the same in any one of the species of Riemannian geometry, i.e., in any of the non-Euclidean geometries and Euclidean geometry, even though these geometries are pairwise incompatible theories. And Riemannian geometry, being the basis of general relativity, is an example of what Feyerabend called a "global" theory.

Discussion

BRODBECK: One of the ways in which we test the general theory of relativity is by the bending of light. When we test it, our notion of the bending of light is not derived from the general theory. We have to make our observations independently of the general theory in order to test it.

FEYERABEND: When we go from classical physics to relativity, what remains the same are the objects. The objects are what they are, only we think different things about them. To a large extent, our operations for getting certain numbers remain the same. The functors which we use and much of the syntactical apparatus may also remain the same. What is different are all the concepts connected with the functors. In classical physics the two spaceships of Professor Grünbaum's example would be supposed to be rigid bodies. In relativity theory, they are not rigid bodies. They would be those kinds of things which in relativity theory correspond to rigid bodies as closely as possible. We have to replace the classical concepts describing these things by relativistic ones if we want to use relativity theory correctly. Thus, velocity will be a new concept; time will be a new concept.

If the concepts are different, how is it possible to have crucial experiments? We require that someone who understands the theory will, under certain circumstances, say that he is wrong. If a proponent of one theory talks to that of another, sooner or later they will discover that they are speaking different languages. But this does not matter because one can learn the other's language, and when he has learned it he will use those terms and then admit that he is wrong. So refutation is possible by using the full-blown theoretical language.

HANSON: It is often said in popular expositions of the correspondence principle that as h-bar goes to zero or as quantum numbers get large, quantum mechanics approximates classical mechanics. In other words, quantum mechanics is very often said to include classical mechanics as a limiting case. Consider the algorithms of classical mechanics and quantum mechanics. It may be that we can derive expressions from the algorithm of quantum mechanics which are syntactically indistinguishable from the expressions of the classical algorithm. That, I take it, is the force behind the claim that quantum mechanics can be taken to subsume classical mechanics as a limiting case. We should not conclude too much from this because the logic of the quantum algorithm is different from the logic of the classical algorithm. In the classical algorithm, the position and momentum operators commute, and the time and energy operators commute, which affects what complete state descriptions of punctiform masses are possible. In the case of quantum mechanics, the operators are not commutative, hence allowing a different set of possible state descriptions.

HILL: I agree very much with what Professor Hanson says. The assertion that when h-bar goes to zero the mathematics of quantum mechanics goes into that of classical mechanics is patently false. I would make the categorical statement that there are no two theories in physics that overlap completely. They all branch. The chocolate-layer-cake theory of physics, in which the higher layers effectively include the lower ones, is wrong. Physics is more like a forest that has a lot of parallel trees that grow up with their branches interlocked and rubbing together. Actually, they are different trees on different standards with some crossbars.

HEMPEL: Paul Feyerabend said quite explicitly that the operations performed by the Newtonian and the Einsteinian in the experiment of Professor Grünbaum's example could very well be the same. These operations should be describable in each theory. Would not there be a residual language in which there were observation terms common to both? This would be the language in which these experimental findings are described.

ACHINSTEIN: According to Professor Feyerabend, if I hold one theory and want to examine another theory, then the only thing I can do is learn the vocabulary of the new theory. Certainly, but it does not follow that retreating to a common vocabulary is precluded. Suppose I believe in radioactivity and you believe in leprechauns. Suppose it follows from my theory that when you put radium in a cloud chamber, beta rays will be emitted and ionize the gas molecules. You think that when you put radium in a cloud chamber, nothing will happen, because cloud chambers are irrelevant to your theory. How do I refute your theory? I don't have to learn all the technical vocabulary of your theory. Your theory predicts that there will be no new visible line in this box. My theory predicts that there will be a visible line in this box. Of course, when I describe the line in my theory, I say that it is an electron track, or I say that it is water droplets which have condensed on ions. But you don't understand that at all. So I retreat to a very primitive level. That is, I describe it as a thin wavy line, and my description does not depend on either of our theories. At this point I can refute your theory without having to learn it.

SALMON: With respect to the first problem about falling bodies, I wonder whether Professor Feyerabend and Professor Hanson would agree that we all mean the same thing by 'stone,' 'ground,' 'earth,' 'dirt,' etc., the very commonplace words that are involved here? And I wonder whether it is relevant to ask, in this context or in the later context of the discussion of the relativistic experiment, any question about whether the hand is a rigid object? I know what a hand is, and I do not need to know whether it is a rigid object or a non-rigid object. Suppose one becomes committed to the view that whenever a new piece of information about a particular kind of object comes along, we change the concept. What we are committed to, then, is a view that all truths are essentially truths of definition. Then it is impossible to use words to describe objects, to find out new things about the objects, and then to use the same words in the same meanings to describe the objects about which we have learned something new. The ostensive vagueness of many observation terms has nothing to do with theory. It has to do with the intrinsic vagueness of ordinary words. We can make these words quite precise without developing any new theories. All of this is theory-laden to an extent that is absolutely negligible and unimportant.

MC MULLIN: I agree with Professor Feyerabend, but now I want to go one stage further. Suppose we have a cosmic theory. This is a theory which applies to everything, and therefore falls under his thesis. The sort of evidence that would be brought in support of such a theory would involve concepts required for the observation statements whose meaning would depend upon the theory. For example, the evidential statements that are brought in support of Newtonian theory and those that are brought in support of relativistic theory would involve concepts whose meaning importantly depends upon the theories themselves. A second question is the more interesting one here. Would it be the case that when a Newtonian and a relativist are reporting the same experiment, the two reports would differ as evidence? Are you saying that the theory-ladenness of these reports affects the way in which they were being brought forward as evidence? By the same report, not only do I mean syntactically the same report, but I mean that the operations on which the report is based will be the same. The difference at the theoretical level will not affect the observation statement as evidence.

FEYERABEND: Whether the operations will be the same is accidental. If the gravitational field of the sun were much stronger, then light rays would be bent to such an extent that we would have to take this fact into account in measuring their angle.

CRAIG: Consider an Einsteinian and a Newtonian, in the case in which each wants to test his theory by means of the advance of the perihelion of Mercury. Do they perform the same experiment?

FEYERABEND: In most cases, yes.

CRAIG: Even when one has very pervasive theories such as relativity theory or Newtonian physics, there is some common ground. This is the only way one can decide between two theories. It is quite true that notions, such as angle, which are often thought of as observational notions, turn out to be theoretical. In fact, an assertion about an angle in relativity theory is tested experimentally in a different way from the same-sounding assertion in Newtonian physics. Statements which sound observational, which sound innocuous, are actually theoretical, and in the framework of different theories lead to different predictions, and have to be tested in different ways.

HESSE: I agree with what Professor Craig has just said. Given simple experimental situations, such as that in the example of Anaximander, there seems to be a kind of neutral language between theorists. But the crucial point is whether it follows from this that there is a layer cake of observation languages going downwards, with each level becoming less theory-laden. It seems that we could always, given the way in which we now report observations, withdraw from the theoretical commitments which are made by these observation statements to something at a lower level. I think that it is false to conclude from this situation that one can get a language which is neutral with respect to the two theories by descending to a sufficiently low-level observation language.

Consider the notion of simultaneity. A Newtonian hopes his observation statements are fairly neutral with respect to his theory. We now believe that most of the relevant observation statements applying to his theory commit him to the view that there is a time slice across the universe so that there is absolute simultaneity between any two points at a given time. This notion was one which was theory-laden in the Newtonian scheme. If, for some reason, a Newtonian had doubts about this question before the advent of special relativity, he could not have reformulated his observation language in terms which did not make this assumption about simultaneity. He would have had to set up some kind of operational definitions about which he and the subsequent Einsteinian could have agreed. But in order to set up this operational definition, he would have to have made some assumptions about the speed of light. He would have to assume that this speed was the same in all directions and for all relative motions. These are assumptions which a Newtonian would have found far more theoretical than the assumption about simultaneity. He could have made the assumptions required to set up operational definitions about simultaneity without presupposing other parts of a relativistic theory. He could not have given the Einsteinian definitions because he would not have had the Einstein theory of light. So it is not at all meaningful to talk about the sort of neutral language which the Newtonian could have withdrawn from, even if the Einstein theory were formulated in such a way that they had something mutual, because the Einstein theory would have had to be used to formulate the new observation language.

GRÜNBAUM: The issue is not whether one could have withdrawn to something which does not involve any part of the theoretical apparatus, but whether when one looks at two theories something is the same in both. In my spaceship example, the Newtonian will use locally synchronized clocks. He assumes that clocks under transport will behave in a certain way. The relativist has a different belief about the behavior of clocks under transport. Therefore he will not synchronize clocks in the same way; rather he will station them in certain places, and then have certain light signals exchanged between them. Although the clocks are not synchronized in the same way, we are using clock readings in both cases. We can specify what is the same and what is different in the two cases. The crucial issue is not whether each theorist can withdraw to some territory which is completely free of any of the commitments of his theory, but what is common to the two theories.

HANSON: There may be an argument to the effect that relativity can have ordinary classical mechanics as a limiting case. It is fundamentally different in the quantum mechanical case.

CORNMAN: Given a difference in meaning, or a change of concept, why must we say that the new and old concepts are incommensurate? Perhaps there is an analogical relationship between the corresponding concepts of successive theories.

SALMON: I have exactly the same point to make. Although there are changes in concepts, which we all admit do occur, incommensurability of concepts is not the result. There is a high degree of commensurability left, and this is how we can talk sensibly.

HANSON: Let us consider again the example of the anomalous advance of the perihelion of Mercury. In the years between 1858 and 1860, Leverrier produced a number of memoirs in order to establish precisely what discrepancy existed between the predictions of unadjusted Newtonian theory for the advance of the perihelion of Mercury and what was observed. This was determined to be 38 seconds of celestial arc per century. There is a rough consilience between this figure and what is actually set out in the papers of Einstein as 42.5 seconds of celestial arc per century. It is perfectly clear that if the instruments of the early twentieth century were transported into the middle of the nineteenth century, the observaions would, in an important sense, have been the same. The configurations within the telescopic view area would have been the same for both. However, if one had undertaken to make the adjustments appropriate for the theory of relativity, these would have been such as to yield a value, based on those same observations, which would have been incompatible with the recorded discrepancy of 38 seconds of celestial arc. Here is a clear case where the fact that a theory has come along which requires a completely different construal of the data means that the observation statements are going to be different in these two cases. This discrepancy supports Feyerabend's case.

BUCK: How we could teach relativity theory to kids in school is in an important sense a question for educational psychology. It is not at all obvious that whether we could or could not do this is a philosophical question.

HEMPEL: The question of how one could teach children relativity or the language of some theory has some philosophical interest. People with strong philosophical consciences think that new theoretical terms are suspect as long as their meaning cannot be fully specified by definition. It seems clear that this cannot be done in any profitable way. The question then arises of how one can teach a person to use concepts without defining them, to understand them satisfactorily, and to use them with great objectivity. The question really acquires considerable interest. I think that the purists are making too strong a demand and an unrealistic one. Here it would be helpful to be able to point out that children could be taught such concepts. Then one could argue that one can come to understand a language without having to make use of explicit definitions in terms of something one antecedently fully understands.

CORNMAN: Let us look at Professor Hempel's schema (2). Schema (2) has no calculus and no correspondence rules. It has internal principles and bridge rules. In the internal principles there are both the theoretical vocabulary and the antecedent vocabulary. In the bridge rules there is a theoretical vocabulary and an antecedent vocabulary. They both have the same vocabulary, and there is no formal way of distinguishing between the two. Now, if there is no way of distinguishing between the two schemata, then what are the consequences for those people who object to schema (1)?

HEMPEL: One objection that I have to what I called schema (1) was the idea that it is not always profitable and illuminating to view a scientific theory as consisting of an uninterpreted calculus plus certain statements which interpret this calculus. I argued that this is no doubt possible; I presented reasons why I thought that it was not always profitable. Nevertheless, it can be very illuminating to have certain specific presentations of this kind. For example, I remarked that Reichenbach has made very illuminating use of this idea in discussing the relation of abstract mathematical geometry to physical geometry. One could think of the formulation of a theory as involving two sorts of specifications: (a), those which describe the theoretical scenario, the new entities and processes which are assumed there; (b), those which say how this is connected with what one has antecedently established, empirical laws, for example. As I pointed out, the bridge principles which establish the connections would necessarily have to make use of whatever theoretical terms the theory introduces and the antecedently available vocabulary. The internal principles would make use of whatever theoretical terms the theory introduces, plus antecedently available vocabulary. Since, in general, the theory will contain antecedently available vocabulary, it is very difficult to make a distinction between the bridge principles and the theoretical ones, except in some pragmatic terms, unless the theoretical principles contain theoretical terms exclusively. The Maxwell equations might have this latter character.

ACHINSTEIN: Here is the value of distinguishing between a concept and a term. One wants to say that some of Bohr's concepts were entirely new, namely the concept of quantization within the atom itself, but the terms he used were not. In the case of entropy one would say that Clausius's concept of entropy was a new concept, even though he could define it by reference to heat and temperature which were antecedently understood. Is your thesis one about concepts, or is it one about terms, or do you not want to distinguish between the two?

HEMPEL: The question here is whether one wants to qualify a term as a "theoretical term not belonging to the antecedently available vocabulary" if in fact it can be explicitly defined in terms of the antecedent vocabulary. I would have originally thought that if we can define all our newly introduced concepts in terms of what we already have, then there really is not a new theory. Now I doubt that, and I think one should say that in this new context there are new theoretical terms, although they can be defined in terms of the available vocabulary. I put this before you tentatively and for discussion.

FEYERABEND: The standard view of scientific theories can be defended under two circumstances which Professor Hanson has enumerated, and it cannot be defended under further circumstances. The first is the case of hydrodynamics. It is a beautifully abstract theory about ideal, not real, fluids. Nevertheless, I would want to use this theoretical scheme for making predictions in a certain domain. Special correspondence rules would have to be added. Because the theory is not about anything in existence, one must modify it so that it may make correct predictions. In this case, a scheme of theory, correspondence rules, observation language is the only way in which we can apply the theory to the observations. Another example is quantization in field theory, which is an inconsistent theory. Here one must employ special correspondence rules in addition to the theory in order to get a consistent observational apparatus, because the theory is inconsistent. If one has a different theory which is neither inapplicable to anything nor inconsistent, then one does not need this scheme. If it is applicable to everything, the theory gives terms for every concrete event there may be. This means it contains a vocabulary and a language which can be taught to people. No correspondence rules are needed; no bridge laws are needed. The theory can be acquired from teaching, not by relating it to some other language.

GRÜNBAUM: What is understood by "a theory that applies to everything there is"? Is it a theory which makes statements that pertain to any space point or any time point? What is there about spatiotemporal pervasiveness that requires this doctrine of conceptual change in a way in which nonspatiotemporal pervasiveness does not require it?

MC MULLIN: Let me try to answer the question for Professor Feyerabend. The kind of theory he wants to restrict himself to, if I am not mistaken, is that very narrow class of theories which affect the meaning of the observation statements upon which they are based. There are very few of those. Relativity theory is one of them. That is, if relativity theory is in fact based upon space-time measurements, and if it is itself a theory of space-time, then the meanings of the evidential statements are going to be affected. One could ask whether or not the way in which a theory affects the evidential statements would change the evidence itself. Professor Hanson said that in the case of the Mercury perihelion, the reformulation of the space-time notion meant that the number would be changed from 38 to 42.5. This is an extreme consequence, namely, the reformulation will change the evidential basis. There will also be a great number of cases where this would not be true. Nevertheless, Professor Feyerabend would want to insist that although the numbers have not been changed, the conceptual basis of the new theory would extend to the evidential statements upon which the theory itself was based. I think those are his universal theories.

ROZEBOOM: I would like to give what I understand it is for a term or a set of concepts to be theory-laden. If my notion is agreed upon, then this has some fairly straightforward implications for Professor Feyerabend's views. Suppose that T is the theory under consideration with which a certain proposition P may be laden. If the concepts in P are laden with theory T, I would understand this to require that P necessitates T. This is the only sense which I can make of the idea that the concepts in P are laden with the theory T.

BUCK: If an observation statement turns out to be false, it was suggested that it counted against the theory and also against the intelligibility and applicability of the terms which the observation statement contained. I suppose that you might build a sense of presupposition to answer to these requirements.

HANSON: I will simply refute this in one step. Suppose that P is an observation statement which refers to double refraction. It will require theoretical backing in terms of Snell's law. This commitment is found in corpuscular Newtonian optics, and also in that theory which in many respects is taken to be the negation of corpuscular Newtonian optics, namely, Huygens's wave optics. Either one of these theories is, to the extent that they involve Snell's law, compatible with the observation statement in question. But it is contradictory to say that the observation in question entails both of these theories.

HILL: It surely is important that no scientist presents raw observations. The observations themselves are always cleaned up before they are presented as data. That is true even in biology. In fact, when you talk about observations you are already talking about cleaned-up things.

HANSON: In virtue of what are they cleaned up? By prior theoretical commitments of some sort?

Discussion

HILL: That is right.

WILLIAMS: It is also important that you not be circular and not clean up the data to yield the point of view that you are trying to demonstrate.

MAXWELL: If Professor Feyerabend rejects Professor Rozeboom's account of what it is to be theory-laden, then it seems to me that it is incumbent upon him to give a clear account of his own.

HILL: Let us say prior conceptions of geometry constitute theory-ladenness. The observer's conceptions are largely determined by his apparatus. He has to understand your apparatus, and use certain ideas in interpreting it. If he does not do that he has no observations.

HANSON: This is shown by the relationship between Hooke's law and the spring balance, and Archimedes' law and the double pan balance. Whatever is observed by virtue of balances of this kind is going to have either Hooke's law or Archimedes' law built into it in some sense.

MC MULLIN: Professor Hill said that in physics it is ordinarily the case that the observation statements which are brought forward are theory-laden in the sense that certain preconceptions are built into them. What is meant by saying that for certain theories the observation statements are theory-laden is the following. Statements are advanced as evidence for a particular theory, but they involve concepts whose meanings will be determined by the theory itself, and this is not ordinarily true. There are not many theories of which this is true. In the case of some of the most important theories in physics, the observation statements will necessarily involve conceptions which, in a certain sense, are being called into question, or are being justified by the theory. This is not quite my previous point; it is a development of it. If, for example, the relativity theorist uses certain observation statements involving the notion of length or the notion of angle, these are theory-laden in the sense that they are themselves being called into question, and a particular answer to the question is being given by the theory which is based on this evidence.

HILL: What has happened today in science is that scientists want even their observations to be theory-laden. They deliberately want them to be theory-laden. There is nobody, practically speaking, who goes out and makes raw measurements anymore. The new concept is that the experiment is not for the purpose of finding out the facts of nature, but for the purpose of testing theories.

CRAIG: There is a great deal of theory mixed up in some way or other with reports about experiments, not only by physicists, but also by biologists or chemists and others. I recall lab experiments in a biology course in college. If you start working with a frog, it is very difficult to find out whether you are working with the liver or some other part of the frog. Where does the liver stop and some other part start? Whenever you start actually reporting about things, you already have theorized to some extent. In this respect physics is not unique. If there is something unique about so-called universal theories in the respect in which they differ from the others, it has still to be brought out. It seems that quite a few different notions of universal theory have been dealt with. One notion is simply that a theory has a wider range of applicability than another. For instance, the theory of spectroscopy has less wide a range than a general theory of optics. Quite often the ranges overlap. It is true that certain theories have a much wider range than others. Everybody would agree that relativity theory and Newtonian mechanics are among these. The mere fact that they have a wider range is of some interest, but not very much. It is more important to try to find out in what respect they are deeper theories than others. This has not been brought out very clearly. I agree that in some sense the general theory of relativity is philosophically much more interesting than a particular theory of spectral lines. However, the word 'universal theory' tends to be misleading in this usage. It tends to lead to the following confusion which may have been in some people's minds. At times 'universal theory' was used almost as a respectable scientific term for old-fashioned metaphysics. The very appeal of such a monolithic theory tends to make us overlook what I think are important differences. If it is to be the truly universal theory, it must account not only for observations in the laboratory, but for the facts of daily life as well. A truly universal theory would have to account for these facts, more or less in the terms we use in describing them now. I think all present theories are a far cry from that.

MC MULLIN: Professor Feigl's diagram of the standard view of scientific theories allows us to see that we are asking two somewhat different questions. One would be the question of discovery; the other would be the question of confirmation or justification. It is clear that this diagram was never intended to be a pattern for discovery. It is intended to be a reconstruction at some point in time. One of the difficulties about it is that there are certain kinds of theories to which it does not apply. For example, hydrodynamics had this character. That is, one had this nice theoretical balloon up in the air, but it was awfully difficult to get the hooks down into the soil of observation, and ultimately they did not get down.

On the other hand, if one looks at many other areas of contemporary physics, for example, quantum physics, there is an attempt, particularly with the use of computers, to build out of observations a set of functional relationships among observations. One gradually works up to a level where there is no theoretical structure at all but simply a mathematical way of relating a set of observations with no further kind of structure. That sort of system would not fit the standard view of scientific theories. In these two examples we could make a critical comment that these are not very good theories. Would one want to say that something like the standard view of theories ought to be applicable to the confirmation structure of a good theory? Let me ask a question of Professor Hanson. If one moves to higher and higher algebraic levels, as in your example of aerodynamics, at what point does this become theoretical and how does it become so?

HANSON: Suppose the standard view of theories were taken to be a normative criterion in virtue of which one evaluates extant theories. Then it is relevant to look at actual theories which have been influential in the history of science. If there are working scientific theories which do not fulfill this criterion, it is worth pointing this out. The number of actual cases which completely fulfill this criterion may be equivalent to the null class. This is a rather misleading criterion, if the scientific theories all around us are illuminated so rarely and so seldom by this particular account.

MC MULLIN: I absolutely agree with you that it is not the case that the standard view applies to all theories. But even in your own writing about this you suggest that it may very well give a direction for a theory at a given time. For example, if one can point out that contemporary aerodynamics lacks this form, this does give a direction for future research in aerodynamics, namely, to develop the theory so that it has that form.

HANSON: Yes, I think that is right. I did grant that point when I characterized ideal fluid mechanics as an ideal language theory, and quantum electrodynamics as a black box theory largely in terms of the conceptual apparatus provided by the standard view. I made my remarks about those theories in virtue of my already having granted a certain analytical utility to this schematization. In addition to noting that function, it is important for understanding what scientific explanation and scientific theory are to note that there is nothing which actually corresponds to this. This bears on the degree to which analysis in the form of philosophy of science can help us to understand great moments in the history of science.

An important activity in the development of experimental laws which might have great significance for the development of the theory is the formal generalizing activity, which one finds in the theory of functions. Here, a number of lower-order algebraic statements can be shown to be subsumed under higher-order statements, where the subsumption is not simply one of conjunction. For any collection of algebraic statements there will not be just one higher-order abstract statement from which they can be generated. This multiplicity of higher-order statements need not all be equivalent each to the other. Your point was that since they would all entail the list of algebraic statements, there must be necessary connections between them. My point was that although these higher-order general claims will have as one of their properties that they entail the lower order statements, they will have other properties as well. Then the choice between the higher order general claims will rest upon other considerations which I call boundary considerations.

ROZEBOOM: To the extent that one simply observes that a set of equations are all instantiations of a more general form, this is simply observing an analytic fact, and I am not objecting to this. Professor Hanson's argument does not do anything to point to how a nondemonstrative inference is made from what one has to a more comprehensive picture of what the world is.

HANSON: The relation between higher- and lower-order claims is analytic, if one considers the more general claim and then generates the lower-order claim. It is not analytic in the other direction. One cannot make an inference from the collection of lower-order algebraic statements to the higher-order general claim.

PART TWO. CURRENT ISSUES AND CONTROVERSIES

On the Relation of Topological to Metrical Structure

In his inaugural dissertation, Riemann emphasized the distinction between the topology of space, which, for him, meant its continuity or discreteness, and its geometry.[1] Taking the topology as essential, he convincingly argued that the concept of a continuous space does not, by itself,
imply a particular geometry, and that therefore the nature of the geometry
of space involves physical considerations. This thesis he expressed by the
assertion that metrical relations are not *implicit* in the concept of a continuum, though they are implicit in the concept of a discretum.[2]

Professor Grünbaum has argued that the ideas of Riemann's inaugural
dissertation establish a much stronger thesis. He takes Riemann to have
established that in a *nondenumerable* and *dense* space, the *self-congruence*
of the metric standard is *conventional*, and to this extent, metrical relations *generally* are conventional. According to this thesis, this means that
they are not *real* relations which may or may not be *discovered* to obtain,

[1] Here and elsewhere, I mean *metric* geometry.

[2] Cf. [4], pp. 424–425: "The question of the validity of the postulates of geometry
in the indefinitely small is involved in the question concerning the ultimate basis of
relations of size in space. In connection with this question, which may well be assigned
to the philosophy of space, the above remark is applicable, namely that while in a discrete manifold the principle of metric relations is implicit in the notion of this manifold,
it must come from somewhere else in the case of a continuous manifold. Either then the
actual things forming the groundwork of a space must constitute a discrete manifold, or
else the basis of metric relations must be sought for outside that actuality, in colligating
forces that operate upon it.

"A decision upon these questions can be found only by starting from the structure of
phenomena that has been approved in experience hitherto, for which Newton laid the
foundation, and by modifying this structure gradually under the compulsion of facts
which it cannot explain. Such investigations as start out, like this present one, from
general notions, can promote only the purpose that this task shall not be hindered by
too restricted conceptions, and that progress in perceiving the connection of things
shall not be obstructed by the prejudices of tradition.

"This path leads out into the domain of another science, into the realm of physics,
into which the nature of this present occasion forbids us to penetrate."

but rather their obtaining or not obtaining must be *stipulated*.[3] On the other hand, if space possesses a *discrete* topological structure, metrical relations are not conventional but real.

The importance of this theory of metrical relations consists in the fact that it forms the basis for Grünbaum's conventionalism. That is, *relative to the contingent truth that space is continuous*, this theory, rather than the thesis that any description may be maintained, given suitable alterations in the relevant laws and auxiliary statements, is held to establish the following: (i) "[T]he alternative between different metrics and hence between their associated metric geometries is one of mere descriptive convenience" ([2], page 242). (ii) Given two descriptions such that II asserts a "unitary change in those functional dependencies or laws of nature which involve variables ranging over lengths [and distances, and] I asserts that all metersticks and extended objects [as well as the distances between them] have expanded by doubling; [I and II] are complete[ly] equivalen[t] or co-legitima[te] . . . with regard to truth value *and* inductive (though not descriptive!) simplicity" ([2], pages 158, 161, 173).[4]

The purpose of what follows is to show that whether or not metrical relations are conventional is independent of the type of consideration that

[3] Thus Grünbaum tells us that ". . . *the very existence* and not merely the epistemic ascertainment of relations of congruence (and of incongruence) among disjoint space intervals AB and CD of continuous physical space will depend on the respective relations sustained by such intervals to an extrinsic metric standard which is applied to them. . . . [And t]he existence of congruence relations among disjoint intervals is . . . a matter of convention precisely to the extent that the self-congruence of the extrinsic metric standard under transport is conventional . . ." ([2], pp. 148–149). And discussing the following statement: "(B) The platinum-iridium bar in the custody of the Bureau of Weights and Measures in Paris (Sèvres) is 1 meter long everywhere rather than some other number of meters (after allowance for 'differential forces')" ([2], p. 101), Grünbaum writes: ". . . (B) is not analytic but is a *stipulative* specification of part of the extension of the spatial equality term 'congruent,' just like the following statement: 'Disjoint space intervals which can each be brought into coincidence with the same unperturbed transported rod are hereby stipulated to be spatially equal' " ([2], p. 271). It should be noted that Grünbaum's theory of metrical relations is held to apply to the congruence of *space-time* intervals as well. Cf., for example, [2], p. 217: ". . . an extrinsic metric standard is self-congruent under transport as a *matter of convention* and not as a matter of spatial fact, although any *concordance* between its congruence findings and those of another such standard is indeed a matter of *fact*. If the term 'transport' is suitably generalized so as to pertain as well to the extrinsic metric standards applied to intervals of time and of space-time respectively, then this claim of conventionality holds not only for time but also *mutatis mutandis* for the *space-time continuum of punctal events!*"

[4] With regard to (i), Grünbaum claims to "have put forward positive structural reasons à la Riemann" ([2], p. 242): while (ii) is held to be "a consequence of a significant property of physical space, i.e., of its being a mathematical continuum of like elements" ([2], p. 180).

is suggested by Grünbaum's theory of metrical relations. Of the two theses which this theory is held to establish, only (i) will be explicitly discussed. The discussion will be further restricted to the space metric of nonrelativistic physics. Thus, there will be no explicit mention of the metrics of time and space-time. These restrictions, however, are *purely* expository, and hence any conclusion reached concerning space is readily extendible to *nonrelativistic* time, and, in the case of relativity, to *space-time*. Similarly, though (ii) will not be explicitly discussed, it follows that if Grünbaum has not established the conventionality of metrical relations, (ii) is as groundless as (i).

I

Our discussion will require the notion of a *distance function* or *metric*. This is simply a mapping which associates with any two points x, y a nonnegative real number, and which satisfies the conditions (a) $d(x,y) = d(y,x)$; (b) $d(x,y) = 0$ if and only if $x = y$; (c) $d(x,y) + d(y,z) \geq d(x, z)$. Given the distance function the class of *straight* intervals[5] is determined. Given any interval [x,y] the class of intervals with which it is congruent—i.e., the congruence class to which it belongs—is given by the distance function. That is, relative to a given distance function, two closed intervals are congruent or belong to the same congruence class if their end points are associated with the same real number. Thus, the distance function or metric associates with each pair of points an interval, and with each interval, a congruence class.

Riemann's assertion that metrical relations are not implicit in the concept of a continuous space may be stated more exactly as follows: If space is continuous, there is an infinite number of possible metrics and, moreover, there is no topologically invariant property (or relation) on which the selection of one of these metrics may be based. Of course, this assertion has nothing to do with the choice of scale, or rather, *given* the selection of an appropriate scale, the assertion still holds. It therefore implies that there are infinitely many possible metrics which differ *significantly* in the sense that any two of these metrics will differ with respect to the congruence class with which at least one interval is associated.

The basis for Grünbaum's claim that in a nondenumerable, dense space, the choice among alternative metrics is conventional consists in this:

[5] Hereafter, by "interval" I shall always mean *straight* interval.

Given any two (nondegenerate) disjoint intervals of a nondenumerable and dense space, whether or not they are congruent depends on the *stipulation* that some (possibly the same) pair of disjoint intervals are or are not congruent, since in this type of space, the congruence of disjoint intervals is not a real relation.[6] The qualification that the intervals be disjoint is necessary, since, given two closed intervals such that one is a proper part of the other, we may say that they are *incongruent*, whatever the form of d. Having noted this, no misunderstanding should result if, henceforth, I omit explicit mention of this qualification.

Finally, the thesis that the metric is conventional is expressible as a consequence of the following two ideas: (1) that there is a plurality of significantly different possible metrics, and (2) that congruence is not a real relation. That is, *given the conventionality of congruence*, (1) implies that the choice among alternative metrics is conventional. Notice that by itself (1) is not sufficient for the conventionality of the metric. (By itself, (1) is at most necessary, since without it the thesis is vacuously satisfied.) This may be made clear if we leave, temporarily, our exposition of Grünbaum's theory and consider, instead, the following criticism which might be raised against it.[7]

Let f represent some admissible property of a physical system S consisting of N particles, where f is some suitable real-valued function of the 6N position and momenta coordinates associated with the state s of S. Then, even if the space and time metrics are held constant, the class G of all similarly suitable real-valued functions g_i which differ from f and from each other in the value associated with at least one s is infinite. Moreover, there is a subclass G^* of G whose members differ *significantly* from f, and which is also infinite. Thus, Grünbaum's claim that the metric is conventional is trivial if it is restricted to metrics which "differ" only insofar as they are based on different scales. And it is false if the metrics are meant to differ significantly, since, by parity of reasoning, it would follow that the congruence of phase space intervals is also conventional. However, given the existence of G^*, all that we are required to admit is that f, unlike any g_i in G^*, reflects changes in the state of S in accordance with the inductively simplest laws.

For this criticism to be valid, it is necessary that the theory be inter-

---

[6] Cf. footnote 3, above.

[7] The criticism which follows is essentially the one which appears in [3], pp. 222ff.

preted as asserting that (1) *alone* (i.e., without (2)) implies the conventionality of the metric. Now when we inquire after the basis for (2), certain considerations may be advanced which suggest a similar interpretation. For example, the conventionality of congruence is in turn based on the fact that space is a continuous manifold. Prima facie this seems to mean that congruence is conventional in *any* nondenumerable and dense structure. But then the congruence of phase space intervals must also be conventional according to Grünbaum's theory, and if so, the theory is clearly false.

From these considerations it follows that the conventionality of the metric cannot be based solely on the possibility of a plurality of significantly different metrics, and that the conventionality of congruence cannot be based solely on nondenumerability and denseness. According to Grünbaum, what is required in addition is that the elements of the continua be intrinsically alike. Now the states of affairs represented by different phase space points *differ* intrinsically. For example, the states of affairs constituting a change in pressure differ with respect to the value of this magnitude. With regard to the spatial case, on the other hand, there is no nonconventional, monadic property of spatial points on which spatial congruence may be based.[8]

Moreover, since it is also the case that in a nondenumerable and dense space congruence cannot be based on cardinality or any other topological

[8] See, for example, [2], p. 15, footnote 7 ([1], p. 415, footnote 7): ". . . while holding for the mathematical continua of physical space and time, whose elements (points and instants) are respectively alike both qualitatively and in magnitude, the thesis of the conventionality of the metric cannot be upheld for *all* kinds of mathematical continua . . ." Also [2], p. 33 ([1], p. 431): ". . . there are continuous manifolds, such as that of colors (in the physicist's sense of spectral frequencies) in which the individual elements differ qualitatively from one another and have inherent magnitude, thus allowing for metrical comparison of *the elements themselves*. By contrast, in the continuous manifolds of *space* and of *time*, neither points nor instants have any inherent magnitude allowing an individual metrical comparison between them, since all points are alike, and similarly for instants. Hence in these manifolds metrical comparisons can be effected only among the *intervals* between the elements, *not* among the homogeneous elements themselves. And the continuity of *these* manifolds then assures the nonintrinsicality of the metric for their intervals." And again: "In the case of space and time, intervals as such are constituted merely by the points and instants *without* as yet involving any metric; but the *metrical* attribute of pressure is indispensable *ab initio* to confer identity on the elements of the continuum of pressures, and—unlike points and instants—the elements of the pressure continuum each have distinctive magnitudes of their own. Hence the difference between the individual magnitudes of the elements of the pressure continuum furnishes the *intervals* of the latter continuum with an intrinsic metric, in contradistinction to the intervals of space and time!" ([2], p. 261). Cf. also the passage quoted in footnote 9, below.

property of spatial intervals, Grünbaum concludes that, in this type of space, congruence is conventional.[9]

By a discrete space Grünbaum understands a finite set of contiguous elements, arranged, in the two-dimensional case, like the squares of a checkerboard, and such that there is no basis for supposing them capable of further subdivision. In this type of space congruence is a real relation, since it may be based on the equipollence of intervals.[10]

II

It is necessary that we be quite clear concerning the meaning of Grünbaum's thesis that space is intrinsically metrically amorphous. What is involved may best be illustrated if we contrast the strong form of this thesis with a weaker version.

In its weak form the thesis is merely a circuitous way of denying that the topology of space implies a unique metric, and hence that metrical relations may be based on topological properties (or relations). Here one means by "space" a manifold of homogeneous elements, having a nondenumerable and dense topological structure. Thus, according to the weak form of the thesis, to assert the intrinsic metrical amorphousness of space is to assert that a unique metric is not capable of being based on the topology of the space manifold.

According to the strong form of the thesis, to assert the intrinsic metrical amorphousness of space is to assert that the metrical structure of any metrical relation, such as the congruence of space intervals, is conventional. That is, according to the strong form of the thesis, the extension of any

[9] ". . . upon confronting the extended continuous manifolds of physical space and time, we see that neither the cardinality of intervals nor any of their other topological properties provide a basis for an intrinsically defined metric. The first part of this conclusion was tellingly emphasized by Cantor's proof of the equicardinality of all positive intervals independently of their length. Thus, there is no intrinsic attribute of the space between the end points of a line-segment AB, or any relation between these two points themselves, in virtue of which the interval AB could be said to contain the same amount of space as the space between the termini of another interval CD not coinciding with AB. Corresponding remarks apply to the time continuum. Accordingly, the continuity we postulate for physical space and time furnishes a sufficient condition for their intrinsic metrical amorphousness" ([2], pp. 12–13; [1], p. 413).

[10] Cf., for example, [2], pp. 153–154: "It is clear that the structure of this granular space is such as to endow it with a transport-independent metric: the congruences and metrical attributes of 'intervals' are intrinsic, being based on the cardinal number of space atoms, although it is, of course, trivially possible to introduce various other units each of which is some fixed integral multiple of one space atom. . . . And the metric intrinsic to the space permits the factual determination of the rigidity under transport of any object which is thereby to qualify as an 'operational' congruence standard."

metrical relation must ultimately rest on an arbitrary stipulation such as the self-congruence of the metric standard.

Now, since the weak thesis excludes the possibility that a continuous space possesses a unique metrical structure, it might seem that it also excludes the possibility that there exists a unique and nonarbitrary partition of the class of all space intervals into distinct congruence classes. That is, it might be assumed that if only the weak thesis were true, then, though a particular partition of the class of all intervals may be stipulated, there is no unique partition to be discovered.[11]

This, however, would be a mistake. For the weak thesis excludes the possibility that spatial congruence has a unique metrical structure, if it is based on some topological property such as the cardinality of intervals. If congruence is not based on one or another topological property, then, so far as the weak thesis is concerned, it is quite possible that there exists a unique and nonarbitrary partition of the class of all spatial intervals into distinct congruence classes.

It is to be observed that this error is equivalent to assuming that the necessity for stipulating the extensions of metrical relations (i.e., the strong thesis) is a direct consequence of the fact that the topology of space does not imply a unique metric (i.e., the weak thesis). But in fact, this inference depends on the additional assumption that any (metrical) spatial relation which cannot be based on the topology is conventional.

In light of the preceding, it should be clear that only the strong form of the thesis will support Grünbaum's claim that the choice among significantly different metrics is merely a matter of descriptive convenience. Hence, the considerations (a) that there is no nonconventional, monadic property of spatial points on which congruence may be based, and (b) that congruence cannot be based on cardinality or any other topological property in a nondenumerable and dense space—if they are relevant at all —must support the strong thesis. Thus, since (a) and (b) are put forward in support of the strong thesis, they must be interpreted as implying that the congruence of spatial intervals is nonconventional only if (a′) as with phase space intervals, it is capable of being based on some nonconventional, monadic property of the points contained in the intervals, or (b′) it is

[11] This, at times, appears to be Grünbaum's view: "Relations of congruence among intervals of space, time, and space-time respectively are specified by equal measures ds_3, ds_1, and ds_4 respectively. And since these respective intervals do not possess intrinsic metrics ds_3, ds_1, and ds_4, the respective congruences among them are extrinsic [i.e., must be stipulated]" ([2], p. 218).

capable of being based on cardinality or some other topological property of the intervals.

Of course, we may grant that there is no nonconventional, monadic property of spatial points on which congruence may be based (a); and we may also grant that congruence cannot be based on cardinality (or any other topological property or relation) in a nondenumerable and dense space (b). For the issue is not whether (a) and (b) are true, but rather, given that they are true, does it follow that in a nondenumerable and dense space, congruence, and hence the metric, is conventional?

The basic difficulty with the condition that spatial congruence, if it is not conventional, must satisfy (a') or (b') is that it omits the most important fact about a spatial interval: that it is connected with the relation of distance. That is, for each class of congruent intervals C, there exists a corresponding distance relation R, composed of the elements in the range of the univocal function f: [x,y] → (x,y), for [x,y] in C. Thus, it would appear that if [x,y] and [w,v] belong to the same congruence class C, (x,y) and (w,v) must belong to the same distance relation R. And therefore, if distance is a real relation, so also must congruence be a real relation.

It appears, to begin with, that Grünbaum's failure to consider the connection of spatial intervals with relations such as distance is due to the fact that an interval is a class. For this makes it seem natural to look to cardinality or proper inclusion for a nonarbitrary basis for comparing intervals, when, as in the present case, the intervals concerned do not conform to the phase space model (i.e., to (a')). At most, however, this makes it appear more plausible that spatial congruence is conventional than that there is something of importance which (a') and (b') omit.

But since Grünbaum believes that (a) and (b) establish the intrinsic metrical amorphousness of space, he has very likely confused the weak and strong forms of this thesis. For the weak thesis is established by the fact that congruence satisfies neither (a') nor (b').

On the other hand, the assumption that slightly modified versions of (a) and (b) establish the conventionality of distance[12] implies that the extension of a contingent relation must be stipulated unless (c') the distance relation of (x,y) is capable of being based on some nonconventional, monadic property of x or y, or (d') distinct distance relations are capable of being distinguished on the basis of the cardinality of the intervals with

---

[12] Cf. [2], pp. 12–13, quoted above in footnote 9, which might plausibly be interpreted as asserting this.

which they are associated. No basis in either science or common sense is possessed by (c') and, if accepted, it would imply the conventionality of contingent relations altogether. And (d') is a totally arbitrary requirement for the distinctness of two relations, since there is no reason why a relation should manifest itself in terms of some property of a class of individuals. Hence we may safely conclude that neither (c) nor (d) affords the slightest support for the claim that distance, and hence the metric, is conventional, or equivalently, the nonconventionality of distance does not require that it satisfy either (c') or (d').

Since distance need not satisfy (d'), it is not at all clear that discreteness implies the nonconventionality of distance. Of course, discreteness does imply that it is possible to distinguish space intervals on the basis of cardinality. But since distance need not bear any interesting relation to the cardinality of the intervals with which it is associated, this fact about discrete spaces is not, by itself, sufficient to establish that distance is not conventional. Notice that it is *not* being suggested that cardinality can bear no relation to distance, only that it *need* not. And if there is no a priori reason for assuming that a given distance relation will only be associated with intervals of the *same* cardinality, it cannot be maintained that discreteness establishes the nonconventionality of distance; nor, therefore, can it be maintained that discreteness establishes the nonconventionality of congruence. This completes our discussion of Grünbaum's theory of metrical relations.

The argument of the preceding section may be summarized as follows: A distinction was first drawn between a *weak* and a *strong* form of the thesis that space is intrinsically metrically amorphous. Next, it was shown that the weak thesis, though acceptable, does not imply the strong thesis, which, on the assumption that space is continuous, Grünbaum's theory of metrical relations is held to establish. It was then shown that Grünbaum's theory must be understood as asserting that the congruence of spatial intervals is nonconventional only if (a') it is capable of being based on some nonconventional, monadic property of the points in the intervals, or (b') it is capable of being based on the cardinality of intervals. Concerning (a') and (b') it was shown, first, given that spatial intervals do not satisfy (a'), (b') is a plausible condition for classes, which, as far as is known, are mere classes; second, if neither (a') nor (b') is satisfied, this establishes the weak thesis. With respect to the first result, it was shown that (a') and (b') omit the connection of spatial congruence with the *relation* of

271

distance, which, if real, implies that congruence is real. With respect to the second result, if the fact that distance fails to satisfy slightly modified versions of (a') and (b') is held to establish the *strong* thesis, consequences follow which are so contrary to plain truth that they need only be made explicit in order to reject the theory from which they follow. Finally, though discreteness establishes the possibility of distinguishing intervals on the basis of cardinality, since there is no a priori reason to suppose that sameness of cardinality will coincide with sameness of distance, it was concluded that discreteness does not establish the nonconventionality of metrical relations.[13]

REFERENCES

1. Grünbaum, A. "Geometry, Chronometry, and Empiricism," in H. Feigl and G. Maxwell, eds., *Minnesota Studies in the Philosophy of Science*, vol. III. Minneapolis: University of Minnesota Press, 1962.
2. Grünbaum, A. *Geometry and Chronometry in Philosophical Perspective*. Minneapolis: University of Minnesota Press, 1968.
3. Putnam, H. "An Examination of Grünbaum's Philosophy of Geometry," in B. Baumrin, ed., *Philosophy of Science: The Delaware Seminar*, vol. 2. New York: Wiley, 1963.
4. Riemann, B. "On the Hypotheses Which Lie at the Foundations of Geometry," in D. E. Smith, ed., *A Source Book in Mathematics*, vol. II. New York: Dover, 1959.

[13] This work was supported by a grant from the Minnesota Center for Philosophy of Science.

272

Asymmetries and Mind-Body Perplexities

"O wad some Pow'r the giftie gie us
To see oursels as others see us!"
from "To a Louse" by ROBERT BURNS

Any satisfactory solution to the mind-body problem must include an account of why the so-called "I," "subjective self," or "self as subject of experiences" seems so adept at slipping through the meshes of every nomological net of physical explanations which philosophers have been able to imagine science someday bestowing upon them.[1] Until this agility on the part of the self is either curtailed or shown to be ontologically benign, not forcing us to attribute inexplicable properties to our self-consciousness or consciousness of self, the mind-body problem is not going to go away. Unless the self itself, however characterized, can be shown to be comfortably at home within the domain of the physical, many of its putative attributes—thoughts, feelings, and sensations—will not seem to be at rest there either.

Nor will it do to attempt to preempt the playing out of these perplexities by launching a frontal attack à la Hume or Ryle on allegedly quixotic views about the nature of the self. The problem I am alluding to does not arise because of quixotic views of the self. It is just the reverse: Philosophers find themselves forced to endorse quixotic views of the self primarily because they systematically fail to show how a human being might conceive of himself as being completely in the world.

Some kind of thoroughgoing physicalism seems intuitively plausible mainly because of a dramatic absence of reasons for supposing that were

[1] I wish to thank Professor Feigl and Mrs. Judith Economos for encouraging me to write up the central ideas in this paper. I have also had the benefit of a number of discussions with Mr. Mischa Penn on these matters. I have no idea how happy they will be with the final result. Cf. the final chapter of Mrs. Economos's "The Identity Thesis" (doctoral dissertation, Department of Philosophy, UCLA, 1967). I am also greatly indebted to my former colleague Charles Chastain, with whom I have discussed this paper in detail. I owe to him the idea for describing Case III in section VI in terms of a Chomsky-like rendition of phrase structure grammar.

we to dissect, dismantle, and exhaustively inspect any other person we would discover anything more than a complicated organization of physical things, properties, processes, and events. Furthermore, as has been emphasized recently, we have a strong sense of many of our mental features as being embodied.[2] On the other hand there's a final persuasiveness physicalism lacks which can be traced to the conceptual hardship each person faces when trying to imagine himself being completely accounted for by any such dissection, dismantling, or inspection. It is not so much that one boggles at conceiving of any aspect of his self, person, or consciousness being described in physicalistic terms; it is rather that one boggles at conceiving of every aspect being simultaneously so describable. For convenience of exposition I shall sometimes use the word "self" to refer to whatever there is (or isn't!) which seems to resist such description. Such reference to a self or aspect thereof will not commit me to any positive characterization of it. Neither will it commit me to the view that one's self remains unchanged from moment to moment or to the view that it doesn't or to the view that it is a thing, process, or bundle of events. What I am committed to is phrasing and unpicking the following problem: If a thoroughgoing physicalism (or any kind of monism) is true, why should it even seem so difficult for me to view my mind or self as an item wholly in the world? And this independently of how I may construe that mind or self: whether as a substance or as a cluster of properties, processes, or events. The paradox becomes this: A physicalistic (or otherwise monistic) account of the mind at the outset seems quite convincing so long as I consider anyone except myself. If, however, physicalism provides an adequate account of the minds or selves of others, why should it not, then, provide an adequate account of the nature of my mind or self so long as I lack any reason to suppose that I am utterly unique?[3] But if I am unable to see how physicalism could account for the nature of my mind or self, why then should it not seem equally implausible as a theory about the mind or self of anyone else, again assuming that I lack reasons for supposing that I am unique? In this way we teeter-totter between the problem of viewing our self as wholly in the world, or physical, and the problem of viewing other people who seem wholly in the world as being somewhat mental. But if the mental is after all physical, why should this be so? Al-

[2] G. A. Vesey, *The Embodied Mind* (London: Allen and Unwin, 1965).
[3] Cf. Paul Ziff's "The Simplicity of Other Minds," *Journal of Philosophy*, 72 (October 21, 1965), 575–584.

though I may not initially believe that in my or anyone else's investigation of the world I or they will find need to riddle our explanations with references to immaterial selves or spirits, it still remains easy to believe that I will never turn up the whole of my self as something cohabiting with items in the natural world. Hence the presumptuousness of assuming I really do find other selves in the world.

II

Descartes claimed that it made sense to suppose the set of limbs called his body and whatever physical thing, gas, air, fire, or vapor (animal spirits), might infuse it were nonentities, but that he would still be left with the need to assert that "nevertheless I am something."[4] And by doing so he called attention in a roundabout way to the seeming difference between whatever is associated with the expression "I" when I use it and whatever else there is which is characterized by my use of (generally physicalistic) descriptions.

In a different metaphysical setting Bishop Berkeley was to write: "But besides all that endless variety of ideas or objects of knowledge, there is likewise something which knows or perceives them; and exercises divers operations, as willing, imagining, remembering, about them. This perceiving active being is what I call MIND, SPIRIT, SOUL, or MYSELF. By which words I do not denote any one of my ideas, but a thing entirely distinct from them, wherein they exist. . . ."[5] This is tantamount to Berkeley having asserted that he does not come upon his mind, spirit, soul, or self as an item of the world in the way in which he is able to come upon cogs or pulleys, dendrites or axons. His claim that we only have notions of the mind, spirit, soul, or self and not ideas (perceptions) of it is another way he has of expressing his belief that there is a basic difference between how it is we can have knowledge of our own mind(s) and how it is we can know knowledge of nature.

And Kant, in spite of his general disaffection with Descartes and Berkeley, echoes to some extent their sentiments concerning the mind when he claims that he "cannot have any representation whatsoever of a thinking

[4] René Descartes, *Meditations on First Philosophy*, pp. 69–70 in *Descartes' Philosophical Writings*, trans. and ed. Elizabeth Anscombe and Peter Thomas Geach (London: Nelson, 1954).

[5] George Berkeley, *The Principles of Human Knowledge* (New York: Meridian, 1963), pp. 65–66.

being, through any outer experiences, but only through self-consciousness."[6]

Furthermore, I believe it can be shown in writings from Fichte to Sartre that a well-advertised view of the self as a free or autonomous subject occurs as a simple corollary to the just discussed claim that whatever its nature the self will not be found to reside as do objects at any spatiotemporal address. As occupant of a more ethereal dwelling the self can hardly be expected to feel constrained by the zoning laws of determinism. (Compare the quotation from Schopenhauer in section VI.)

In a contemporary vein Professor Herbert Feigl's view[7] that even with the weapons of a "Utopian neurophysiology" at our disposal the (admittedly suspect) argument from analogy for the existence of "raw feels" in others would not be obsolete but, indeed, necessary is still another way of claiming that the "subjective" selves of others are beyond the pale of physical descriptions. This conclusion need not, yet may, be arrived at by way of the belief that it is difficult to make sense out of one's own "raw feels" being located in the net of physical descriptions ("physical$_2$ descriptions" in the terminology of Sellars and Meehl[8]). This is not the same as, but is a companion to, the view that one's self seems to slip through the net.

Strictly speaking, Professor Feigl's identity thesis commits him to the claim that mental states are wholly characterizable in terms of features within the nomological net (of physical$_2$ descriptions). This should cast the admittedly controversial "argument from analogy" into disuse. But I believe that his desire to retain that argument in his repertoire of inferences can be appreciated not as a blatant inconsistency (which it seems to be) within his physicalistic theory, but as an honest acknowledgment that to date there remains something fishy about viewing one's own self and

[6] *Critique of Pure Reason*, trans. Norman Kemp Smith (London: Macmillan, 1929), p. 332.

[7] Cf. Herbert Feigl, "The 'Mental' and the 'Physical,' " in H. Feigl, G. Maxwell, and M. Scriven, eds., *Minnesota Studies in the Philosophy of Science*, vol. II (Minneapolis: University of Minnesota Press, 1958), pp. 370–497; see especially pp. 429–430. The utility of some version of the argument from analogy within the framework of Feigl's physicalistic theory became more apparent to me during a number of discussions with Professors Herbert Feigl, Paul Meehl, and Grover Maxwell, during a Colloquium on Mind-Body Problems sponsored jointly by the Minnesota Center for the Philosophy of Science and the UCLA Philosophy Department, March 1966.

[8] See P. E. Meehl and W. Sellars, "The Concept of Emergence," in H. Feigl and M. Scriven, eds., *Minnesota Studies in the Philosophy of Science*, vol. I (Minneapolis: University of Minnesota Press, 1956), pp. 239–252.

hence other selves or other selves and hence one's own self as items within the net. This in spite of the fact that physicalism may seem in most other respects impeccable. In short, Feigl's espousal of the argument from analogy is a way of admitting that something very like the paradox stated at the outset of this essay exists.

Thomas Nagel in his recent article "Physicalism" writes: "The feeling that physicalism leaves out of account the essential subjectivity of psychological states is the feeling that nowhere in the description of the state of a human body could there be room for a physical equivalent of the fact that I (or any self), and not just that body, am the subject of those states."[9] No doubt (as Nagel himself intimates) such puzzlements are to some extent reflected in (perhaps in some sense caused by?) the peculiar linguistic role played by expressions such as "I" ("now," "this," and so on) or what have been called egocentric particulars (by Russell),[10] token reflexives (by Reichenbach),[11] indicators (by Goodman),[12] and indexicals (by Bar-Hillel).[13] Even so, what then needs to be shown is that the pragmatic conditions underlying the difference in use between the indexical "I" and nonindexicals do not add up to a metaphysical difference between whatever the indexical "I" denotes when it is used and the sorts of things which the nonindexicals might refer to or characterize. Only after this is done will it be easy to concur with Russell's claim concerning egocentric particulars "that they are not needed in any part of the description of the world, whether physical or psychological."[14]

In brief, I believe that a major temptation to reject a physicalistic theory of mentality, or any monistic doctrine, and by default flirt with some variety of Cartesianism or epiphenomenalism derives from the as yet inadequately assessed asymmetry between (a) how I am able to view myself as a potential object of investigation (within a spatiotemporal setting) and (b) how at first sight it seems one would be able to investigate virtually anything else including (supposedly) other people within such a setting. Given this asymmetry it is cold comfort to be told that my sensations and

[9] In *Philosophical Review*, 74 (July 1965), 354.
[10] Bertrand Russell, chapter VII, "Egocentric Particulars," in *An Inquiry into Meaning and Truth* (New York: Norton, 1940).
[11] Hans Reichenbach, *Elements of Symbolic Logic* (New York: Macmillan, 1960), pp. 284–287.
[12] Nelson Goodman, *The Structure of Appearance* (Cambridge, Mass.: Harvard University Press, 1951), pp. 290–295.
[13] Y. Bar-Hillel, "Indexical Expressions," *Mind*, 63 (1954), 359–379.
[14] *Ibid.*, p. 115.

feelings may be identical with certain brain processes in the way that a lightning flash is identical with an electrical discharge or a cloud is identical with a mass of tiny particles in suspension.[15] Such comparisons may serve to assuage whatever logical qualms had been felt concerning the compatibility of an identity statement ("Sensations are identical with brain processes") with the supposedly synthetic empirical character of the mind-body identity thesis. (For we have learned that although a lightning flash is identical with an electrical discharge we had to make empirical discoveries to disclose it.) But as long as we seem systematically unable to view our own mind or self as something which can be wholly investigated in the way in which lightning flashes or electrical discharges or, as it seems, other people can be wholly investigated, illustrations involving lightning flashes, electrical discharges, and the like will seem less than illustrative. It is for this reason that the seeming duality of the phenomenal and the physical does not constitute an analogue to the "complementarity" involved in the Copenhagen interpretation of quantum mechanics. For both particles and waves are, in some sense, equally at home in or "out of" the world.

The invisible bull in the china shop of the physicalist's analogies is the ominous absence of whatever those arguments might be which would show one that his own self is as wholly amenable to physical investigation as are either clouds or molecules or lightning flashes or electrical discharges. The identity analogies usually engaged in the service of physicalism involve only identities between entities rather obviously susceptible to eventual specification and characterization by expressions which conveniently locate them within a spatiotemporal framework and describe them in physicalistic ways. The question of whether my mind or self is wholly amenable to even roughly this sort of description is one of the major points at issue. It is not sufficient to argue that if other minds seem to consist of nothing other than that which can be physically located and characterized then my mind must be too, unless I suppose it is unique; for the failure to suppose it's unique can be utilized to show that other minds cannot be accounted for in a purely physicalistic way.

If the diagnosis above is correct, any solution to the mind-body problem must proceed through (at least) two stages: At the first stage what must

[15] Cf. J. J. C. Smart's "Sensations and Brain Processes," in V. C. Chappell, ed., The Philosophy of Mind (Englewood Cliffs, N.J.: Prentice-Hall, 1962). Such analogies are, of course, scattered throughout the writings of proponents of the identity thesis.

be overcome is a natural resistance to viewing one's own mind or self as something which can be wholly investigated in a way in which other people and things can be imagined as being wholly investigated by one's own mind or self. I shall refer to the difficulties encountered at this first stage as the *Investigational Asymmetries Problem*. Once such difficulties have been dissolved one may go on to attempt to answer the question of whether one's mind (and hence other minds) which is amenable to such investigation can best be characterized after such an investigation as "a certain kind of information processing system," as "a coalition of computerlike routines and sub-routines," or instead as "a certain type of entelechy" or as "a certain sort of vital force" and so on. I shall refer to the difficulties encountered at this second stage as the *Characterization Problem*.

I mention entelechies and vital forces in passing because I wish to emphasize that a solution to the *Investigational Asymmetries Problem* does not settle in favor of physicalism the question of whether physicalism is true. The extent to which this latter question remains unanswered is the extent to which a theory such as vitalism could blossom from our investigation of nature in general. For example, it might seem reasonable to conclude on the basis of current physical theory that there are entities (say entelechies) inexplicable within the framework of that theory. (Compare Hans Driesch's vitalistic conclusions insofar as they were based on his investigation of the development of sea urchins and not based on his investigation of Hans Driesch.[16])

Also, as I have already intimated, the problem of the first stage is not just a problem for a physicalist view of the mind.[17] (I shall, however, generally treat this problem as a problem for physicalism since I currently view this as the most persuasive monism abroad in the land. But see my final spooky footnote.) Suppose we wish to ask, sensibly, whether my mind, self, or consciousness is identical with some entelechy or vital force. Then, too, we must first establish that my mind, self, or consciousness is

[16] Hans Driesch, *The History and Theory of Vitalism* (New York: Macmillan, 1914). Vitalistic metaphysics did not generally depend on puzzles about the self for its antimaterialistic conclusions. These conclusions were usually based upon seemingly inexplicable but publicly observable features of animals or people such as self-adaptive behavior. This is one reason why the counterexamples which cybernetic machines provided to the claim that nonmechanistic explanations (involving entelechies, etc.) were needed to account for self-adaptive behavior did not settle certain basic mind-body problems. That is to say, puzzles about the self could be utilized on behalf of a mind-body dualism whether or not cybernetic models made reference to entelechies unnecessary in explaining behavior.

[17] Cf. Thomas Nagel's "Physicalism."

the sort of thing which is amenable to the investigations we use for finding out about entelechies or vital forces. It is obvious, for example, that even if Descartes had been willing to contend that he could imagine entelechies or vital forces as being nonentities he would *still* have thought himself left with the need to assert that "nevertheless I am something." The "residue" of self or the I which remains once one has doubted away the existence of all physical and/or vital things or features is precisely that which seems intuitively so implausible to identify with any physical and/or vital thing or feature. So too, the dualism of the "knowing subject" and the "objects of knowledge" so prominent in a variety of idealist writings can be argued for quite independently of how nature in general is conceived—whether, say, in panpsychic or materialist terms.

An unsettling feature of most altercations concerning the mind-body relationship during recent years is that the disputants have often (a) ignored the necessity for passing through what I have called the first stage or (b) prematurely argued about the details of the second stage possibly in the hope that once these were worked out this might settle the perplexities encountered at the only dimly defined first stage, or (c) restricted their attention to asymmetries closely akin to yet not fully reflecting the *Investigational Asymmetries Problem.* These asymmetries are closely associated with the "other minds" problems, but they are not identical with it. In what follows I shall concentrate on the first stage, and propose a solution to the *Investigational Asymmetries Problem.* I shall set aside for the most part the issue of *Characterization.* This issue at the moment, I believe, can be best dealt with by developing and assessing analogies between minds and machines. But first some remarks on (c).

III

In the context of current controversies concerning the problem of "other minds" much attention has been given to the asymmetries[18] expressed by the claims (A) that first-person psychological statements when honestly proffered are incorrigible and that third-person psychological statements are generally corrigible, and (B) that in order to know about my own thoughts, feelings, sensations, and so on, I need know nothing about my own neurophysiology whereas if physicalism were true you could

---

[18] For summaries of these asymmetries and numerous references to further discussions of them, see Jerome A. Shaffer's "Recent Work on the Mind-Body Problem," especially pp. 3–5 under the heading "The 'Asymmetry' of Mental Reports," in *American Philosophical Quarterly*, 2 (April 1965), 1–24.

be certain of my thoughts, feelings, and sensations only by knowing something about my neurophysiology. The overlap between the problem of other minds and the problem of the mind-body relationship is that where there seem to be radically different ways of knowing about my own as opposed to your thoughts, feelings, and sensations, there is some reason to suppose that the sort of things I know about on the one basis, my own thoughts, feelings, and sensations, cannot be identical with any things of the sort I know or find out about on the basis of the other. The semantically unpalatable view that when I say "I am in pain" and when I say "you are in pain" I mean two different things by "pain" retains an edge of reasonableness only because it's not wholly unreasonable to deny that my sensation, say, could be identical with any brain process. For it seems I need know nothing about my brain processes in order to know that I am in pain, whereas all that I can ever know in order to know (if I can know) that you are in pain is something like a brain process (together with behavior). This line of reasoning, of course, often leads to the claim that I don't really know anything about your mind at all. So too, if first-person psychological statements when honestly proffered really are incorrigible, then how could the items which they are statements about (thoughts, feelings, sensations) be identical with the items which statements about neurophysiological events processes are about? For these latter statements are generally thought to be not incorrigible. In brief, as long as these asymmetries persist, there may be ways of arguing that physicalism is not home safe.

Claims (A) and (B) are at best crude paraphrases of richly textured positions which are celebrated enough to need no detailed recounting here. In some important respects progress has been made in clarifying the exact nature of these asymmetries and the extent to which they jeopardize a physicalistic interpretation of mentality. Two of these respects should be briefly discussed:

I think recent writers have convincingly argued that whatever the nature of the asymmetry with respect to me vis-à-vis my own mind and vis-à-vis someone else's mind, it does not consist simply in the capacity to frame incorrigible psychological statements pertaining to my own case, as distinct from at best corrigible statements with respect to other people. For it seems conceivable, though perhaps surprising, that I might, with good reason, be persuaded to doubt and relinquish honestly proffered first-person reports of my own thoughts, feelings, and sensations. One of the most recent proponents of this claim, Paul Meehl, has sketched a persuasive case[19] of a person becoming convinced that he is not experiencing a "visual raw feel of red" although it seems to him that he is. The person is brought to the point of believing that his own honestly offered report on his current experience may be inappropriate due to his overwhelming conviction that the neurological theory which tells him he should be experiencing something other than what he has said he is, is true. It's not simply a case where a person comes to *feel* that his first-person psychological statements are in error. He may, in fact, persist in *feeling* they are correct. It's rather a case where the person has sound theoretical backing for believing himself mistaken. (Compare some of the claims set forth in Richard Rorty's imaginative article "Mind-Body Identity, Privacy, and Categories"; see section 5.[20]) Consequently if the asymmetry between what and how I can know about my psychological self and what and how I can know about the psychology of others which poses a problem for physicalism had consisted simply in the asymmetry between first-person psychological statements which were seemingly incorrigible and third-person psychological statements which were seemingly corrigible, then the problems encountered at what I have called stage one of any solution to the mind-body problem would have been solved.

I also think that J. J. C. Smart et al.[21] have undermined the assumption that physicalism can be refuted simply by proving that I may know that I am in a certain mental state without knowing anything at all about my

[19] Paul Meehl, "The Compleat Autocerebroscopist: A Thought-Experiment on Professor Feigl's Mind-Body Identity Thesis," in Paul K. Feyerabend and Grover Maxwell, eds., *Mind, Matter, and Method: Essays in Philosophy and Science in Honor of Herbert Feigl* (Minneapolis: University of Minnesota Press, 1966). Cf. Meehl's remarks on privacy in this article, p. 134. He begins by saying: "It is agreed that no other person is the locus of my raw-feel events. This simple truth can be reformulated either epistemically or physiologically, as follows: a. A raw-feel event x which belongs to the class C_1 of events constituting the experiential history of a knower K_1 does not belong to the class C_2 of a different knower K_2. b. The tokening mechanism whose tokenings characterize the raw-feel events of organism K_1 is wired "directly" to K_1's visual cortex, whereas the tokening mechanism of K_2 is not directly wired to the visual cortex of K_1." If I had chosen to phrase one of my asymmetries above explicitly in terms of privacy or privileged access, I would then have included an adumbration of Meehl's analysis from which the quotation above is excerpted. I agree with the essentials of his treatment but would simply add that seeing why it is that privileged access to our own mental state does not refute physicalism does not show us why it seems we are unable to treat ourself as just another item of the physical world.

[20] *Review of Metaphysics*, 19 (September 1965), 24–54.

[21] "Sensations and Brain Processes"; Thomas Nagel, "Physicalism." Also see Smart's "Materialism," *Journal of Philosophy*, 60 (October 24, 1963), 651–662.

neurophysiology. They argue as follows: Just as I may know something about Cicero without knowing that what I know is also true of Tully without thereby threatening the identification of the person Cicero with the person Tully, so too I may know something about my own psychological states or processes without knowing that what I in effect have knowledge of is the same thing you have knowledge of through knowing about my neurophysiological states or processes. Thus if the asymmetry between how and what I can know about my own psychological states and what and how I can know about the psychological states of others which poses a problem for physicalism had turned out to be simply the asymmetry between needing no neurophysiological knowledge in my own case and much neurophysiological knowledge in the case of others (or this asymmetry plus the first-mentioned one), then again there would be reason to suppose that the problems encountered at stage one of any solution to the mind-body problem would have been dissipated by recent writings.

Along these lines Professor Feigl reports[22] that Bruce Aune had suggested (in conversation) that because of the "referential opacity" we do not at first realize that in talking about raw feels you are "really" (also) talking about certain (configurational) aspects of the cerebral states or processes. Feigl thinks that Aune's suggestion implies that by introspection we can do a crude sort of neurophysiology! He goes on to say "perhaps, if you try hard 'three times before breakfast' (Alice in Wonderland) you'll manage to believe this." But apart from reservations one might have concerning this approach, I think that Smart's remarks and Aune's suggestion at least point out that it's not a conclusive objection to the identity thesis of mind and body simply to show that I can know about my own psychological states without apparently knowing neurophysiology whereas you can know about my psychological states only by knowing about my neurophysiology. Furthermore, it is not at all clear to me that Aune's suggestion implies that introspection is a crude sort of neurophysiology. However it is we obtain information about Tully, there is no reason to suppose that this amounts to a crude version of however it is we obtain information about Cicero.

But I am not summarizing these views with which I am in general sympathy in order to defend or develop them. Instead what I wish to argue is

---

[22] In a mimeographed outline called "Crucial Issues of Mind-Body Monism" distributed at the University of Minnesota–UCLA Joint Colloquium on Mind-Body Problems held at UCLA, March 1966.

that even if the incorrigibility claims made on behalf of first-person psychological reports could be undermined, and even if it could be established that "talking about raw feels" might really amount to "talking about certain (configurational) aspects of the cerebral states or processes," physicalism is not free from trouble. A reasoned resistance to it would remain. For the feeling would linger that wherever and however I might investigate the physical universe I could never come across the whole of the self which I am. In particular, I would never come across the self or aspect of it which was doing the investigation. Hence discussions of the asymmetries mentioned above do not really tune in on a basic mind-body perplexity. They do not, in short, exhaust the *Investigational Asymmetry Problem* as I have stated it. It is, of course, helpful to be shown that first-person psychological statements are not incorrigible simpliciter. And it is clarifying to see that even if my mental state is identical with a certain physical state it does not follow by Leibniz's law that if I know I am in that mental state then I know I am in that physical state (since the context is intensional). But we are not thereby informed how it is that we could ever view ourself as a purely physical being. For in order to do this it certainly seems that I must be able, at least in principle, to see myself simply and wholly as one among many physical things in a physical universe. But this is precisely what remains so very difficult to do. And given that this is difficult to do, one is disinclined to accept the claim that descriptions of brain states are *in fact* descriptions of mental states. One might even suggest that the difficulty is of such magnitude that it is more appropriate to claim that it hardly makes sense to propose that talk about mental states might really be talk about neurophysiological states in the way that talk about Cicero is really (often) talk about Tully. (Compare "talk about $\sqrt{-1}$ might really be talk about the wind in the way that talk about Cicero is really talk about Tully." But why should anyone ever believe this?) *So the problem I wish to focus on is not simply that my self seems so private to me and hence could not be a physical object of scientific investigations carried out by others, but rather that it seems in some part so unpublic to me, and hence cannot be viewed by me at any given time as an item wholly susceptible even in principle to scientific investigations by me.* (We might call this the problem of empirically "underprivileged access" to ourselves.) But if my self could never be wholly public to me in the way that cogs or pulleys, dendrites or axons seem to be, it is easy to be persuaded that it is not really wholly public to anyone else either.

Hence a thesis such as physicalism, which certainly ought to be committed to the view that my mental states are public in virtue of their being physical states or processes which are incontestably public, still seems implausible.

So what I now hope to show is that the asymmetry between how I am able to investigate myself (and thereby the subject of my thoughts, feelings, and sensations) and how it is I can investigate what I regard as other selves and other things within some spatiotemporal scheme is structurally similar to other ontologically benign asymmetries. By seeing why it is that these analogous asymmetries fail to thrust upon us any dualistic ontology of things, processes, or features, I think it will be shown that there is no need to suppose that the *Investigational Asymmetries* underlying the mind-body problem force upon us a dualistic ontology of things, processes, or features. If this is correct we shall pass through what I called the necessary first stage of any solution (and hence any physicalistic solution) to the mind-body problem. What will remain of the mind-body problem will be the *Problem of Characterization*, or the problem of providing an adequate inventory and anatomy of those features which we, in fact, find other persons to possess. Given the notable absence of any (current) arguments on behalf of vitalism or kindred anti-physicalist doctrines, the inventory and anatomy of other persons at present is heavily weighted in favor of physicalism.

IV

Although I regard each of the following cases to involve asymmetries analogous to the *Investigational Asymmetry*, the first will seem somewhat removed from it and as it stands is more problematic and perhaps less ontologically benign than the other cases. I include it in spite of some unsettled opinions about it mainly because it provides some indication of the variety of ways in which an *Investigational Asymmetry Problem* may be stated. If I am correct, it can be used to illustrate the manner in which problems concerning the mind-body relationship have a bearing on certain problems in linguistic theory: namely, the problem of disambiguation and the formulation of an adequate speaker-hearer model. But this I shall only hint at and not develop.[23] The second case bears more directly on the issues at hand, and has in slightly different forms appeared in the

---

[23] Cf. my *Lockean Linguistics* (in preparation).

writings of others (for example Wittgenstein[24] and Ruyer[25]). As I shall try to show, the wrong conclusions are generally drawn from this sort of case. The third case and a curious corollary to it are, I trust, wholly on target.

Case I: The My Meanings Problem

Speaker-hearer asymmetries. When I say "I'm going to the bank," you (the hearer) may have to "disambiguate" my utterance. You may need to interpret whether I'm going to the river bank or to a bank where one deposits money. But I do not, in the normal case, need to disambiguate for myself my own utterance. And not only do I not need to do so, in the usual case I could not do so. We can, of course, imagine a speaker going through the motions of doing this. For example, we can imagine Professor Chomsky asserting that "Flying planes can be dangerous" and then asking himself whether he meant "Flying" to be construed as a verb or as an adjective. But here we have only imagined someone going through the motions of disambiguating an utterance. Given that the speaker is actually making an assertion, it is absurd to suppose he should have to figure out for himself what he has asserted at the time of asserting it. (That someone may later have to disambiguate his own utterance for himself—coming across it in a diary or because he has a strange memory such that he can hardly remember the last two words he has spoken—is a logically possible case which need not be discussed once it is acknowledged.) Hence in the usual case we find that there is an asymmetry between the speaker vis-à-vis his utterance and the hearer vis-à-vis the speaker's utterance. One could say, following Ziff,[26] there is an asymmetry between the encoding process involved in producing an utterance and the decoding process involved in interpreting that same utterance. This fact I shall redescribe for the purpose at hand by saying that for the usual case we as speakers are unable to make our utterances items for public interpretation by ourself after the manner in which we find, as hearers, the utterances of others to be public items for interpretation by ourself (or anyone else excluding the speaker).

---

[24] Ludwig Wittgenstein, *Tractatus Logico-Philosophicus*, trans. D. F. Pears and B. F. McGuiness (London: Routledge and Kegan Paul, 1961), 5.621–5.641.

[25] R. Ruyer, *Néo-Finalisme* (Paris: Presses Universitaires de France, 1952).

[26] This asymmetry was first called to my attention by Paul Ziff. Cf. Charles E. Osgood and Murray S. Miron, eds., *Approaches to the Study of Aphasia* (Urbana: University of Illinois Press, 1963), especially "A Mediation-Integration Model," by Charles Osgood, pp. 95–101.

Why, in general terms—not in terms of any specific information processing system—is this so?

Let us try to imagine its being otherwise. In what sense could my utterances be my utterances if I had to interpret them in a speaker-hearer context in the way a hearer does, if the problem of disambiguating my own utterance arose as naturally for me vis-à-vis my utterance as it does for a hearer vis-à-vis my utterance? There would, of course, still be an output (syntactic, lexicographic) which was mine, in a sense, but in what sense or to what extent would it be my assertion? To put it in a slightly different way, in what sense could my linguistic output be treated as a specific locutionary act with a specific illocutionary force (following Austin[27])? For example, in what sense could I in saying "I am going to the bank," where I too need to disambiguate what I have said, be making a statement, if I was meaning to say something, if I know what I was saying when I was saying it, and so on? For to say that I would need to know what I was meaning to say in order to perform a certain illocutionary act is the same thing as saying I would not need to disambiguate my own utterance in the way that someone else would. For a condition for saying something and meaning it is simply that one is choosing to utter those phrases which will get his meaning across, and to say that one is doing this is to say that one does not need to disambiguate what he means, for what he is saying is being determined by what he means.

Suppose we tried to break down the asymmetry between a speaker's stance toward his utterance and the hearer's stance toward that same utterance. What would this involve? It would involve attempting to treat the locutionary meaning and illocutionary force as a hearer's input. But how could this be done if what the speaker is saying is being determined by what he means? How can he reasonably be put in the position of having to ask "What does what I am saying (which is being determined by what I mean to say) mean?"

So suppose I utter "I am going to the bank" and someone else one minute later utters "I am going to the bank." The two utterance tokens are tokens of the same utterance type, but the problem of disambiguating for me arises only in the case of the utterance token which was not mine. In other words, I am unable to view my utterance as an item for my own interpretation in the way I must treat the other speaker's utterance as an

---

[27] J. L. Austin, *How to do Things with Words* (Cambridge, Mass.: Harvard University Press, 1962).

item demanding interpretation by me. There is, it seems, a systematic difference between any of my ambiguous utterances and any of anyone else's ambiguous utterances uttered as potential items in interpretation by me. But could such differences provide a basis for some kind of dualism with respect to the nature of my utterances as distinct from the utterances of others? Would, for example, such differences justify the claim that the nature of my utterance is utterly unlike the nature of utterances which I hear, since the latter can need interpretation by me, but my own do not need interpretation by me. (Consider: "My utterances could never be identified with the sorts of things which utterances needing interpretation consist in.") Of course not. So long as what I am saying is determined by what I mean, and what someone else is saying is being determined by what that someone else means, there is bound to be an asymmetry. In the usual case, I can never be a full-blown hearer, as it were, of my own utterance at the time of the production of that utterance, and so, if you produce the same utterance that I do at roughly the same time, I will stand toward your utterance in a way I do not stand toward mine. But that does not mean that there is anything peculiar about my utterance as distinct from any other utterance I may hear which stands in need of interpretation. To think so would be to overlook the fact that my output is for everybody else except myself like everybody else's verbal output vis-à-vis me. (As I will claim later, my self also is to everyone else except myself like everybody else's self vis-à-vis me.) The temptation to imbibe a dualistic ontology with respect to the nature of utterances can be seen to derive wholly from the harmless though ultimately exciting fact that I cannot be the receiver and sender of a particular message simultaneously. Consequently whenever I am sending a message, there will be at least one particular utterance in the world, the meaning of which I seem to have a privileged access to, namely, the utterance I am uttering. To wish to be either in as unprivileged a position with respect to the meaning of the utterance I am uttering, or in as privileged a position with respect to the utterances of others, is to wish that there was no such thing as the difference between a speaker and a hearer, or an encoder and a decoder of utterances.

But the bearing which an asymmetry of the speaker-hearer sort has on the problem of the self, as well as the metaphysical harmlessness of its character, can be made clearer and more convincing by considering another case.

Case II: The My Eyes Problem

How can I tell what both my eyes look like (at one time)? Not in the same way I can tell what someone else's eyes look like, not simply by observing them. Only by looking in mirrors, or at photographs, or at movies of me, or by asking others to tell me what my eyes look like.

There are two ways of finding out what a person's two eyes look like at one time: (1) a way of finding out about the eyes of others, and (2) a way of finding out about my own. A familiar division. Here we have a kind of other minds problem in reverse. I can know by directly looking at them what other people's eyes look like, but I can never know what my own eyes look like by looking at them—except with the aid of mirrors, photographs, and so on. (I shall hereafter rule out the latter.)

I can imagine what it would be like to be in a position to see what anybody else's eyes look like, but I cannot imagine what it would be like to be in a position to see with my present eyes what my present eyes look like. At least I cannot imagine being in a position to see what my eyes look like, without, say, imagining something like the case where I have my current eyes removed and replaced by a different pair of eyes. But this changes the case. By "my eyes," I mean to mean "the eyes which I now possess in my body and which I now see through."

But is there any reason to suppose that the general characteristics which my eyes have differ in kind from the sorts of features which other people's eyes have and which I can see that they have by looking at them? That is to say, are we in the least bit tempted here to propagate a double ontology concerning eyes: (1) the sorts of features my eyes have and which from my point of view seem nonvisible and (2) the sorts of features (or looks) everybody else's eyes have and which I am aware of whenever I look at them? Are we to imagine that there is more (or, better still, less) to other people's eyes than meets my eye since I have no reason to suppose I am unique, and, seemingly, every reason to suppose there is something in eyes which cannot be investigated since I cannot investigate my own?

Consider the complications which would arise if we tried to refute the testimony of mirrors, photographs, and other people as wholly adequate for our own case. We would have to assume that though we know what everyone else's eyes looked like, and know that they looked just as they looked in the mirror, ours do not look as they look in the mirror to everyone else. Ours, we might insist, have nonvisible features. If we were to do this we would also have to assume that the looks of our eyes differ from

289

any other part of the body with respect to their reflections in a mirror. We know our hands look as our hands look in the mirror; we know our stomachs look as our stomachs look in the mirror; and so on. But eyes, well, no, or maybe, or we can't tell whether they do look as they look in the mirror or that they are even the sorts of things of which it makes sense to say "they look a certain way," and so on. We would have to believe that everybody else lies with respect to our eyes, and that mirrors "lie" with respect to our eyes, though they "tell the truth" with respect to every other reflectable feature of us.

I do not conclude that my eyes look as they look in the mirror because I adopt a simple "reverse" argument from analogy: "Since other people's eyes look like what they look like in the mirror, therefore my eyes must look like what they look like in the mirror." Rather, it is because it would take an immensely complicated and implausible theory to try to explain my eyes not looking as they look in the mirror, given that my hands do, given that my feet do, given that other people's eyes do, given that I have no reason for thinking other people are lying when they tell me what my eyes look like, and so on. Lacking a special theory for my own case, I not only accept what other people say about the looks of my eyes, and what is shown in the mirror, I have excellent reasons for accepting this. And certainly I do not conclude that my eyes are, say, "featureless." This latter absurdity, however, is one we would be bossed into accepting were we to decide to "start from our own case" and reason by analogy to the nature of the eyes of others. We would be forced to submit to the conclusion, even in the face of other faces, that the eyes of others have no visible features, for our own eyes seem to us to have none. But we never seemed pressed to such calamitous conclusions, and this is because we have a perfectly good explanation of why we could never be in a position to see the features of our own eyes in the way we are in a good position (potentially) for seeing the features of anyone else's eyes. And the explanation is that in order for me to see my eyes with my eyes, my own eyes would have to be in two (actually four) places at once: in front of themselves to be looked at as well as at the point from which they are being looked at. Note too the conceptual absurdity involved in supposing we know how eyes (in general) look by "starting from our own case" and then reasoning by analogy that other eyes are as ours seem to us, i.e., featureless. What possible sense could be given to the claim that our eyes seem to us to have or lack features of any sort whatsoever if we suppose there are no mirrors about etc.?

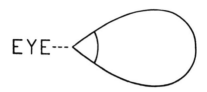

EYE---

Wittgenstein remarked in the *Tractatus*: "For the form of the visual field is surely not like this,"[28] that is, like the diagram, which connects up with his earlier remarks at 5.631: "There is no such thing as the subject that thinks or entertains ideas." "If I wrote a book called *The World as I Found It*, I should have to include a report on my body, and should have to say which parts were subordinate to my will, and which were not, etc., this being a method of isolating the subject, or rather, of showing that in an important sense there is no subject, for it alone could not be mentioned in that book."[29] Surely Wittgenstein here has his eye on what I have called the *Investigational Asymmetries Problem*. In other words, just as the eye does not, cannot, see itself in its own visual field, so too, the self will never, in its inventory-taking of the world, find itself in the world in the manner in which it finds other people and things. But Wittgenstein wrongly concludes from this that the self ("the subject") in "an important sense" does not exist. What I am arguing is that there's no more reason to suppose the self does not exist because it is unable to observe itself than there is reason to suppose eyes have no visible features since they are unable to observe them for their own case. Compare: The speaker's meaning does not exist for utterance tokens because only items which require some degree of interpretation or disambiguation can have meaning, and a speaker doesn't interpret or disambiguate his own remarks! Note that Wittgenstein anticipates this move but seems willing to accept the consequences: "You will say this is exactly like the case of the eye and the visual field. But really you do *not* see the eye."[30] Of course. But we really *do* see eyes, and (1) have an explanation of why it is we cannot observe our own, and (2) have no reason to assume our case is unlike the case of other eyes which we do observe. But if our having eyes does not guarantee that we ourselves will be able to inspect their looks, their visible natures, why should we suppose that our being or having selves should guarantee that we will be in a position fully to inspect their natures, and so on? The way back into the world for the self which seems to itself not to be there is

[28] *Tractatus*, p. 117.
[29] *Ibid.*
[30] *Ibid.*

291

simply its coming to realize that it is to other people what other people are to it.

The case of "my eyes," though it is a kind of other minds problem in reverse, about which I'll have something to say later, is exactly parallel to that problem of the self which has concerned me here. For the problem of convincing myself that the looks of my eyes are (more or less) exactly like the looks of other eyes is parallel to the problem of becoming convinced that the nature of my self is (more or less) exactly like the selves of others which, it seems, can be exhaustively described by reference to their behavior and physiology.

But one last case will help us to consolidate and further clarify the conceptual theses advanced in the first two cases. It might be noted that nowhere in our third case is any mention of a living organism involved. This should serve to erase any suggestion that the form which the *Investigational Asymmetry Problem* takes is peculiar to sentient agents.

Case III: The Self-Scanning Scanner Problem

Suppose we have a nonconscious scanning mechanism, call it SM_1, which is able to scan what we shall call its communication cell, CC_1, somewhat after the manner in which current computing machines are able to scan symbols in their communication cells. We shall imagine CC_1 to be a cell of rather flexible size. It will be able to expand and contract. We shall suppose that SM_1 could scan CC_1 for the appearance of symbols, or the presence of objects, say a bug or a watch or a feather. Let us also imagine that SM_1 could scan other scanners, SM_2, SM_3, . . . , SM_n, all of which would differ from SM_1 only in that they'd be smaller than SM_1 during the time at which they were being scanned. Whenever a scanner such as SM_1 scans another scanner, the scanner being scanned will have to shrink suitably in order for it to appear in CC_1. (Later we shall relax the restriction that any object being scanned by SM_1 must appear in SM_1's CC_1. This will lead us to temper some of the following claims.) We shall also suppose that SM_1 could scan other scanners while they were in the state of scanning things. Scanners SM_1, SM_2, . . . , SM_n will be similar in all interesting respects: in design and structure, and in the sorts of inputs, outputs, and so on which are possible for them. So now let us suppose that SM_1 is able to perform what we shall call a "complete scan" of the workings of SM_2, while SM_2 is scanning its communication cell CC_2 in which there appears some symbol or thing. Thus the nature of SM_2, its program, its actual operation while scanning CC_2, will be made available to SM_1 in the form of descriptions. Each scanner we shall suppose to be endowed with certain pattern recognition or generalization capacities. For example, each scanner will be able to recognize various instances of triangles as triangles, apples of different sizes and color, all as being apples, and so on. Thus a scanner will be equipped to answer simple questions about how a certain item is classified. Let us imagine that such information could be stored as an entry to a list contained in some storage system which is an appendage of SM_1. Let us call this list SM_1's "World List." And let us call the list of all possible World Lists (of scanners SM_1, SM_2, . . . , SM_n) "The World List." So now let us imagine that SM_1 goes on to scan SM_3, SM_4, . . . , SM_n while each is in a state of scanning a symbol or a bug or a watch or a feather. Thus a description of each scanner scanning would be potentially available to SM_1 and could be stored in SM_1's storage system on its "World List." And if we imagine the universe in which SM_1 exists as being a universe in which there are only other scanners SM_2, SM_3, . . . , SM_n and a few objects and events—feathers, bugs, watches, scanners scanning—then we can imagine SM_1 being able to describe virtually everything in its universe. That is, it could in principle scan almost everything in its universe and store a description of each item on its World List. But obviously there is going to be one description which SM_1 will never be able to insert on its own World List: namely, any complete description of SM_1 while it is in a state of scanning. SM_1 is, of course, not able to obtain information about itself while scanning in the same way that it is able to gain information about SM_2, SM_3, . . . , SM_n. In order for SM_1 to obtain information about itself scanning an X, say, in the way that it obtains information about SM_2 scanning an X, SM_1 would have to be in two places at once: where it is, and inside its own communication cell.

Nevertheless, a description of SM_1 in a state of scanning will be available to the World List of some other scanner—say SM_{27}. Hence, such a description could appear on The World List. All descriptions of scanners in a state of scanning could appear on The World List. So if we think of scanners SM_1, SM_2, . . . , SM_n as all being of a comparable nature, and the descriptions of them in a state of scanning as depicting that nature, then we can see that SM_1 in a state of scanning is the same in nature as the other scanning scanners it scans, though it will never be able to locate a description which depicts its nature of other scanners scanning which

293

are descriptions on its (SM_1's) World List. In other words it would be utterly wrong to conclude from SM_1's failure to find a description of its own state of scanning on its World List (or for SM_1 somehow to report on the basis of this) that SM_1 possessed features different in nature from the features which other scanners scanning had and which could be revealed to SM_1. For this would be to suggest that a comparison between SM_1's features and the features of other scanners made available to SM_1 while scanning had been made and that radical differences had been found. Although scanner SM_1 is in principle unable to construe itself on the model of other scanners in a state of scanning in the restricted sense that it cannot construct a list of information about itself comparable to the lists of information it can compile about other scanners, there is no reason for it to report or for us to suppose that some kind of scanning dualism is in order. In other words, there is no reason to assert that SM_1 while in the state of scanning differs in nature from the sorts of features it finds other scanners to possess while in a state of scanning X. Hence, there is also no reason to suppose that there is more to other scanners in a state of scanning than that which could be revealed to SM_1.

An alternative and somewhat more formal way of stating some of the points above together with further ones suggested by them is to construct a "grammar" with which we can generate all possible semantically well-formed descriptions of any scanning mechanism, SM_i, which is in a state of scanning. Our grammar will be a "device" for parsing or analyzing any SM_i in a state of scanning into its constituent parts. Following, metaphorically and incompletely, Chomsky's account (in *Syntactic Structures*[31]) of phrase structure grammars, we can conceive of a thing-structure "grammar" for describing nonabstract objects, SM_i's, in a state of scanning as being "a finite set F of instruction formulas" of the form $X \rightarrow Y$ interpreted: "rewrite X as Y." Thus given our "grammar" (Σ, F) we can "define a *derivation* as a finite sequence of strings, beginning with an initial string of Σ, and with each string in the sequence being derived from the preceding string by application of one of the instruction formulas of F." Those derivations which cannot be rewritten by any further application of the rules F will be called terminated derivations.

So now let SM_1s be the symbol for a given scanning mechanism in a state of scanning. Given our previous account of the nature of SM_i's, their

---

[31] The Hague: Mouton, 1957; see especially pp. 26–30.

flexible communication cells, the variety of objects they might scan, etc., we can ask: What according to our "grammar" should be the permissible combinations of constituents for SM_is's? And the answer seems to be that our grammar should consist of, and certainly not violate, the following rules:

(1)   (i) $SM_is \rightarrow S_i + CC_i$
       (ii) $CC_i \rightarrow ($ $)_i +$ _____, SM_is, $Object_i$, etc.

where each rule $X \rightarrow Y$ of (1) involves a left to right rewrite, and where a given SM_is is the symbol for a scanning mechanism in a state of scanning; S_i is the symbol for a scanner; CC_i is the symbol for SM_is's communication cell; ( )$_i$ is the symbol for the space in that communication cell; _____$_i$ is the symbol for the absence of any object in CC_{ii}; and for any object in SM_i's communication cell other than an SM_i we shall use italicized terms such as bug, feather, watch. (Restrictions on SM_is in (ii) will be discussed in due course.)

Now consider:

(2) SM_is
       $S_1 + CC_1$ (i)
       ( )$_1 +$ _____$_1$ (ii)

where the second line of (2) is derived from the first by rewriting SM_1s in accordance with rule (i) of (1), and the third line is obtained by rewriting $S_1 + CC_1$ as ( )$_1 +$ _____$_1$ in accordance with rule (ii) of (1). We can also represent (2) by means of a tree diagram:

(3)

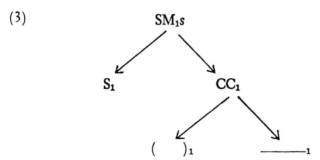

$S_1 + ($ $)_1 +$ _____$_1$ may be thought of as a terminated derivation in our "grammar" since, given our restricted vocabulary, we have no way of rewriting either S_1, ( )$_1$, or _____$_1$. This terminal line we may interpret as describing a possible SM_is with the nature of its constituents fully analyzed by the tree with which the terminal string is associated. If we were to substitute *feather* for _____₁, however, we would get a terminal line showing SM_1s as a scanning mechanism scanning its communication cell in which there was a feather, etc.

Interesting problems arise, however, when we ask what, if any, constraints should be placed on the occurrence of other SM_is's in SM_1's communication cell. Certainly for any $SM_is \neq SM_1s$ it is permissible to substitute SM_is (say SM_2s) for _____₁ in tree diagram (3). Thus we could have:

(4)

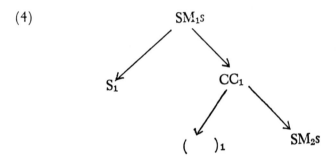

where $S_1 + (\quad)_1 + SM_2s$ would not be a *terminated* derivation since SM_2s could itself be parsed. Hence the complete tree, call it (4'), might be:

(4')

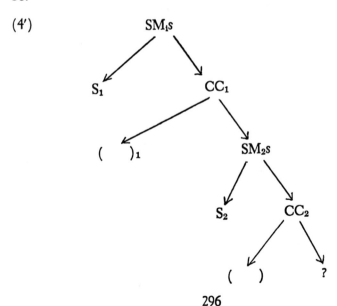

where if ? is replaced by _____$_2$ or *feather* we would have a *terminated* derivation; whereas if another SM_1s, say SM_3s, were substituted for ? we would have an *unterminated* derivation, etc.

But now we must ask: Can we allow an $SM_1s = SM_1s$ to recur as the label of some right-most node in a tree which finds SM_1s as the label of the top-most node? Certainly one conclusion of our less-formal previous discussion would seem to preclude this. For it was explicitly stated that SM_1 could not occur in its own communication cell. And since "contained in" denotes a transitive relation SM_1s could not occur in any communication cell, say SM_2s's, if the scanning mechanism to which that communication cell belonged occurred in SM_1's communication cell, etc.

There is, so far as I can tell and as I shall subsequently show, no way of relaxing this restriction without falling into contradiction. And given our thing-structure "grammar" we are now in a position to say, in a rather novel, and I think interesting way, exactly why this is so. We can do this by comparing the role of the SM_1s in our grammar to the role of S = sentence in an actual phrase-structure grammar. Suppose we have:

(5)

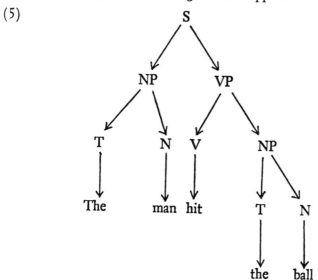

where S = sentence, NP = noun part, VP = verb part, etc., and The + man + hit + the + ball is a *terminated* derivation. (For rules parallel to our (1) above and further details of phrase-structure grammars see Chomsky, *Syntactic Structures*.) What is of primary interest to us in (5) is whatever restrictions are to be placed on the recurrence of S. S is a category symbol for sentences in the same way SM$_i$s is a category symbol for scanning mechanisms in a state of scanning. Clearly there is no reason why S should be prohibited simpliciter from recurring at nodes of either right- or left-branching paths descending from S. For such a recurrence would be a convenient way of indicating a sentence which had as one of its constituents an embedded sentence. (Cf. Chomsky's Aspects of the Theory of Syntax.[32] Note, however, in Syntactic Structures reference to two distinct sentences, S$_1$ and S$_2$, does not occur in (Σ, F) type grammars. The nature of this constraint need not trouble us here.) On the other hand it is obvious that some restrictions must be placed on the embedding of sentences within sentences. And clearly one restriction, the restriction that is of most interest here, must be that a sentence cannot embed itself. For what would this involve? It would involve the recurrence of S, call it S', such that S' = S. But this would be tantamount to allowing S with itself embedded as one of its constituents to be equivalent to S, which is a contradiction. (Compare: The man hit the ball $\neq$ The man hit The man hit the ball the ball. And a still different yet analogous example suggests itself: the difficulty of a painting of a painting where the first painting = the second painting. I might paint a painting of "Guernica" and preserve all the dimensions of "Guernica" in doing so. So might Picasso; but not while painting "Guernica.") Hence S can recur in a tree only if it is understood that the initial string S $\neq$ the recurrent string S. Part of a sentence, or another sentence, can, of course, occur in a sentence as one of its constituent parts, e.g., the main clause of a sentence can occur in the subordinate clause of the same sentence as in "The man who told me that the man hit the ball," etc. (Compare: Any SM$_i$s might scan part of itself.)

So now let us suppose, as seems reasonable, that total consciousness of one's self would be analogous to a scanner completely scanning itself. Further let us suppose that SM$_i$s's are describable in terms of a thing-structure "grammar" which has a formal equivalent in a phrase-structure grammar (of the sort mentioned above). Then the following conclusion emerges: *Total consciousness of one's self is formally equivalent to a self-embedded sentence; any self-embedded sentence involves a contradiction, for no whole sentence can be one of its own constituents; hence any example of total consciousness of one's self involves a contradiction.* And the corollary of this which interests us is: *For one's self to observe one's self*

[32] Cambridge, Mass.: MIT Press, 1966.

as a wholly physical object would be an example of one being totally conscious of one's self, which is a contradiction.[33]

We have reached this conclusion by a most curious route; namely by extending our (allegedly) ontologically benign asymmetries so that they now include the asymmetry between any whole sentence S vis-à-vis its constituents and certain sentences $S \neq S_i$ vis-à-vis S_i's constituents.

It might seem, however, that some of our conclusions above would be left begging for premises were we to relax our restriction that for SM_1 to scan an object, that object must appear in SM_1's communication cell. Instead we might allow SM_1 to scan any objects which, while remaining wholly external to SM_1, come within what we may call the "scope" of SM_1's scan. Further suppose that any scanner scanning an object internalizes some sort of image of that object which it uses as the basis for its derivation of a description of the object being scanned. Now let us imagine the case where SM_1s is being scanned by SM_2s and where SM_1s is able to scan the image of itself appearing in SM_2s. It might then be argued, would not this be a way of SM_1s wholly inspecting itself? In a sense, of course. But not in a sense which abrogates the sort of asymmetry which has been troubling us. For SM_1s can only store a description of itself by way of an examination of the image it obtains from the image of itself which appears in SM_2s. Thus SM_1s can confront the whole of itself only in the sense that I can, through my current eyes, see those eyes, namely by looking in a mirror. But there is no way that SM_1s can confront the whole of itself which does away with the asymmetry between its confrontation with the whole of itself and its confrontation with the whole of some other SM_is.

So too we can easily imagine SM_1s at time t_1 scanning, together with a feather, say, a description of itself at t_1. Thus:

(6)

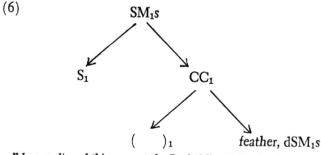

[33] In a reading of this paper at the Rockefeller University (February 15, 1968), Professor Robert Nozick (now back at Harvard) suggested that my conclusion would hold

where $dSM_1s = $ a token of the tree diagram (6).

But if we suppose that in order to be informed that a description actually describes a given object the description must in some sense be seen to apply to or be true of that object, then we cannot imagine SM_1s being able (1) to scan a description of itself scanning, and (2) recognize the applicability of the description to itself, if recognizing the applicability of the description involves making SM_1s wholly available to SM_1s. But of course, SM_1s could do this in the case of SM_2s. Hence the troublesome asymmetry persists.

The case of the scanning mechanism is very much like a case of a periscope which is able to sight other ships, even parts of the ship to which it belongs, but which is unable to place itself in its own crosshairs (compare R. Ruyer's *Neo-Finalisme*[34]).

The import of such examples for the mind-body problem should by now be transparent: The difficulty in construing our self at any given moment as an item wholly susceptible to third-person physicalistic and behavioristic descriptions is comparable to the difficulty a periscope would face in attempting to place itself between its own crosshairs.

V

On the basis of the foregoing cases, I think it is correct to conclude that the following statements are true: (1) My (potentially ambiguous) utterance tokens are identical in kind with members of the class of utterance tokens requiring disambiguation even though I am, as encoder of my utterances, never in a position of needing to disambiguate them. (2) Even though I can never be in a position to look at them, the visual appearances which my eyes possess are identical in kind with members of the class of visual features to which the visual features of the eyes of others belong, which features are revealed to me simply by looking at other eyes (sans mirrors, photographs, etc.). (3) Whenever scanner SM_1 is in a state of scanning its communication cell, it is in a state identical in kind with the sorts of states other scanners in a state of scanning are in, and which are revealed to SM_1 whenever other scanners scanning are placed in SM_1's communication cell. (A periscope$_1$ is identical in kind with members of

only if the mind were finite, and that if it could be represented by an infinite sentence total self-embedding would be possible. A number of us including myself found this a very interesting suggestion, though none of us seemed wholly to understand it.

[34] See footnote 23.

the class of periscopes$_2$ . . . $_n$ any one of which periscope$_1$ can place be-tween its crosshairs.)

The three statements above might all be called "metaphysically neu-tral." For even though it can be shown that X, say, is identical in kind with Y's, it leaves open the question what kind of things Y's are. Yet such a "metaphysically neutral" identity statement is, I believe, exactly what must first be shown to be true if a theory such as physicalism is to seem justified. The statement which the analogies above are designed to support is this: "I am identical in kind with what I find other people to be" where by "what I find other people to be" is meant as they are (or might be) re-vealed to be on the basis of empirical investigation, etc. (which could in principle be found out by me). This latter qualification must be made in order to distinguish our case from the case where someone accepts the claim that he is identical in kind with other people but asserts that other people all have private selves not amenable to empirical investigation where this is derived from his belief that he is a self not amenable to such investigation. That is, I am assuming that what we find other people to be will no longer be colored by the assumption that since I cannot wholly investigate myself, then since there is no reason to assume I am unique, there is something for all people which does not yield to empirical investi-gation either. In other words, "what I find other people to be" will be construed as "what I find them to be sans use of the just mentioned as-sumption." This again, as argued earlier, does not load the dice in favor of physicalism. It simply seems that way, since it is obvious at the moment that unless we resort to such an assumption there is no reason to suppose other people are not wholly physical beings.

It is easy to saddle oneself unwittingly with the view that if I am an object wholly amenable to scientific investigation, I had better be able to imagine myself as being an object which I myself could wholly investigate. This, I have argued, is an absurd demand. And for the same reason that it's absurd to demand that I be able to see my own eyes if I am to credit them with the same sorts of features I ascribe to the eyes of others. Yet the self-centered insistence that if I am an item susceptible to empirical in-vestigation I ought to be able, at least in principle, to carry out the com-plete investigation is quite understandable. For what this really adds up to is the insistence that anyone, myself or others, be able to demonstrate to *my satisfaction* that I am such an item. And the insistence that it be possible to demonstrate to my satisfaction that I am such an item slips into seeming comparable to the demand that I be in a position to appreciate the demonstrations. (Compare: I must be the sort of creature which can never die, since I cannot imagine being in a position to observe my own death. And since I'm not unique, perhaps we are all immortal.)

But given that what I have called the "metaphysically neutral identity statement" is true, the main obstacle is removed from the path leading to the conclusion that I am exactly like the sort of thing I find other organisms to be, even though I can never examine myself fully in the way that I could at least in principle examine other organisms. And in the absence of uncovering in our investigations of others any evidence or reason to support the existence of immaterial entelechies, psychoids[35] (in Meehl's sense), Cartesian egos, or what not, there is every reason to suppose that physicalism will do. If any kind of dualism is to show its mettle, it must now do so from a third-person standpoint. It will have to be solely from an investigation of other people or things that we find reason to suppose the existence of Cartesian egos, entelechies, or psychoids.

Furthermore, once it is realized why it is we are never in a position fully to investigate ourselves, it should seem less counterintuitive to suppose that as we find out more and more about the neurophysiology of other people we will be really finding out more about the nature of their (and hence our) subjective minds. Let us suppose that at any given time we can only investigate ourselves up to a certain point, before we, as investigator, are unable to treat ourselves as the object of our investigation; where, say, some information-processing center in the brain, some subregion of the cortex, is unable to process information about its own information processing. Let us call this point P. Let us further suppose that we are able to investigate others beyond point P, to point BP. And let us assume that what we find out at BP is similar in kind to what we find at P. What might at first sight seem odd is that what we find out by advancing to BP in the case of others should in effect be an indirect explanation of the nature of whatever there is to our own minds which we seem utterly unable to investigate in the way in which we can investigate ourselves up to P. It is easy to assimilate the rest of our mental life to this seemingly intractable residue of self and then comment on the amalgam by saying, "But I have knowledge of my self which consists of sensations, of thoughts, and of feelings—simply by having thoughts, feelings, and sensations." And what we find when we advance to point BP in the case of others will intuitively

---

[35] See Paul Meehl, "The Compleat Autocerebroscopist," pp. 120–127.

seem to have nothing to do with the clarification of our own self and what we take to be its various aspects, namely thoughts, feelings, and sensations. But the answer to the person who feels the necessity for saying something like this is simply to point out that all he knows about thoughts, sensations, and feelings is that he knows what it is to have them, what it is to have a sensation, or a thought, or a feeling. He does not thereby know how to sketch a complete picture of the nature of whatever it is he is having. To be an item which is having experiences does not guarantee a knowledge of their nature anymore than to be an item with an anatomy guarantees an awareness of the nature of that anatomy. (Compare the defect in "I know what a roller-coaster ride consists in by having a ride on a roller coaster.")

VI

The initial plausibility of nonphysicalism according to my story turns out to derive in large part from the fact that I cannot be in two places at the same time, and the tendency to wish to be in as good a position to investigate myself (scientifically) as I am in (at least potentially) with respect to others. Curiously enough the failure to be able to be in two places at once in this case forces us to feel that we, as minds, or consciousnesses, are not in any place at all. As Schopenhauer was to write:

. . . the body is an object among objects, and is conditioned by the laws of objects, although it is an immediate object. Like all objects of perception, it lies within the universal forms of knowledge, time and space, which are the conditions of multiplicity. The subject, on the contrary, which is always the knower, never the known, does not come under these forms, but is presupposed by them; it has therefore neither multiplicity nor its opposite unity. We never know it, but it is always the knower wherever there is knowledge.

So then the world as idea, the only aspect in which we consider it at present, has two fundamental, necessary, and inseparable halves. The one half is the object, the forms of which are space and time, and through these multiplicity. The other half is the subject, which is not in space and time, for it is present, entire and undivided, in every percipient being.[36]

And it is this sense of nonlocation which tends to reinforce the view that the mind is only contingently connected with the physical. But the foregoing arguments, if correct, should rid us of any temptation to adopt either view.

[36] Arthur Schopenhauer, *The World as Idea*, I, 2, pp. 5–6, in Irwin Edman, ed., *The Philosophy of Schopenhauer* (New York: Modern Library, 1928). Cf. George Pitcher's

Keith Gunderson

So too, if I am really never in a position to exhaustively inspect the self which I am, to include it in my inventory of physical facts garnered through physical investigations, since I would have to be investigator and object of investigation simultaneously, we might expect this fact to reflect itself in our language. And this, I believe, is precisely what underlies the odd referential status of the indexical "I." (Recall Professor Ryle's discussion of the "systematic elusiveness" of 'I'[37] where part of his strategy was to show that the evasive reference of the indexical did not foist Cartesianism upon us.) What I have, in effect, tried to show is that the puzzle surrounding "I" is a manifestation of a deep dissatisfaction with either physicalism or any monism, a dissatisfaction which is rooted in the seeming unpublicness of our self. That "I" when used in an utterance refers only to the user and not to anyone else and is in this sense for each token occurrence private, or nonpublic, is parallel to the fact that I cannot make myself an object for public inspection by me and still remain myself. That I cannot refer wholly "outward" by using "I" reflects the fact that I cannot investigate myself as a wholly public object. By saying that the use of "I" in an utterance is restricted in that it can never be used to refer wholly "outward," I wish to call attention to the fact that even in cases where I use "I" in an utterance to specify a certain person (obviously myself) who was or will be, "I" will necessarily specify as well the person I am at the moment of utterance. Consider: I say (while looking at an old photograph) "Here I was in 1950." In such cases the "I" still serves to specify the speaker (whatever else it might do). (Note: It is precisely this fact that "I" is in any utterance token speaker-specifying which leads to the oddity Hintikka[38] points out in the case where a person attempts to convince someone he doesn't exist by saying "I don't exist.")

This feature of the first-person pronoun indexicals is the linguistic frosting on the factual cake: the ingredients of the latter being my (or anyone's) inability to treat myself (himself) wholly as another person or thing. If I were able on occasion to treat myself wholly as another person or thing, then "I" would not be, could not be, restricted in the manner specified above. The reflection of this fact underlying this restriction on

discussion of this passage in his *The Philosophy of Wittgenstein* (Englewood Cliffs, N.J.: Prentice-Hall, 1964), p. 147.

[37] In *Concept of Mind* (New York: Hutchinson's University Library, 1955), pp. 195–198.

[38] Jaakko Hintikka, "Cogito, Ergo Sum: Inference or Performance?" *Philosophical Review*, 71 (January 1962), pp. 3–32.

"I" would, of course, be plowed under were we to imagine our language sans egocentric particulars (following Russell). Yet in agreement with Russell we can see why in one sense "I" is not needed in our descriptions of the world. For once we are aware that the unpublic nature of my self vis-à-vis me does not show my self to be unpublic simpliciter, each of us is then in a position to see and describe ourselves by way of an acceptance of how others see and describe us—and others do not (cannot), of course, utilize "I" in their descriptions of us.

The solution to stage one of the mind-body question, on the analysis above, turns out to be nothing more than coming to accept that we will never be in as good a position to investigate ourselves as we are to investigate others, and seeing why this makes no more difference (ontologically) than the fact that a submarine's periscope cannot locate itself in its own crosshairs makes an ontological difference between the nature of the periscope doing the sighting and the things it can sight. If this is true, then what remains to the mind-body is working out the details of characterization which so far, at least, seems snugly confined to the province of biology, chemistry, and physics.

The Schopenhauerian self which could not locate itself as just another member in the sprawly family of physical objects is comparable to a periscope which is safe from the torpedoes of its own submarine because it can never locate itself in its own crosshairs. But just as such a periscope is only safe from its own torpedoes and not the torpedoes of others, so too a Schopenhauerian self (or Cartesian ego) is only safe from spotting itself in the physical world. It is not safe from being spotted there by others.

VII

Finally, I wish to sketch the way in which I believe it possible to derive from the foregoing solution to the problem of the unpublic self a corollary solution to the problem of other minds insofar as that problem has seemed to place a stumbling block in the path leading to a consistently formulated physicalism. This corollary seems to me to deserve more elaboration than I can give it here, but at least the strategy involved can be adumbrated.

It was claimed earlier that the "My Eyes" problem is (1) exactly parallel to the problem of the self as herein described; and (2) exactly the reverse of the other minds problem. In the one case I lack a kind of knowledge of my self which I possess with respect to others. In the other

305

case I lack a kind of knowledge of others which I seem to have with respect to myself. Now if I have shown that the asymmetry involved in the case of the unpublic self does not constitute a threat to physicalism, or to any other monism, there is hope that the asymmetry involved in the case of other minds does not constitute a threat to physicalism either. That is, if the problem of "My Eyes" does not force dualism upon us, and if that problem is the reverse of the other minds problem, then perhaps the *solution* can simply be reversed and yield a solution to the other minds problem. In other words, although the problem of the unpublic self and the problem of other minds are to some extent separate investigational asymmetries, they are similarly asymmetric. They have a common source and structure.

Let me now explain what I mean by this. I shall do this by showing in more detail how one's being limited to being in one place at one time gives rise to the other minds problem. The best way to proceed may be to begin with a case where there is a simulation of a solution to the other minds problem. By inspecting the simulation we may be able to uncover the conditions which would in actuality have to obtain in order to prevent an other minds problem from arising.

The Cinematic Solution to the Problem of Other Minds

What I shall refer to as *the cinematic solution to the problem of other minds* is really a simulation of a situation in which aspects of someone else's mental life are made fully available to me in the way my own are. This epistemological shiftiness is encountered in those movies where a director wishes to provide the moviegoer with a sense of being fully acquainted with, say, the visual experience of one of the characters on the screen. Suppose, for example, the director wishes to make (or seem to make) available to me the visual experience vC of some character C as C walks down the garden path. How best to effect this? Not, surely, by simply flashing on the screen the scene of C walking down the garden path. Why not? Because what we seem to see or are made familiar with, come to know in such a case, is not C's visual experience vC, but our own visual experience vO which includes the view of C walking down the garden path. What is needed to avoid this, and to effect a sense of view C's vC, is some way of exporting our own perspective from the situation, and importing C's perspective into it. But how can this be done, or even appear to be brought about? The way to do this, and the way any movie director with slight cunning will do this, is to create the illusion that we are where C is. And this is most simply (and in fact necessarily) brought about by showing little or nothing of C's body as C walks down the garden path. The camera, in other words, is able to produce a sense of moving down the garden path without showing someone moving down the garden path. In brief, we are provided with C's perspective, the visual experience expected for C when walking down the garden path, by removing C's body from our own perspective. (Of course, selected parts of C's torso may remain in view without destroying the illusion.) In such a case it is not untoward to say we seem to see C's seeing. Consequently we are provided with a simulated solution of the other minds problem as it arises for us vis-à-vis C's visual experience.

A suspiciously whimsical example perhaps, but one with a nonwhimsical aspect. The connection between the *cinematic solution to the other minds problem* and the solution I have suggested to the problem of the unpublic self should be clear. Both demand what cannot in practice be brought about: being in two places at once. This is made even more apparent when it is pointed out that the problem of the unpublic self has a similar pseudo-solution in the case of movies. The solution is, simply, to create the illusion that there is no investigational asymmetry, or that I stand to myself in the way I stand to others, by flashing my body on the screen along with the bodies of other people. Hence, it would be as if I were to seem to see myself walking down the garden path. Note: Although it does not seem possible to have one's perspectives and see them too, even this might be simulated in the case where (a) I see my body on the screen, and then (b) the *cinematic solution to the other minds problem* is brought about for me vis-à-vis my own body as viewed on the screen! In other words, I am made to share the perspective of myself by the removal of most or all of the body from the screen with which I have identified myself. What this selected short subject suggests is that one cannot envision a situation in which the problem of the unpublic self is in practice solved for a person without the problem of other minds arising for that person vis-à-vis himself.

To encapsulate: The other minds problem as I have been construing it is really a problem in sharing perspectives. And to share perspectives with another would entail being in two places at once. (Note: What others have called egocentric particulars, etc., William James labeled "words of

307

perspective."[39]) We never do share perspectives, so in fact there always is, in one sense, a problem of other minds. We must in the case of coming to know the nature of someone else's visual experiences, for example, rely on reasonable inferences based on what we observe, discover, and so on with respect to that person (where he stands, whether he is an organism of such-and-such a sort, and similar matters). But once the nature of the other minds problem is made clear, it should be easy to see why it does not constitute a threat to a consistent physicalism. For it is based on an investigational asymmetry which is as ontologically benign as the one underlying the problem of the unpublic self. Anyone else's visual experiences, for example, which include a perspective of the world, will never be made available to me through behavioral or neurological information. But from this it does not follow that there is anything nonphysical involved in, say, anyone's visual experiences. Although we will not find someone else's perspective by inspecting his brain, there need be and is no reason to suppose that there is some nonphysical feature involved in that person's mental life.

I'm inclined to think that the so-called grain argument attributed to Sellars and recently discussed by Meehl rests on the mistaken assumption that physicalism involves commitment to the view that the (non-"gappy"-grained) perspective which someone has of the world could, in principle, be found in his ("gappy"-grained) brain. And it may well be that some physicalists have themselves thought this. But this is the same as supposing that the best way for a movie director to provide us with C's vC while walking down the garden path would be to flash on the screen a detailed cross section of C's brain as he walks down the garden path plus a (Utopian) neurophysiological interpretation of it. Such factors could never yield even a simulation of our seeing C's seeing. For C's brain, neurophysiological interpretation and all, would still remain mere items embraced in our own perspective.[40]

---

[39] William James, *The Varieties of Religious Experience* (Modern Library Edition), p. 192. In remarking on beliefs of Buddhists or Humians he writes: "For them the soul is only a succession of fields of consciousness: yet there is found in each field a part, or sub-field, which figures as focal and contains the excitement, and from which, as from a centre, the aim seems to be taken. Talking of this part, we involuntarily apply words of perspective to distinguish it from the rest, words like 'here,' 'this,' 'now,' 'mine,' or 'me'; and we ascribe to the other parts the positions 'there,' 'then,' 'that,' 'his' or 'thine,' 'it,' 'not me.' But a 'here' can change to a 'there,' and a 'there' become a 'here,' and what was 'mine' and what was 'not mine' change their places."

[40] The problem of other minds, which I have tried to rephrase as a problem of sharing perspectives, was clearly construed that way by G. T. Fechner in his *Elemente der*

A Remark on ψ-φ Correlations

It is no doubt apparent that nothing has yet been said about correlations between psychological states and physiological states—a topic which has certainly been at the forefront of many contemporary treatments of mind-body perplexities. The reason this has not been dealt with (except indirectly) is that if my diagnosis of the mind-body problem is correct, it need not be dealt with at all in discussing the *Investigational Asymmetry Problem*. If, as I have contended, the temptation to reject physicalism (in the absence of reasons for hypostatizing vital forces, etc.) derives wholly from *Investigational Asymmetries*, then given that my foregoing arguments are correct we are freed from such temptation, for the asymmetries have been rendered harmless. Thus, in the absence of reasons for accepting vitalism or like doctrines, physicalism "wins the game" by forfeit.[41] So-called ψ-φ correlations, no doubt, have contributed and will contribute to solutions to what I have called the *Problem of Characterization*. But before there is a solution of the *Investigational Asymmetries*, these correlations cannot even with the aid of Occam's razor trim the psychic sideburns from the face of physicalism.

Psychophysik (Leipzig: Breitkopf & Härtel, 1907). He worked up to this construal by first remarking: "When anyone stands inside a sphere its convex side is for him quite hidden by the concave surface; conversely, when he stands outside, the concave surface is hidden by the convex. Both sides belong together as inseparably as the psychical and the bodily sides of a human being, and these also may by way of simile (*vergleichweise*) be regarded as inner and outer sides, but it is just as impossible to see both sides of a circle from a standpoint in the plane of the circle, as to see these two sides of humanity from a standpoint in the plane of human existence." And he goes on to write: "The solar system seen from the sun presents an aspect quite other than that which it presents when viewed from the earth. There it appears as the Copernican, here as the Ptolemaic world-system. And for all time it will remain impossible for one observer to see both systems at the same time, although both belong inseparably together, and, just like the concave and convex sides of a circle, they are at bottom only two different modes of appearance of the same thing seen from different standpoints." And finally he claims: "The difference of standpoint is whether one thinks with one's brain or looks into the brain of another thinker. The appearances are then quite different; but the standpoints are very different, there an inner, here an outer standpoint; and they are indescribably more different than in the foregoing example (i.e., the circle and the solar system), and just for that reason the difference of the modes of appearance is indescribably greater. For the double mode of appearance of the circle, or of the solar system is after all only obtained from two different outer standpoints over against it; at the centre of the circle, or on the sun, the observer remains outside the line of the circle, or outside the planets. But the appearance of the spirit to itself is obtained from a truly inner standpoint of that underlying being over against itself, namely the standpoint of coincidence with itself, while the appearance of the bodily self is obtained from a standpoint truly external to it, namely, one which does not coincide with it."

[41] But a question too crazy and deep for me to consider here has begun to haunt me: In the last analysis might physicalism and panpsychism turn out to be the same doctrine?

Psychological Determinism and Human Rationality: A Psychologist's Reactions to Professor Karl Popper's "Of Clouds and Clocks"

In the Second Arthur Holly Compton Memorial Lecture, engagingly titled "Of Clouds and Clocks," Sir Karl Popper addresses himself to a long-familiar problem about psychological determinism, indicated by the lecture's subtitle, "An Approach to the Problem of Rationality and the Freedom of Man."[1] The lecture treats of several interconnected themes, ontological, historical, and methodological. I want to emphasize that the present paper is in no sense an "attack" on the lecture as a whole, which abounds with the usual Popper stimulation and perspicuity, and from which I have learned much. Some of the interpretation (e.g., the indeterministic features of classical physics) is beyond my competence even to discuss, let alone criticize. What I consider herein, qua philosophically oriented psychologist, is only one specific thesis, to wit, *psychological determinism is incompatible with human rationality*. The core idea here, in spite of the new aspects illuminated by Popper, is an old one, no doubt familiar in one form or other to almost any undergraduate philosophy major. (I recall first hearing it, when a sophomore, forcefully presented and ingeniously defended by Professor Alburey Castell. I did not buy it then, and I find that, some thirty years having passed, I cannot buy it now. But I trust that my reasons are somewhat better today than they were in 1939, as I am now more cognizant of the genuine puzzles and paradoxes involved.)

AUTHOR'S NOTE: I am indebted to the Louis W. and Maud Hill Family Foundation and to the Carnegie Corporation of New York for support through summer appointments as professor in the Minnesota Center for Philosophy of Science.

[1] K. R. Popper, "Of Clouds and Clocks: An Approach to the Problem of Rationality and the Freedom of Man," the Arthur Holly Compton Memorial Lecture presented at Washington University, April 21, 1965 (St. Louis: Washington University, 1966). All quotations of Popper's language are from this source.

It is important to be clear about three matters right off: First, the thesis is ontological, not epistemological, and I therefore bypass evidential questions, freely invoking "what Omniscient Jones knows," "what a Utopian physiologist would say," "what is actually going on," "what is true concerning the state of Nature." The thesis is a claim about incoherence in a deterministic ontology; it says that if all human thought and action were completely determined, then it *could not be* rational. That kind of question can of course be examined without reference to the evidential issues of how we could find out that determinism was false, or how we could ascertain that we are rational (?!).

Second, I do not attempt a defense of complete psychological determinism, partly because its truth or falsity would not bear on its consistency with rationality, but also because I am not myself a convinced determinist, and consider the substantive issue in doubt on present evidence.

Third, being a psychologist I am naturally suspect and vulnerable to a kind of *ad hominem* complaint that "You of course defend determinism because of trade-union interests, thinking that your scientific and clinical jobs require an implicit faith in the ultimate strict orderliness of all psychological processes." For me, at least, this is not true. I anticipate that no development of the behavior sciences will eliminate their current stochastic features, and I am not aware of any research programs that would have to be abandoned as fruitless if an element of radical indeterminism were postulated. For example, it seems fair to say that the greatest degree of behavioral prediction and control achieved thus far by psychologists is found in the work of Skinner and his disciples.[2] Aside from the fact that these modest triumphs of "behavioral engineering" are quantitatively

[2] In recognizing the unblinkable fact that Skinner's epoch-making book *The Behavior of Organisms* (New York: Appleton-Century-Crofts, 1938) gave rise to a technology of behavior control which has, to an unprejudiced mind, no real competitors, I do not commit myself as regards its long-term *theoretical* adequacy. As would be true for most of my philosophical readers, I have grave reservations about Skinner's account of language, in his *Verbal Behavior* (New York: Appleton-Century-Crofts, 1957), vigorously attacked by N. Chomsky, "Review of *Verbal Behavior*, by B. F. Skinner," *Language*, 35 (1959), 26–58. But see K. MacCorquodale, "On Chomsky's Review of Skinner's *Verbal Behavior*," *Journal of the Experimental Analysis of Behavior*, 13 (1970), 83–99. I also believe the Skinnerian group underestimates the importance of genetic factors —and resulting real taxonomic entities—in behavior disorder, and hence they underrate the importance of formal diagnosis, although there is nothing about the theoretical position that requires this attitude. Finally, as a psychotherapist I have reservations about the adequacy with which Freud's constructs are translated into behaviorese by J. G. Holland and B. F. Skinner in *The Analysis of Behavior* (New York: McGraw-Hill, 1961), a programmed text which I recommend to readers approaching this sub-

311

tighter when the subjects are pigeons pecking keys than when they are humans speaking words—let alone philosophers engaged in criticism—whether or not one sees the laws as deterministic depends upon the level of analysis. The main dependent variable studied is rate of responding, as represented by the slope of a cumulative response record. The conceptual and mathematical relationships between this "operational" variable and the underlying probability-of-responding—the relation between a finite relative frequency and a "propensity"—have never been precisely explicated by the operant behaviorists. Usually they can go along quite well with their work without a rigorous explication of it. But when radical determinism is under discussion, we need more than a mere showing that a response-curve slope is highly manipulable. Whether or not a rat, pigeon, or human emits a certain response during a small interval Δt, and whether the response has such-and-such narrowly specified topographic, durational, and intensive properties, are *not* under complete experimental control, but remain probabilistic only. Besides, the various kinds of human psychological activity differ in how "clocklike" versus "cloudlike" they are, and Popper could quite properly argue that any showing that there is quasi-clocklike orderliness in a well-conditioned eye-blink reflex is only faintly relevant to the question "How clocklike are political theorizing, mathematical invention, and philosophical criticism?"

With these disclaimers made about what I am not attempting to do, what I shall attempt is to criticize Popper's view that *if* human thought and behavior were completely determined, *then* they could not be rational. If I understand him rightly, he believes that if strict determinism were

ject matter for the first time. Another good introductory presentation can be found in B. F. Skinner, *Science and Human Behavior* (Cambridge, Mass.: Harvard University Press, 1951). See also, but requiring varying amounts of technical preparation, B. F. Skinner, *Cumulative Record* (New York: Appleton-Century-Crofts, 1961); T. Verhave, ed., *The Experimental Analysis of Behavior: Selected Readings* (New York: Appleton-Century-Crofts, 1966); T. Ayllon and N. Azrin, *The Token Economy: A Motivational System for Therapy and Rehabilitation* (New York: Appleton-Century-Crofts, 1968); L. Krasner and L. P. Ullmann, eds., *Research in Behavior Modification* (New York: Holt, Rinehart and Winston, 1965); L. P. Ullmann and L. Krasner, eds., *Case Studies in Behavior Modification* (New York: Holt, Rinehart and Winston, 1965); R. Ulrich, T. Stachnik, and J. Mabry, eds., *Control of Human Behavior* (Glenview, Ill.: Scott, Foresman, 1966); B. F. Skinner, *The Technology of Teaching* (New York: Appleton-Century-Crofts, 1968); C. B. Ferster and B. F. Skinner, *Schedules of Reinforcement* (New York: Appleton-Century-Crofts, 1957); W. K. Honig, ed., *Operant Behavior: Areas of Research and Application* (New York: Appleton-Century-Crofts, 1966); and A. C. Catania, ed., *Contemporary Research in Operant Behavior* (Glenview, Ill.: Scott, Foresman, 1968).

true, we could not, in any genuine sense, give reasons, be influenced by reasons, engage in critical thought, etc., and that the validity or invalidity of arguments could not influence the course of human happenings. The line of his argument can best be seen from a few representative quotations. Popper writes:

[Quoting Compton] "If . . . the atoms of our bodies follow physical laws as immutable as the motions of the planets, why try? What difference can it make how great the effort if our actions are already predetermined by mechanical laws . . .?"

Compton describes here what I shall call 'the nightmare of the physical determinist.' A deterministic physical clockwork mechanism is, above all, completely self-contained: in the perfect deterministic physical world there is simply no room for any outside intervention. Everything that happens in such a world is physically predetermined, including all our movements and therefore all our actions. Thus all our thoughts, feelings, and efforts can have no practical influence upon what happens in the physical world: they are, if not mere illusions, at best superfluous by-products ('epiphenomena') of physical events. [pages 7–8]

I believe that the only form of the problem of determinism which is worth discussing seriously is exactly that problem which worried Compton: the problem which arises from a physical theory which describes the world as a *physically complete* or a *physically closed* system. By a physically closed system I mean a set or system of physical entities, such as atoms or elementary particles or physical forces or fields of forces, which interact with each other—and *only* with each other—in accordance with definite laws of interaction that do not leave any room for interaction with, or interference by, anything outside that closed set or system of physical entities. It is this 'closure' of the system that creates the deterministic nightmare. [page 8]

For according to determinism, any theories—such as, say, determinism —are held because of a certain physical structure of the holder (perhaps of his brain). Accordingly we are deceiving ourselves (and are physically so determined as to deceive ourselves) whenever we believe that there are such things as arguments or reasons which make us accept determinism. Or in other words, physical determinism is a theory which, if it is true, is not arguable, since it must explain all our reactions, including what appear to us as beliefs based on arguments, as due to *purely physical conditions.* Purely physical conditions, including our physical environment, make us say or accept whatever we say or accept; and a well-trained physicist who does not know any French, and who has never heard of determinism, would be able to predict what a French determinist would say in a French discussion on determinism; and of course also what his indeterminist opponent would say. But this means that if we believe that we have accepted a theory like determinism because we were swayed by the logical force of certain arguments, then we are deceiving ourselves, according to physical determinism; or more precisely, we are in a physical condition which determines us to deceive ourselves. [page 11]

For if we accept a theory of evolution (such as Darwin's) then even if we remain sceptical about the theory that life emerged from inorganic matter we can hardly deny that there must have been a time when abstract and non-physical entities, such as reasons and arguments and scientific knowledge, and abstract rules, such as rules for building railways or bull-dozers or sputniks or, say, rules of grammar or of counterpoint, did not exist, or at any rate had no effect upon the physical universe. It is difficult to understand how the physical universe could produce abstract entities such as rules, and then could come under the influence of these rules, so that these rules in their turn could exert very palpable effects upon the physical universe.

There is, however, at least one perhaps somewhat evasive but at any rate easy way out of this difficulty. We can simply deny that these abstract entities exist and that they can influence the physical universe. And we can assert that what do exist are our brains, and that these are machines like computers; that the allegedly abstract rules are physical entities, exactly like the concrete physical punch-cards by which we 'program' our computers; and that the existence of anything non-physical is just 'an illusion,' perhaps, and at any rate unimportant, since everything would go on as it does even if there were no such illusions. [page 12]

For obviously what we want is to understand how such non-physical things as *purposes, deliberations, plans, decisions, theories, intentions,* and *values,* can play a part in bringing about physical changes in the physical world. [page 15]

Retaining Compton's own behaviorist terminology, Compton's problem may be described as the problem of the influence of the *universe of abstract meanings* upon human behavior (and thereby upon the physical universe). Here 'universe of meanings' is a shorthand term comprising such diverse things as promises, aims, and various kinds of rules, such as rules of grammar, or of polite behavior, or of logic, or of chess, or of counterpoint; also such things as scientific publications (and other publications); appeals to our sense of justice or generosity; or to our artistic appreciation; and so on, almost *ad infinitum.* [page 16]

I believe these quotations suffice to give the essential argument, which purports to show that complete psychological determinism, arising on the basis of complete brain-process determinism ("mind is a function of brain"), renders genuine rationality and purposiveness impossible, and "giving of reasons" a spurious idea or an inefficacious irrelevancy. I turn now to my analysis and criticism of that contention.

There is, at least prima facie, a certain oddity about the position of those who wish to reject psychological determinism on the ground that it precludes human rationality, since part of their reason for insisting upon a "something else" which is not the mere workings of the cerebral machinery is the obvious fact that our conduct *is* causally influenced by "the giving of reasons." If I want to control your behavior with regard to a certain decision, it is true that I may proceed by various kinds of irrational appeals (e.g., after the manner of Hitler); but it is also true that I may proceed by giving you what I believe to be good reasons for your behaving in the way that I desire. In fact, if I know you fairly well and believe you to be a highly rational man, I may well operate on the assumption that the *most* effective way to control your behavior is to present you with good reasons. Thus we have a situation in which the idea of *control*, or the determination of one event (your action) by means of introducing another event (my giving you good reasons), with reliance upon a kind of regularity ("Jones is influenceable by good reasons," roughly), is combined with the idea of *rationality*. Some hold that these two ideas cannot be thus conjoined in discoursing about human conduct, because, they say, "causes" cannot be "reasons." It is this alleged truism, frequently asserted without further justification, that I wish first to examine.

Let us consider a simple arithmetical example as a "pure case," one in which rational inference plays an absolutely crucial role, in which the inference is one of deductive necessity, and in which (precisely because of this deductive necessity) the behavior is determined as completely as we determine the behavior of macro-objects in ordinary physics. Let us suppose I am a practical jokester of philosophic bent. I put a jar down on the table in front of you and allow you to inspect it at leisure. I swear an oath on the Bible (let us suppose you know me to be a pious Christian believer) that I am not a magician, that there is nothing phony about the construction of the jar, and that I am not going to lie to you or engage in any kind of legerdemain. We presuppose that you take these things as true, and that you believe them with as much certainty as we can generally reach about any empirical matter. While you observe me, I now place five pennies one by one in the jar, counting out loud, and put on the lid. Then I hand you two pennies and invite you to place them also in the jar. After you have replaced the cover, I then say the following: "I want you to be-

lieve for ten seconds that there are now eight pennies in the jar. No harm
will be done to anybody by your believing this, and I don't require that
you assert it. So you don't even have to tell a white lie for the short run, if
that would bother your conscience. I'm going, however, to attach a psy-
chogalvanometer to your palms, and then I shall point to the figures 'six,'
'seven,' 'eight,' 'nine' on the blackboard one by one, and the instrument
will reveal which numeral corresponds to your actual momentary belief
about the contents of the jar. You understand I am only asking you to believe the proposition (that $N = 8$) for ten seconds, so you don't have to
worry about developing bad arithmetical habits, or becoming psychotic
through chronic reality distortions. Now, if you are able to believe for ten
seconds that there are now eight pennies in the jar, I will give ten thousand
dollars to your favorite charity. Surely you can have no moral objections or
psychological fears about this procedure."

Now it is perfectly obvious that under these circumstances the experimental subject would very much like to entertain the proposed belief for
ten seconds, but if he is a sane man, acquainted with the rules of arithmetic, it would be literally impossible for him to do so. I do not mean he
would have a hard time "willing to do so," or that he would be "rationally
reluctant to do so." In terms of the utilities involved, it would in fact be
rational of him to make up his mind (in the pragmatic metalanguage) to
believe this false sentence for ten seconds, but the fact is that it would be
impossible. You could afford to wager as much money on the outcome of
this experiment as you can on the outcome of a neurologist's tapping the
patellar tendon to elicit the knee jerk, or on the color and size of a negative
afterimage. Yet it is equally obvious that this behavior control, which is as
deterministic as anyone could desire, involves a rational process, namely,
the process of mental addition obeying the rules of arithmetic, as a crucial
feature. By placing five pennies in the box and having the subject place 2
pennies in the box, I *determine* his belief that it now contains 7 pennies,
and I render it *impossible* for him to believe that there are 8. There is, I
submit, little or no more "play" in this system than there is in the elicita-
tion of a reflex from a spinal animal or the putting of a sugar lump into
solution. If the conditions stipulated are fulfilled, I would lay as large a
wager on the outcome of any one of these experiments as on any other.

There are, it seems, two opposite dangers to beware of in discussing
causes and reasons in relation to human behavior. The first danger, which
is not likely to be made by anyone who is philosophically sophisticated, is

to conflate causes and reasons; but there is an opposite danger, one which we sometimes find in philosophically sophisticated persons, to conclude that since causes and reasons are not the same sort of entity, there cannot be any intimate connection between them, so that "explaining someone's behavior" must either be a causal-analysis enterprise or a reason-providing enterprise, but no single instance of behavior-explaining can be both. This radical separation of discourse about causes from discourse about reasons is in my view mistaken when the domain of explananda is human conduct, even though I admit (nay, insist) that the words 'cause' and 'reason' designate utterly different sorts of being. I grant the premise, that the terms 'cause' and 'reason' refer to nonoverlapping classes of designata. But I deny that from this premise we can validly infer the usual conclusion, to wit, that to provide a causal account of a person's behavior is inconsistent with giving an account in terms of his reasons. If this is paradoxical, I can only argue that it is not contradictory, and hope that its paradoxical flavor will be dissipated by sufficient immersion in my examples.

The view that I wish to develop is that while causes and reasons are utterly different sorts of things, and while in an important sense we can say that causes are "in the world" whereas reasons are not "in the world," nevertheless the *giving* of reasons, the *holding* of reasons, the *stating* of reasons, the *tokening* of reasons, the *belief* in reasons, are all psychological events, and as such are very much "in the world," and part of the chain of causality. I wish to maintain further that such psychological events have a *content*, the character of which cannot be fully set forth without employing the categories of logic. Hence, in formulating the causal laws of behavior, at least at the molar level, regarding the influencing of behavior by the tokening of reasons, the question whether or not a certain proposition or belief or sentence is a *good reason* is psychologically relevant. This is a question which can be put without conflating causes and reasons, because while a reason is not an event "in the world," the giving of a reason (or the believing of a reason, or the accepting of a reason) *is* a psychological event and *is* "in the world."

Let us take an example of simple purposive behavior to examine in this light. I mail a letter to a hotel in New York City for the purpose of arranging a room reservation because I am planning to attend a convention there. My plan to attend the convention is a good reason for sending a letter. There is a quasi-lawlike statement relating the sending of letters and the establishment in New York of a room reservation in one's name, which, while it is not a fundamental nomological, can either be made into a nomological by a suitable *ceteris paribus* clause or formulated as a statistical generalization of high p. We then have a kind of "causal law" (belonging in the domain of sociology), which relates one event, the mailing of a letter, to a subsequent event by alleging a causal connection between the two. Now this statistical generalization (or derived nomological, presupposing the *ceteris paribus* fulfilled) is known to me. And taken together with my intention, it provides a good reason for mailing the letter.[3] In causal terms "having-the-intention-cum-believing-the-law" is a composite inner (mental) event or state that acts as an efficient cause of my letter-mailing behavior.

To say, "A reason caused my behavior" is perhaps a harmless ellipsis, but strictly speaking, it involves a confusion of the two realms which we must be careful to avoid. What we should rather say, so as to steer safely clear of this confusion, is, "The tokening of a reason was the psychological cause of my behavior." Or, lest even this formulation be taken wrongly, we could say, "The tokening of a sentence S which expresses a proposition p, where p is a good reason for action A, was the psychological cause of my emitting action A." So, even in this simple example we have at least four linkages or "connections" to consider and distinguish: First, mailing letters is a cause of room reservations expressible as a hypothetical 'If one mails a letter, he gets a room'; second, this causal relation is, *in the realm of inference*, a good reason for mailing a letter, granted the premise that one wants a room reservation; third, my tokening of this good reason functions as a psychological cause of my performing an act whose description occurs in the antecedent statement of the hypothetical (a relation in pragmatics which, in general, is a characteristic of purposive behavior); fourth,

---

[3] See C. G. Hempel, "The Concept of Rationality and the Logic of Explanation by Reasons," pp. 463–486 in his *Aspects of Scientific Explanation* (New York: Free Press, 1965); G. H. von Wright, "The Logic of Practical Discourse," pp. 141–167 in R. Klibansky, ed., *Contemporary Philosophy* (Florence: Nuova Italia, 1968); A. Pap, *An Introduction to the Philosophy of Science* (New York: Free Press, 1962), pp. 263–267; a very stimulating analysis and criticism of "the view that meaningful human actions are not amenable to causal, scientific explanation" is May Brodbeck's "Meaning and Action," *Philosophy of Science*, 30 (1963), 309–324. Because her mode of resolution is incommensurable with mine, and repudiates the mind-body identity thesis presupposed in Professor Popper's formulation of "Compton's problem"—see her "Mental and Physical: Identity versus Sameness," in P. K. Feyerabend and G. Maxwell, eds., *Mind, Matter, and Method* (Minneapolis: University of Minnesota Press, 1966), pp. 40–58—I have not found it feasible under space limitations to integrate Professor Brodbeck's discussion into my paper.

an external observer would in turn have a good reason for expecting me to mail this letter, and that good reason would be *his* understanding of the psychological causal law which says that, *ceteris paribus*, if a rational person wills the consequent of a causal law which the person believes, he tends also to will the antecedent. (The *ceteris paribus* clause must, of course, include such qualifiers as 'absent countervailing means-end structures' and the like.) In this analysis I have not, I trust, anywhere conflated causes with reasons. Yet I have explicitly recognized that a critical element in what makes certain kinds of mental events causally efficacious is that they are tokenings of sentences which, in the realm of logic, constitute valid reasons.

Consider next the case of a simple desk calculator. In order for it to compute sums accurately, its internal structure must have some kind of isomorphism with decimal arithmetic. Thus, the machine is constructed so that after a wheel has turned through ten positions, this physical fact causally produces a one-position displacement in the next adjacent wheel, i.e., the wheel which "corresponds" to the next integer to the left. The machine behaves rationally, in that it makes legitimate or valid transitions in the arithmetical language game. If it were not constructed in the way it is, or in some alternative way preserving the necessary machine-arithmetic correspondence, it would not be able to do this. We telephone the company and ask for a repairman to be sent out when the machine begins to make counter-arithmetical transitions, i.e., it "makes mistakes" and "gives the wrong answers." (Even a philosophy professor normally finds these locutions quite natural under such circumstances.) Such a desk calculator is clearly a "clock" rather than a "cloud," but as it gets old and worn out and becomes a little more cloudlike, it also becomes more irrational, i.e., slippage in the gears leads it to make arithmetical mistakes. We can carry the analogy further, still remaining at the level of a mere desk calculator rather than the big modern computers. What the machine will do with the numbers we punch in depends upon our giving it instructions, which is (formally) comparable to the "intention" or "mental set" adopted by a human being as he listens to us giving reasons. It is no objection to this analogy that the machine does not have conscious intentions, because it is imperative to distinguish the components of sentience and sapience,[4] and what we are concerned with in the present section is wheth-

---

[4] H. Feigl and P. E. Meehl, "The Determinism-Freedom and Body-Mind Problems," in P. A. Schilpp, ed., *The Philosophy of Karl Popper* (LaSalle, Ill.: Open Court, forth-

Paul E. Meehl

er determinism is in any way incompatible with that aspect of sapience which we call 'rationality.' (Even in the human being there is, of course, plenty of evidence to say that sapience can occur, and sometimes in very complicated forms, in the absence of reportable phenomenal events. The well-known examples of unconscious literary composition or scientific problem-solving, not to mention the quite complicated content of means-end connections involved in psychoanalytic mechanisms, suffice to show this.)

With the kind permission of Professors Schilpp and Freeman, I would like to quote here a passage from the forthcoming contribution by Professor Feigl and myself to the Schilpp volume on Sir Karl Popper's philosophy.[5]

Returning to the question of the sense in which a physicalistic account in brain-language is "complete" *even though it does not say all that could be said*, we suggest the following as a first approximation to an account which, while maintaining the distinction between logical categories and the categories of physics or physiology, nevertheless insists that a physicalistic micro-account is nomologically complete. We have a calculus, such as arithmetic or the rules of the categorical syllogism. We have a class of brain-events which are identified by appropriate physical properties—these, of course, may be highly "configural" in character—at, say, an intermediate level of molarity (i.e., the events involve less than the whole brain or some molar feature of the whole acting and thinking person, but are at a "higher" level in the hierarchy of physical subsystems than, say, the discharge of a single neuron, or the alteration of microstructure at a synapse). Considered in their functioning as inner tokenings—that is, however peripherally or behavioristically they were originally acquired by social conditioning, considering them as now playing the role of Sellars' mental word[6]—there is a physically-identifiable brain-event b_M which "corresponds" (in the mental word sense) to the subject-term in the first premise of a syllogism in Barbara. There is a second tokening event b_P

coming); H. Feigl, The "Mental" and the "Physical": The Essay and a Postscript (Minneapolis: University of Minnesota Press, 1967); P. E. Meehl, "The Compleat Autocerebroscopist: A Thought-Experiment on Professor Feigl's Mind-Body Identity Thesis," in Feyerabend and Maxwell, eds., Mind, Matter, and Method.

[5] Feigl and Meehl, "The Determinism-Freedom and Body-Mind Problems."

[6] W. Sellars, "Empiricism and the Philosophy of Mind," in H. Feigl and M. Scriven, eds., Minnesota Studies in the Philosophy of Science, vol. I (Minneapolis: University of Minnesota Press, 1956), p. 328. Also in Science, Perception, and Reality (New York: Humanities, 1963); W. Sellars, "Thought and Action," in K. Lehrer, ed., Freedom and Determinism (New York: Random House, 1966); R. Chisholm and W. Sellars, "Intentionality and the Mental," in H. Feigl, M. Scriven, and G. Maxwell, eds., Minnesota Studies in the Philosophy of Science, vol. II (Minneapolis: University of Minnesota Press, 1958), Appendix, pp. 507–539.

which is a token of the type that designates the predicate-term of the conclusion; a brain-event b_S which corresponds to a tokening of the type that designates the subject-term of the conclusion of the syllogism; and finally a brain-event b_C corresponding to the copula. (These expository remarks are offered with pedagogic intent only. We do not underestimate the terrible complexity of adequately explaining the words 'correspond' and 'designate' in the immediately preceding text.)

A physically-omniscient neurophysiologist [=Omniscient Jones estopped from meta-talk about logic] can, we assume, identify these four brain-events b_M, b_P, b_S, b_C on the basis of their respective conjunctions of physical properties, which presumably are some combination of *locus* (where in the brain? which cell assemblies?) and *quantitative properties* of *function* (peak level of activation of an assembly, decay rate, pulse-frequency of driving the next assembly in a causal chain, mean number of activated elements participating). For present purposes we may neglect any problem of extensional vagueness, which is not relevant to the present line of argument, although it is of considerable interest in its own right.

Our physically-omniscient neurophysiologist is in possession of a finite set of statements which are the nomologicals (or quasi-nomologicals) of neurophysiology, which we shall designate collectively by L_{phys} [=neurophysiological laws]. He is also in possession of a very large, unwieldy, but finite set of statements about structure, including (a) macrostructure, (b) structure of intermediate levels, e.g., architectonics and cell-type areas such as studied microscopically in a brain-histology course, and (c) micro-structural statements including micro-structural statements about functional connections. We take it for granted that "learned functional connections" *must* be embodied in micro-structure (although its exact nature is still a matter for research) since there is otherwise no explanation of the continuity of memory when organisms, human or animal, are put into such deep anesthesia that all nerve cell discharge is totally suspended for considerable time periods, or when normal functional activity is dramatically interrupted by such a cerebral storm as a grand mal seizure induced in electroshock treatment. Thus the class of structural statements S_t includes two major sub-classes of statements, one being about the inherited "wiring diagram" of a human brain, and the other being the acquired functional synaptic connections resulting from the learning process.

Our omniscient neurophysiologist can derive, from the conjunction $(L_{phys} \cdot S_t)$, a "brain-theorem" T_b, which, to an approximation adequate for present purposes, may be put this way: Brain-state theorem T_b: "Whenever the composite brain events $(b_M b_C b_P)$ and $(b_S b_C b_M)$ are temporally contiguous, a brain-event $(b_S b_C b_P)$ follows immediately." This brain-theorem is formulated solely in terms of the states b_i which are physicalistically identifiable, and without reference to any such meta-concept as class, syllogism, inference, or the like. The derivation of T_b is one of strict deducibility in the object-language of neurophysiology. That is,.

neurophysiology tells us that a brain initially wired in such-and-such a way, and then subsequently "programmed" by social learning to have such-and-such functional connections (dispositions), will necessarily [nomological necessity] undergo the event $(b_S b_C b_P)$ whenever it has just previously undergone the events $(b_M b_C b_P)$ and $(b_S b_C b_M)$ in close temporal contiguity.

But while for the neurophysiologist this brain-theorem is a theorem about certain physical events and *nothing more*, a logician would surely discern an interesting formal feature revealed in the descriptive notation —the subscripts—of the b's. It would hardly require the intellectual powers of a Carnap or Goedel to notice, *qua* logician, that these brain-events constitute a physical model of a sub-calculus of logic, i.e., that these physical entities $[b_M, b_P, b_S, b_C]$ "satisfy" the formal structure of the syllogism in Barbara, if we interpret

$b_M =$ tokening of middle term $\qquad b_S =$ tokening of subject term
$b_P =$ tokening of predicate term $\qquad b_C =$ tokening of copula

The "brain-theorem" T_b can be derived *nomologically* from the structural statements S_t together with the microphysiological law-set L_{phys}, given explicit definitions of the events $[b_M, b_P, b_S, b_C]$. These explicit definitions are not the model-interpretations, nor are they "psycholinguistic" characterizations. We can identify a case of b_P by its physical micro-properties, *without knowing that it is a tokening-event*, i.e., without knowing that it plays a certain role in the linguistic system which the individual who owns this brain has socially acquired. But brain-theorem T_b has itself a *formal* structure, which is "shown forth" in one way, namely, by the syntactical configuration of the b-subscripts [M,P,S,C]. In this notation, "which subscript goes with what" is determinable, so long as the events b_i are physically identifiable. There is nothing physically arbitrary in this, and there is nothing in it that requires the physically-omniscient neurophysiologist to be thinking about syllogisms, or even, for that matter, to know that there is any such thing as a syllogism. Although again, it goes without saying that he himself must reason logically in order to derive the brain-theorem. But he does not have to meta-talk about rules, or about his own rule-obedience, in order to token rule-conformably in his scientific object-language, and this suffices to derive T_b.

One near-literal metaphor which we find helpful in conveying the essence of the "syllogistic brain-theorem" situation, as we see it, is that the sequence of brain-events $(b_i b_j b_k)$ $(b_j b_k b_l)$ . . . *embodies* the syllogistic rules. Their defined physical structure plus the physical laws of brain-function causally necessitate that they exemplify syllogistic transitions, a fact revealed when the notation designating them is considered in its formal aspects. In the usual terminology of thinking processes and logic, the brain-theorem T_b says, in effect, that the existence of a formal relation of deducibility (truth of logic) provides, in a brain for which the theorem obtains, the necessary and sufficient causal condition for a factual transition of *inference* (a mental process). This assertion may appear to "mix the languages," to "commit the sin of psychologism," to "conflate causes with reasons"; but we maintain that none of these blunders is involved. It is a *physical* fact that a certain *formal* relation is physically embodied. If the formal features of the initial physical state were otherwise, the ensuing physical result would have been otherwise. Hence the physical embodiment of the formal relation—a *fact*, which is "in the world" as concretely as the height, in metres, of Mount Everest—is literally a condition for the inference to occur.

I need hardly say that the idea that strict rationality in a deductive-inference situation is not only compatible with determinism but at the common-sense level requires it—"If I am 100% rational, I will be *unable* to deny conclusions strictly implied by premises"—is hardly a new insight on my part, and I have not felt it useful to canvass the philosophical or psychological literature for citations. Since the first draft of this paper was written, two explicit statements on this point have been brought to my attention, one by Ruth Macklin, in an illuminating paper entitled "Doing and Happening," where we read:

The problem of trying to make this distinction [between things a person does and things that happen to him] hold for all cases becomes even more complex when we consider mental acts such as believing, thinking, and wanting. Although choosing, deciding, and forming intentions appear to be mental acts in the sense that they seem to be clear cases of something a person does, what about believing? Does a person choose to believe the things he believes? Or to think the thoughts he thinks? Does he have control over his beliefs in the psychological sense that he can, in fact, avoid believing that p in cases where evidence in favor of the truth of p is overwhelming? If he cannot control his beliefs in such cases, are we to say that believing that p is not something which that person does, but rather something that happens to him? This result is obtained by using an analogue of the physiological control criterion which may be somewhat infelicitously termed "the mental control criterion." It does seem counter-intuitive to claim that believing is not something that someone does; yet it is not clear that either the mental control criterion or another appeal to linguistic usage will answer the question satisfactorily. We do sometimes say, "I cannot help believing that," or "Try as I might, I cannot believe that," indicating that the ability to choose or control our beliefs is open to question. This problem can be met, in part, by making the further distinction between deliberate and non-deliberate doings, and between believings that are reflective and those that are not. Hence, application of the mental control criterion would result in the position that some types of believing are

Paul E. Meehl

not things that one does, but rather things that happen to one. Perhaps, then, the criterion should be rejected. But on what grounds? Presumably, on the grounds that it conflicts with our intuition that believing is always something persons do. Of course, there is still another alternative, namely, that the distinction between what a person does and what happens to him is inapplicable to mental acts such as believing. On this view, it is inappropriate to claim that believing is either something that one does or something that happens to him.[7]

The other quotation, as succinct and explicit a statement of my position as one could easily find anywhere, goes back to 1905, in Max Weber's critique of Eduard Meyer's methodological views.

The error in the assumption that any freedom of the will—however it is understood—is identical with the "irrationality" of action, or that the latter is conditioned by the former, is quite obvious. The characteristic of "incalculability," equally great but not greater than that of "blind forces of nature," is the privilege of—the insane. On the other hand, we associate the highest measure of an empirical "feeling of freedom" with those actions which we are conscious of performing rationally—i.e., *in the absence of physical and psychic "coercion," emotional "affects" and "accidental" disturbances of the clarity of judgment,* in which we pursue a clearly perceived end by "means" which are the most adequate in accordance with the extent of our knowledge, i.e., in accordance with empirical *rules.* If history had only to deal with such rational actions which are "free" in this sense, its task would be immeasurably lightened: the goal, the "motive," the "maxims" of the actor would be unambiguously derivable from the means applied and all the irrationalities which constitute the "personal" element in *conduct* would be excluded. Since all strictly teleologically (purposefully) occurring actions involve applications of empirical rules, which tell what the appropriate "means" to ends are, history would be nothing but the applications of those rules. The impossibility of purely pragmatic history is determined by the fact that the action of men is *not* interpretable in such purely rational terms, that not only irrational "prejudices," errors in thinking and factual errors but also "temperament," "moods" and "affects" disturb his freedom—in brief, that his action too— to very different degrees—partakes of the empirical "meaninglessness" of "natural change." Action *shares* this kind of "irrationality" with every natural event, and when the historian in the interpretation of historical interconnections speaks of the "irrationality" of human action as a disturbing factor, he is comparing historical-empirical action not with the phenomena of nature but with the ideal of a purely rational, i.e., absolutely

[7] Ruth Macklin, "Doing and Happening," *Review of Metaphysics,* 22 (1968), 246–261, quoted from 257–258. See also the same author's "Action, Causality, and Teleology," *British Journal for the Philosophy of Science,* 19 (1969), 301–316.

purposeful, action which is also absolutely oriented towards the adequate means.[8]

I have no doubt made my task somewhat easier, as Sir Karl might object, by confining myself to that restricted form of rationality involved in thinking syllogistically. But while the choice of such an example simplifies the problem, I cannot think that the use of such an example is tendentious or prejudicial. And this is the more so since Sir Karl, whose position is under examination here, is such a firm and articulate opponent of the idea that there exists such a thing as "inductive logic." Without making overmuch of the Reichenbachian dichotomy between the context of discovery and the context of justification (a dichotomy which is not clearcut, but still useful for many purposes) I would suggest that those portions of what is ordinarily considered subsumable under "inductive inference" that involve the testing of hypotheses and generalizations can be given a syllogistic form (e.g., modus tollens), so that the preceding syllogistic example can function as a paradigm case for dealing with Sir Karl's position in this respect; and that what is not so formulated can either (a) be viewed as satisfying or not satisfying some overarching methodological prescription in the pragmatic metalanguage (and hence examinable in an essentially syllogistic way, i.e., we inquire whether a given bit of concrete pragmatics of inference is or is not in accord with the methodological prescription), or (b) really properly relegated to "context of discovery" in the strong sense of the phrase. Since Sir Karl himself explicitly repudiates the problem of the psychology of discovery as not belonging to the logic of the matter, the question how (historically, psychogenetically, sociologically) a particular scientist comes to hit upon a theory or an experimental arrangement is not relevant to our present issue. The possibility of novelty, of genuine "creation" by the scientist or artist, does involve this context-of-discovery question, and Sir Karl adduces it as a further objection to determinism; but that is outside the scope of this paper.[9]

One's philosophical (and perhaps, more importantly, one's "personal, human, existential") discomfort about determinism in relation to the possibility and limits of human rationality is, I suggest, often exacerbated by

[8] M. Weber, "A Critique of Eduard Meyer's Methodological Views" (1905), in Max Weber, *Methodology of the Social Sciences*, ed. E. A. Shils and H. A. Finch (New York: Free Press, 1949). I am indebted to my Law School colleague Professor Carl Auerbach for calling this reference to my attention.

[9] See, however, Feigl and Meehl, "The Determinism-Freedom and Body-Mind Problems."

two mental habits in our thinking about the psychological causation of beliefs and related cognitive processes (e.g., perceiving, inferring, recalling). The first is our habit of associating the idea of psychological causation primarily with the nonrational or irrational class of causes, such as unquestioned beliefs carried over from childhood, political manipulation of mass opinion through propaganda, personal idiosyncratic prejudices having a variety of origins, unconscious determinations of the Freudian type, and the like. One has the impression, in talking with either philosophers or "plain men," that when asked to contemplate the possibility that their own behavior may be completely (or even largely) determined by antecedent causal conditions, they tend to think immediately of such factors as the fried eggs they had for breakfast, or the prejudice they learned from their Norwegian grandmother against the Danes, or the subtle influence of TV advertising upon consumer choice, rather than such psychological facts as the fact that they have been presented with certain evidence in the form of statements from reliable sources, or have been subjected to criticism in the course of discussion with a colleague, or have made certain observations in the laboratory. Thus it seems that our tendency to polarize "human reason" over against "psychological causality" infects our thinking by influencing the very examples that occur to us in this connection, so that we tend to think of psychological causality solely in terms of the kinds of causes which normally are used to explain a piece of human irrationality. No doubt the influence of Freud and Marx, at least among educated persons, is important here, inasmuch as they both stressed the "hidden, nonrational" forces that play a greater role in the molding of man's opinions than had been formerly supposed.

The second habit is our tendency to connect such "psychological determiners" as motives, affects, training, group pressures, psychoanalytic identifications, and the like almost entirely with a person's *particular substantive (object-language) opinion*, forgetting to take into account the fact that these same classes of psychological determiners also exert a powerful causal influence upon his *meta-talk* and *meta-thought*. That is, we readily recognize the possibility that I am a zealous pacifist or jingoist because I have a strong unconscious father-identification and my father was a pacifist or jingoist, as the case may be. But we neglect the possibility (thankfully, one realized in at least an appreciable minority of human beings!) that I am committed to certain overarching procedural or methodological principles, such as rationality, critical discussion, and the examination of contrary evidence, as a result of my psychological history. These overarching methodological habits are, of course, subject to learning by reward or father-identification or peer-group conformity pressures, and it is a mistake to assume that only our specific-issue opinions are "psychologically caused." In common life, we have occasion to characterize persons according to their meta-talk dispositions. We may say of a certain individual, "Well, he's usually a very rational fellow on most subjects, and he tries hard to be fair-minded and to see the other fellow's viewpoint; but you'd better not get him on the race question, because there he goes haywire." What do we mean by a remark of this kind? We mean that this person is one who has developed very strong pervasive "meta-habits" of the kind we call 'rational,' so that we expect that these overarching considerations, e.g., his self-concept of being a rational person, and his sincere desire to find out the truth about matters to which he addresses himself, his compulsion to derive implications from hypotheses and compare them with facts, will in general be controlling with respect to his processes of thinking about particular substantive matters; but that this overarching monitor or control system, which causes him to think rationally in general, is not sufficiently strong to countervail the influence of a very strong emotional commitment on this particular substantive issue of the race question. (Each of us has on occasion in his smaller way to make Henry Clay's famous decision on whether we would rather be right than be President!)

It may be objected that in these remarks I have fallen into the confusion I earlier renounced, to wit, mixing the category of "causes" with the category of "reasons." If so, it has been through some subtle philosophical mistake, because in writing the foregoing I had this distinction constantly in mind. But I have permitted myself such locutions as "thinking rationally," or "countervailing influence against rationality" and it is obligatory upon me to explicate the cause-reason relationship indicated by such language.

What, then, is the situation here, as regards the relation of causes to reasons, when we inquire concerning a particular individual's thinking about a certain subject whether it was mainly influenced by "rational" or "irrational" factors? For ease of exposition and to avoid nonrelevant issues in psycholinguistics (but, I hope, without loss of generality), I confine myself to thinking processes sufficiently "symbolic-linguistic" in nature to make appropriate the notion that the individual tokens a sentence. I do not mean thereby to prejudge the question whether all forms of intentionality involve sentence-tokening; but that sort of intentionality which is (debatably) present when I simply image a state of affairs, or when I have an unworded "expectation" that is rudely disappointed (as in Russell's well-known example of a nonlinguistic belief, that of experiencing surprise upon finding another step on the stairs when I thought I had come to the bottom even though I was not consciously "thinking about it" at all) is too dubiously "rational" to be of use for our analysis. I have no stake in asserting that such non-worded, inchoate expectations *cannot* (on some suitable reconstruction) be considered rational; and in fact I tend to believe some of them should be called so. But since the detailed reconstruction is not available, and since some readers would disagree with me, I avoid these marginal cases. The issue is the compatibility of rationality and determinism, and it seems unlikely that the alleged incompatibility would show up in the case of dubiously intentional (mental) acts of a nonsymbolic, nonlinguistic sort, but for some strange reason be lacking in the clear-cut linguistic case. And since the linguistic case is the one which has been subjected to more adequate philosophical analysis, it is the only one I shall consider here.[10]

Consider an example in which I begin without any personal involvement or emotional prejudice one way or the other. I am an educated man but a non-mathematician, and I have no particular philosophical leanings about the idea of infinity. In a semipopular volume on mathematics, I come across the question whether there is a largest prime. I think to myself, "That's an interesting question; it never occurred to me; I shouldn't be surprised one way or the other; and I couldn't care less." I read through Euclid's short and easy proof, which I find convincing. From that moment onward, I firmly believe that there is no largest prime. This would seem to be a rather clear-cut case of practically 100 percent rationality. Now in what sense, if any, can we speak of the "causes of my belief" in the infinity of primes being "rational causes," without conflating the distinct categories of *cause* and *reason*?

It seems to me that there is no great mystery here, that the correct analysis is quite straightforward, and not even paradoxical. In reading through the proof, I token consecutively the sentences which, according to the syntactical rules of the language I speak, constitute (in the realm of logic) a

---

[10] For a fascinating—although, in the end, somehow unsatisfying—analysis of the concept *rational* as applied to imaginary dance-language of bees, see J. Bennett, *Rationality* (London: Routledge and Kegan Paul, 1964).

formally valid proof of the infinity of primes. These sentences express propositions which, in the realm of logic, constitute "valid grounds for believing the conclusion." That is, the propositions which these sentences express are *good reasons*. What makes them good (deductive) reasons is, of course, that the conclusion can be reached by a finite number of steps, each step being taken in accordance with the transformation rules of the language. The tokenings ($=$ my *thinking the sentences*) are not, strictly speaking, reasons. The tokenings are psychological causes, that is to say, they are events which go on in my mind (or, if physicalism is true, we can also say "in my brain").[11] What makes them effective causes is the fact that my brain is wired (or, more accurately, we would have to say wired plus programmed) to make language transitions in accordance with certain syntactical transformation rules. When I token a sentence S_1 and a sentence S_2, tokenings of sentences related such that the sentence S_3 logically follows from S_1 and S_2 in accordance with the transformation rules of the language in which I have learned to "think," I am strongly disposed (to the extent that I am a rational man) to token S_3. What is the mystery about this? Except for vast differences in complexity, how does this differ from the fact that my desk calculator has a mechanical construction such that the movements of its gears "obey the laws of arithmetic"? Elliptically, it is therefore unobjectionable, provided there is no danger of confusion, to say that I am "caused to believe in the infinity of primes by valid reasons." But this locution should probably be avoided in the interests of clarity. The reasons are not causes, but the tokenings of the sentences which express the reasons are causes. If my terminal tokening of Euclid's conclusion is in fact psychologically produced not by admiration of Euclid (for some students the letters Q.E.D. could stand for "quod Euclid dixit"), or by the fried eggs I had for breakfast, but by the fact that my brain has been programmed to token sentences in accordance with transformation rules, then my belief in the infinity of primes is "rationally determined."

I have in this analysis made free use of the notion of an inner "tokening" as if it were an obvious and clear idea, which it admittedly is not. Unfortunately this is one of those concepts in whose consideration philosophy unavoidably overlaps with one of the empirical sciences, namely, psycholinguistics; and psycholinguistics is a science presently in a primitive state of development so that it cannot provide clear-cut, well-established laws

---

[11] See Meehl, "The Compleat Autocerebroscopist," pp. 108, 160–163.

(or, hence, adequate implicit definitions of its theoretical entities) for the use of the philosopher interested in semantics. It seems clear that the mental word need not possess a complex internal structure capable of correspondence, picturing, or "isomorphism" with the external nonlinguistic event that it designates. For example, a single noise, not capable of division into parts or components which have a separate "meaning," may, to an Eskimo, be equivalent to an English language statement, "There is today a great deal of snow, of a slushy variety." No psycholinguist or philosopher can, in the present state of knowledge, specify in detail what are the necessary and sufficient physical and psychological conditions for an Eskimo to token this word internally. Fortunately, it is not necessary for the philosopher to rely heavily upon technical psycholinguistics in discussing Popper's thesis. All that we need suppose is that there occurs a certain kind of physical event in the brain, whatever its "internal" nature, that has the required relation to an external event such that it is entitled to be called a tokening of a sentence, when that sentence expresses a proposition designating the external event. In the case of an English speaker, the resources of his language are such that he must say "There is a great deal of snow, of a slushy type." Whereas for an Eskimo speaker it may be sufficient to say "glop," which means the same thing to him as the more complicated expression means to an English speaker. The point for our purposes is that *whatever* physical event in the brain has come (by social learning) to possess this kind of statistical correspondence to slushy snow, perceptions normally produced by slushy snow, expectations about correlates of slushy snow, etc., constitutes the physical tokening of the proposition. The occurrence of slushy snow is physically describable. The occurrence of a particular token event is also physically describable. (I set aside ontological dualism for the time being; but since determinism and not materialism is in issue, I do not believe that prejudices our present discussion. Would not Popper's objection hold against a deterministic dualistic interactionism, if it holds against a deterministic identity thesis?) Examining the role of a given kind of tokening event in the total tokening system of an individual belonging to a particular culture, we can (in principle) determine the external event to which it "refers" (insofar as it refers precisely, which it almost never does). Having determined that, we understand the "meaning" of the sentence which the individual tokens. And if he tokens that sentence when the external event does not occur, then we say that he "tokens falsely."

Of course we know that the formulation of semantic rules on the basis of studying a natural language ("English as she is spoke") will always involve a certain element of arbitrariness, because we may or may not choose to embody certain statistically deviant locutions in the rules. This will depend upon their statistical rarity, in large part, although not wholly. (Cf. the dictionary-maker's problem of deciding between "second usage" and "erroneous usage.") Thus Carnap[12] says that after observing the verbal behavior of people who speak a particular language B and noticing that 98 percent of the time they use the word 'mond' to mean the moon, but 2 percent of the time use 'mond' to refer to a kind of lantern, we are free to formulate the semantics of language B either to include this special and rare usage or not. That is, the formulation of a pure semantics, like any idealization or rational reconstruction, is something done with *attention to* (and *on the basis of*) the empirical statistics of descriptive semantics and pragmatics, but cannot usefully aim to *exactly correspond* to the latter. (If it did, as a desideratum, it would not be rule-construction, since a rule is, by definition, something capable of being violated.)

In the present state of knowledge of how the human brain mediates behavior and experience (on *either* a dualistic or "identity" view of the mind-body problem) it is pointless to speculate philosophically about possibilities as regards detail. Nor can one anticipate with any confidence just where on the cloud-clock continuum various kinds of psychological processes will be located when psychophysiology has reached a relatively advanced state. When the contemporary psychologist deals with rational behavior, such as that of a logician performing the task of classifying a simple syllogism as formally invalid, he proceeds in the same way as the philosopher or the layman does, namely, at what is usually called the "molar" level of analysis.[13] That is to say, in forecasting Professor Popper's verbal

[12] R. Carnap, *Foundations of Logic and Mathematics*, vol. 1, no. 3 of *International Encyclopedia of Unified Science*, ed. O. Neurath (Chicago: University of Chicago Press, 1939), pp. 6–7.
[13] E. C. Tolman, *Purposive Behavior in Animals and Men* (New York: Appleton-Century-Crofts, 1932), pp. 3–23 and passim; E. C. Tolman, *Collected Papers in Psychology* (Berkeley: University of California Press, 1951); C. L. Hull, *Principles of Behavior* (New York: Appleton-Century-Crofts, 1943), pp. 19–21; B. F. Skinner, *The Behavior of Organisms* (New York: Appleton-Century-Crofts, 1938), pp. 3–6, 33–43; R. A. Littman and E. Rosen, "Molar and Molecular," *Psychological Review*, 57 (1950), 58–65; H. A. Murray, *Explorations in Personality* (New York: Wiley, 1938), pp. 55–58, 96–97; K. MacCorquodale and P. E. Meehl, "Edward C. Tolman," in W. K. Estes et al., *Modern Learning Theory* (New York: Appleton-Century-Crofts, 1954), especially pp. 218–231; and the operant behaviorists generally, cited in footnote 2 *supra*.

response to a syllogism which commits an Illicit Distribution of the Major, we do not—in part because we *cannot*, at the present state of knowledge of brain function—mediate this prediction in the language of neurophysiology. Instead we rely upon dispositional properties (setting aside for the moment whether these are probabilistic or nomological) of the "whole organism," the man Popper, whom we know to be sane, sober, attentive to his task, and who has a history of having been educated in formal logic. The fact that our predictions of his verbal behavior are mediated on this molar basis gives rise to an interesting problem in the methodology of those sciences that deal with the behavior of human beings, namely, to what extent must the behavior scientist employ the concepts of the logician?

It is important to make some distinctions here which are sometimes overlooked. A logical category, such as "valid syllogism," can be involved in the psychologist's task in three different ways. First, the psychologist wants to proceed rationally in his own scientific thinking, i.e., it is necessary that he himself as a knowing organism *exemplify* or *be obedient to* the laws of logic. That is to say, in his own (object-language) discourse he must avoid committing fallacies. Second, we recognize that a considerable amount of scientific writing and discussion involves, in addition to object-linguistic assertions describing observations or propounding theories, processes of rational criticism. Here the psychologist moves periodically into the metalanguage as he engages in such processes as evaluating experimental evidence, examining the theoretical derivations offered by himself and others, carrying on rational inquiry about the internal consistency of a system of theoretical propositions, and the like. So far as I can discern, in these two respects the behavioral scientist's "use" of logic does not differ in any essential way from that of the botanist or the astronomer. All scientists must think logically, whether in the object language (substantive derivations and classifications) or in the metalanguage (criticism, evaluation, and research strategy). But it seems that insofar as the psychologist treats of human cognitive processes at the molar rather than at the neurophysiological level, he is forced to employ the concepts of logic in a third way, a way that is unique to social science as a subject matter. The reason for this peculiarity of psychology, sociology, economics, etc., as Professor Popper would be the first to emphasize, is, put most simply and directly, that plants and stars do not think, but human beings do.

This undisputed fact (did even John B. Watson *really* doubt it?) about the special nature of the psychologist's subject matter gives rise to the paradox that *concepts customarily regarded as metalinguistic unavoidably appear in the psychologist's object-linguistic discourse whenever he is attempting to mediate predictions about rational human behavior.* It might be supposed that more adequate behavioristic formulation of rational behavior could, in principle, dispense with the employment of such metalinguistic concepts. Quite apart from the fact that this hope refers to a Utopian state of behaviorism, whereas we all admit that the psychologist, like the layman or the philosopher, can successfully mediate high-probability predictions given the present non-Utopian state of the psychology of cognition and psycholinguistics, I must further point out that it is far from obvious that even in a Utopian state of these molar disciplines it would be theoretically possible to dispense with the logician's metalanguage in giving a psychological description or causal analysis of rational human behavior. Since the possibility of their permanent indispensability at the molar level of analysis is the alternative most favorable to Professor Popper's position, let us scrutinize the consequences of this alternative in more detail.

Consider again the simplified, idealized example of a logician being confronted with a syllogistic argument containing an Illicit Major. We set aside the empirical problems involved in ascertaining those aspects of the individual's emotional and motivational state which are relevant to his momentary disposition to think rationally. That is, we assume that the "test conditions" for activating his disposition to classify an argument as Illicit Major are known to be momentarily fulfilled. The usual gambit for an arch-behaviorist who aims at eliminating any vestige of "mentalistic concepts" from his descriptive or theoretical language is to invoke the individual's learning history and to infer from it that the logician has a very strong "habit" of responding with the phrase "Illicit Major" when he is presented with a certain stimulus, say, the formally invalid syllogism appearing in three lines on a clearly printed page. Leaving aside the current controversies in psycholinguistics (which call into serious question the theoretical adequacy of any "stimulus-response" model of verbal processes) let us proceed on the (probably false) assumption that such an analysis could be given and satisfactorily corroborated. That is, let us assume that the molar behaviorist can make good on his claim of accounting for the logician's current disposition to token the metalinguistic expression

"Illicit Major" as a response to the printed tokens of such a formally in-valid syllogism. The question now arises, how is the stimulus class to be characterized in molar language? As is well known, there are terrible diffi-culties involved in the whole problem of pattern recognition, such that no one has as yet provided an adequate theoretical model of the necessary structures which, if we were in possession of it, would enable us to con-struct and program a computer to duplicate even fairly simple visual pat-tern-recognition functions of the human brain.[14] We shall set this whole class of difficulties aside also, and assume as an oversimplified situation that the typeface, size, spacing, etc., are physically identical with those which have been presented to the logician in his previous experiential history.

But even these idealizations and oversimplifications do not, it seems to me, get rid of the behaviorist's fundamental problem. We know that it is possible to present the logician with *any* syllogism having the requisite formal structure of an Illicit Major, and confidently predict that his re-sponse will be a tokening of 'Illicit Major,' or some equivalent thereof. (Let us set aside the problem of what are "equivalent responses" and con-fine our attention solely to the problem of identifying the stimulus class.) Now the fact that the presented visual stimulus, a syllogism on the printed page, need not be physically identical in terms of mounds of ink with any stimulus previously presented to the logician is not in itself a serious ob-jection to the behaviorist's analysis, it being admitted on all sides that some underlying concept of stimulus equivalence or stimulus generaliza-tion will be required by any adequate molar theory (since from the sheer standpoint of their physics no two stimulus inputs, at least among those occurring in ordinary life, are strictly identical). That is, what Skinner in *The Behavior of Organisms* calls "the generic nature of the concepts of stimulus and response" is taken for granted by psychological theorists of many different persuasions. The scientific problem here is not (at the molar level) to derive or explain the basic phenomenon of stimulus equi-valence or stimulus generalization, which is rather taken as a rock-bottom fact, a basic postulate in any molar behavior theory, and presumably finds its own explanation in turn at another level of causal analysis, i.e., at the neurophysiological level. The problem at the molar level, once having in-cluded some suitable theoretical postulate regarding stimulus equivalence

[14] K. Sayre, *Recognition: A Study in Artificial Intelligence* (Notre Dame, Ind.: Uni-versity of Notre Dame Press, 1965).

or stimulus-generalization gradients, is that of *formulating*, in the descriptive language which we employ to characterize the stimulus side, *what the common property of the stimulus inputs which belong to a stimulus-equivalent class must be*. Or, speaking not in terms of strict stimulus equivalence but rather in terms of the stimulus-generalization gradient, the problem is one of formulating the relevant features of the physical dimensions which constitute the input variables with respect to which the generalization gradient is to be plotted as a hyper-surface in a stimulus hyper-space. To avoid the mathematical complexities involved here, we shall simplify further by speaking of stimulus equivalence rather than stimulus-generalization gradients, i.e., we shall dichotomize the syllogism inputs into illicit and licit distributions. We shall also neglect the fact that even for a logician there might be certain formal presentations which would be more "seductive" in leading him to misclassify the syllogism as valid when it is actually fallacious.[15] Our problem then becomes, how do we characterize the stimulus input in a molar-psychological formulation of the stimulus side, of a verbal habit whose response side consists of the tokening 'Illicit Major'?

Now it is a truism, routinely pointed out to students in an elementary logic course, that the fallacious character of such a syllogism is revealed by its form alone, so that one can identify an Illicit Major even if the terms (other than the logical constants) are terms whose meaning is not known to the classifier. For that matter they could be neologisms which have no meaning, in anybody's natural or artificial language. So that when we present to the logician a syllogism which says, "No glops are klunks; all klunks are fabs; ergo, no glops are fabs," we will be perfectly confident that he can respond with the tokening of 'Illicit Major' in spite of the fact that the terms 'glop,' 'klunk,' and 'fab' are novel to him. And we have this confidence because we know that, for a logician, the defining property of an Illicit Major is its possession of a certain syntactical form, i.e., all syllogisms of this form are stimulus-equivalent to him as determiners of the verbal response 'Illicit Major.' The possession of this syntactical form is both a necessary and a sufficient condition for the logician to token the fallacy's name.

---

[15] R. S. Woodworth and S. B. Sells, "An Atmosphere Effect in Formal Syllogistic Reasoning," *Journal of Experimental Psychology*, 18 (1935), 451–460; for a summary of this and related studies, see generally R. S. Woodworth, *Experimental Psychology* (New York: Holt, 1938), pp. 810–817.

It might be argued by the staunch behaviorist that he can describe the syllogism-input stimulus class without making use of the logician's concept *Illicit Major*. Now no one wants to maintain that the behaviorist must employ the logician's terminology, i.e., he need not employ the actual expression 'Illicit Major' to mediate his prediction. But does this get around the behaviorist's difficulty? I do not think it does. After all, the metalinguistic expression 'Illicit Major' is introduced by the logician through explicit definition and, therefore, is in principle eliminable from *his* discourse as well. But the point is that when the behaviorist attempts to really deliver the goods on his claim to be able to characterize the stimulus side of the logician's disposition, he will find himself unavoidably driven to set forth the formal (syntactical) characteristics of the adequate stimulus class; and these characteristics will (if the logician is really a logical logician!) be identical with the defining syntactical property which the logician expresses by the shorthand phrase 'Illicit Major.'

It would therefore seem more honest for the behaviorist to admit from the start that he employs the syntactical category referred to by the metalinguistic expression 'Illicit Major,' and that being the case, he might just as well include the logician's phrase in his scientific vocabulary and be prepared to utilize it in object-language derivations.

Should this distress him? I think not. Consider an analogous infrahuman example. An animal psychologist is studying form discrimination in the monkey, say in a Skinner-box situation, where the discriminative stimulus is the presentation of a visual pattern on an illuminated screen and the response alternatives are to depress one of two levers. During the training phase of the experiment, the monkey is reinforced if he presses the right-hand lever in the presence of an isosceles triangle composed of three straight lines, and he is also reinforced for pressing the left-hand lever if the visual stimulus is a single circle of approximately the same size as the triangle. After this discriminative control is thoroughly established, the experimenter then presents the monkey with a visual pattern consisting of three small circles in a triangular arrangement and lacking symmetry of placement such that the circle which constitutes the top vertex is displaced leftward from the midline between the other two. So the novel visual stimulus on test trials consists not of an isosceles triangle formed out of three visible straight-line segments, as in training trials, but rather of the three vertices only, arranged so that they define a scalene triangle, and the vertices are circles which, although of smaller size, are, as circles, members of the same geometrical class as the original training stimulus for pressing the *left*-hand lever. It is likely that if these circles were made very large and close to collinear, the monkey would respond to them as approximately stimulus-equivalent to the original circle; if they are made very small, almost points, he may or may not respond; and with a considerable departure from collinearity he will (one hopes) respond to them as triangular, in spite of the fact that the physical lines connecting the vertices of this triangle are missing.

Suppose the psychologist, by trying various combinations after such original training, finds that the probability or strength of the response disposition to the right versus left lever depends in a complicated way upon (a) the absolute size of the original negative circle, (b) the absolute sizes of the test circles used as vertices, (c) the ratio between these circular areas, (d) the degree of departure from collinearity of the three points in the test trial, (e) the distances between the vertices in relation to the angles of the test triangle, and so forth. Let us imagine that (whether on theoretical grounds or by a blind, curve-fitting process) the investigator succeeds in constructing a complicated configural function which relates the response strengths of the two lever-pressing operants to these geometrical features. That is, he writes response-strength equations or probability equations of the type $P_R = F (L_1, L_2, R_1, R_2, \theta)$ and $P_L = G (L_1, L_2, R_1, R_2, \theta)$. Obviously, in explaining what these variables are, our psychologist employs concepts of analytic geometry and trigonometry, i.e., he has to explain that the variable θ which appears in these response-strength functions refers, say, to the smaller of the two angles between the lines connecting the vertices of the circles on the test trial, and so forth. Thus he employs, in his descriptive discourse characterizing the stimulus side, an interpreted formalism, i.e., physical geometry.

Now suppose either a hard-nosed behaviorist or an anti-behaviorist philosopher with intent to gore the behaviorist's ox were to object to this procedure by saying, "But, my dear fellow, you said that you were a behaviorist; that is to say, you alleged that stimuli and responses, which are mere physical energy inputs and effector events, would constitute your subject matter. Now I find you forced to employ a set of nonbehavioral concepts—namely, those of geometry. Furthermore, you do not employ them merely in the sense that you use your knowledge of geometry in designing the apparatus or what not. No, you employ them in a *substantive* way—that is to say, you use concepts from a nonbehavioral formal discipline as an essential part of the language with which you characterize the monkey's stimulus input. This seems to me to be contradictory to your expressed behaviorist aim."

I cannot imagine anyone voicing this complaint, and if anyone did, I cannot imagine any behaviorist taking the objection seriously. Of course he must at times employ mathematical formalism in other ways than as transformation rules in making theoretical derivations. Physical objects exemplify formal properties, and these properties are behaviorally relevant. It is just not possible to characterize certain stimulus inputs if one is precluded from employing the language of geometry. One way of viewing this is that we generally take the physical language, whether the ordinary physical-thing language or the theoretical language of physical science, as including certain portions of the languages of formal disciplines. That is, we do not forbid the physicist to write a Riemann integral, or the descriptive statistician to write down the expression for the gamma function, on the grounds that physics and descriptive statistics are supposed to deal with physical things which are "in the world," arguing that hence these sciences may not employ abstract or formal categories such as those found in mathematics or in an uninterpreted calculus. We are less accustomed to think of the formal features of a printed syllogism as a kind of "geometrical configuration," although Carnap made the point explicitly in his great work of 1934:

Pure syntax is thus wholly analytic, and is nothing more than combinatorial analysis, or, in other words, the geometry of finite, discrete, serial structures of a particular kind. Descriptive syntax is related to pure syntax as physical geometry to pure mathematical geometry; it is concerned with the syntactical properties and relations of empirically given expressions (for example, with the sentences of a particular book).[16]

It is just as possible to construct sentences about the forms of linguistic expressions, and therefore about sentences, as it is to construct sentences about the geometrical forms of geometrical structures. In the first place, there are the analytic sentences of pure syntax, which can be applied to the forms and relations of form of linguistic expressions (analogous to the analytic sentences of arithmetical geometry, which can be applied to the relations of form of the abstract geometrical structures); and in the second place, the synthetic physical sentences of descriptive syntax, which are concerned with the forms of the linguistic expressions as physical structures

[16] R. Carnap, Logical Syntax of Language, trans. Amethe Smeaton (New York: Humanities, 1937), p. 7.

(analogous to the synthetic empirical sentences of physical geometry, see ∫25). *Thus syntax is exactly formulable in the same way as geometry is.*[17]

The sentences of syntax are in part sentences of arithmetic, and in part sentences of physics, and they are only called syntactical because they are concerned with linguistic constructions, or, more specifically, with their formal structure. Syntax, pure and descriptive, is nothing more than the mathematics and physics of language.[18]

The point is made repeatedly and with beautiful clarity in several papers by Wilfrid Sellars, although I shall content myself with only two brief quotations from his early (and insufficiently noticed) "Pure Pragmatics and Epistemology." Discussing the necessity of a pure pragmatics that avoids the philosopher's sin of psychologism, Sellars writes:

Today, then, the analytic philosopher establishes his right to attack psychologism with respect to a given concept if he is able to show that it is capable of treatment as a concept the nature and function of which is constituted by its role in rules definitive of a broader or narrower set of calculi. The issue was joined first over the concepts of formal logic and pure mathematics, and it can be said with confidence that the attack on factualistic and, in particular, psychological accounts of these concepts rest on solid ground. Logic and mathematics are not empirical sciences nor do they constitute branches of any empirical science. They are not inductive studies of symbol formation and transformation behavior. (And if, at a later stage in our argument, we shall find *formal* science dealing with language *facts*, it will not be because logic is discovered by a more subtle analysis to belong to empirical science after all, but rather because of a less naive analysis of the relation of language to fact.) This first battle was won because of the development of pure syntax. The concepts of formal logic and pure mathematics were clarified through being identified with concepts which occur in the formation and transformation rules definitive of calculi. These rules constitute a logic of implication and deducibility. In this stage of the battle against psychologism, an apparently clear-cut distinction arose between *symbol-behavior* and *formal system*, a distinction sometimes summed up as that between *inference as fact* and *deducibility as norm*.[19]

And later in the same article he says:

On the other hand, if we are asked, "Isn't it absurd to say that syntactical properties do not apply to symbol behavior?", we should find it extremely difficult not to agree. How, indeed, can we characterize an *infer-*

[17] *Ibid.*, pp. 282–283.
[18] *Ibid.*, p. 284.
[19] W. Sellars, "Pure Pragmatics and Epistemology," *Philosophy of Science*, 14 (1947), 181–182.

ence, for example, as valid, unless it makes sense to attribute syntactical properties to symbol-behavior in the world of fact? If we say that syntactical properties belong in the first instance to expressions in a calculus or language which is a model or norm for symbol behavior, can we then go on to say that in the second instance they belong to language as behavioral fact? But to say this would be to put metalinguistic predicates into the object-language. Is there, then, no way out of our dilemma? Must we hold either that syntactical predicates are object-language predicates, or that syntactical predicates are not applicable to language as behavioral fact? Perhaps we can find a way out by drawing a distinction between language as behavior (that is, as the subject-matter of empirical psychology), and language behavior *to the extent that it conforms, and as conforming, to the criteria of language as norm;* or, in the terminology we shall adopt, between language behavior qua behavioral fact, and language-behavior qua tokens of language as type.[20]

A difficult question which arises in connection with microanalyses of systems that perform logical and mathematical operations is the following: Suppose we deal with such a system, one which is clocklike rather than cloudlike, and we present a detailed causal analysis of the workings of the mechanism, including of course those *structural* and *configural* characteristics of the machine by virtue of which it "mirrors" or "embodies" logical and mathematical rules. If we do this microanalytic job adequately, it seems that we have performed the task of causal analysis, and yet we seem to have *left something out of our account,* namely, that which the mechanism is "achieving" or "doing." This puzzle arises in the philosophical analysis of conduct at least as far back as Plato (in the *Phaedo*) and continues to bother us today.

It seems not to be a mere matter of omitting adequate description of how the parts are arranged. It is obvious that one cannot be said to "describe" an ordinary desk calculator if he merely *lists* the parts, as such-and-

---

[20] *Ibid.,* pp. 184–185. No philosopher or psychologist concerned with the "rules-and-facts" problems of semiotic can afford to leave Sellars's contributions unread or unstudied. See especially his "A Semantical Solution of the Mind-Body Problem," *Methodos,* 5 (1953), 45–85, and "Intentionality and the Mental [correspondence with Professor Roderick Chisholm]," in Feigl, Scriven, and Maxwell, eds., *Minnesota Studies in the Philosophy of Science,* vol. II, Appendix, pp. 507–539. See also "Epistemology and the New Way of Words," *Journal of Philosophy,* 44 (1947), 645–660; "Realism and the New Way of Words," *Philosophy and Phenomenological Research,* 8 (1948), 601–634; "Mind, Meaning, and Behavior," *Philosophical Studies,* 3 (1952), 83–94; "Some Reflections on Language Games," *Philosophy of Science,* 21 (1954), 204–228; "Empiricism and the Philosophy of Mind," in Feigl and Scriven, eds., *Minnesota Studies in the Philosophy of Science,* vol. I; and "Empiricism and Abstract Entities," in P. A. Schilpp, ed., *The Philosophy of Rudolf Carnap* (LaSalle, Ill.: Open Court, 1964).

so gears, levers, cams, cogwheels, and the like, even if he also gives a description of how a gear "works" (i.e., how it acts upon another gear in terms of the laws of mechanics) but omits to specify how the gears are physically arranged in the calculator. So, "calculating purpose" aside, it is clear that no one can claim to have provided a complete physical description of the machine if he leaves out an account of the arrangement of its parts, "how it is all put together." Let us suppose such a complete physical description to have been given. But let us suppose that the knower who carries out this "internal" analysis in terms of the principles of mechanics is a Martian visitor who uses a binary or duodecimal number system. (Or, even if he used ours, he might be unacquainted with the particular sign vehicles [= numerals] which we employ to designate the natural numbers.) That is to say, he has his own mathematical equipment (which is necessary for him to be able to solve the equations of mechanics involved in describing the inner workings of the machine); but he is not in possession of the rules of translation between his number system and ours, and therefore he might (conceivably) be forced to treat the Arabic numerals which are stamped on the keys, and which pop up in the register dials, as uninterpreted forms. If the calculator is structurally intact and functioning "properly," the rules of decimal arithmetic are perfectly embodied in the machine's structure, so that the operations of arithmetic are in perfect isomorphism with certain corresponding changes of state of the machine. The machine is—in the technical sense of the logician—a (physical) *model* of the interpreted calculus *arithmetic*. Thus, for example, punching the key marked '1' in the extreme right-hand column and then punching the key marked '+' is a physical operation sequence corresponding to the abstract specification "taking the successor of an integer." It is evident that the Martian *could*, in principle, possess a mathematics adequate for a science of mechanics that would provide a "complete causal analysis" of the functioning of the machine, and *not* thereby (necessarily) understand the correspondence between the machine's structure (and structure-determined functional properties) on the one hand, and the Earthlings' numerical system on the other.

There is a sense in which, when the Martian has given his structural and functional analysis of the workings of the mechanism, he has "said everything that can be said," in the sense that nothing is "left out" of the causal analysis. But there is another sense, which is equally important, in which the Martian has *not* "said everything that can be said" about the proper-

341

ties of the machine, because he has not said that the machine "does arithmetic," or, less teleologically, that the machine's wheel movements constitute a model of a decimal arithmetic (= the wheel movements and positions satisfy the postulates of arithmetic). It is partly a matter of semantic convention how we choose to employ the locution 'saying everything that can be said.' But it is not purely conventional that one can distinguish between the following two kinds of *text* (I speak of 'text' because I want to emphasize that the following distinction is not a distinction of "mere descriptive pragmatics"):

1. A text is stated which consists of a conjunction of sentences exhaustively descriptive of the physical structure-dependent properties of the machine, and which suffice to entail all true statements about the machine's dispositions.

2. A conjunction of sentences (1) is stated, *together with all theorems which flow as consequences of the conjunction (1) given certain definitions.*

Now what *is* conventional or stipulative about the locution 'saying everything that can be said' is, of course, the possibility of stipulating that an individual who asserts the postulates also implicitly asserts the theorems. If anyone wishes to adopt this locution for certain purposes of logical analysis, I shall not complain of it. The fact remains that a text may contain the postulates without the theorems, or it may contain both the postulates and the theorems. And it is a familiar truth that while one in some sense "implicitly holds" the consequences of his postulates, in the sense that if he is consistent and rational he *ought* to believe the theorems that follow from them, the limitations of the finite intellect are such that we often do not hold all of the theorems which flow from our postulates because, for example, nobody has as yet succeeded in showing whether a certain well-formed formula is a theorem, or a counter-theorem, or even whether it is decidable. E.g., we do not know whether Fermat's Last Theorem is true or false; and we know that no one has as yet presented a valid proof of it, or a proof of its undecidability; so that it is somewhat misleading to say that we "believe it" or "hold it" or "know it to be true," supposing that Omniscient Jones knows it to be true, i.e., to be a consequence of the postulates of arithmetic. It is not, I think, an excessive reliance upon the usages of vulgar speech to ask that metalanguage stipulations avoid needlessly paradoxical consequences, such as that I am bound to hold that my late grandmother believed the number of her noses to be $-e^{\pi i}$.

Now it might be said that whereas the Martian would lack (better, could lack) our semantics for interpreting the numerical sign vehicles that pop up in the dials of the desk calculator, and if he were a particularly rigid or stupid Martian he might not develop insight into the translatability of the physical properties of the machine into a number system which was in turn transformable into his own, he could, nevertheless "predict everything about the machine's behavior," because we have just assumed that he gives a complete mechanical-causal analysis of its micro-structural (and, as a consequence, micro-function) properties. And it seems evident that one who understands everything that happens in the causal order about any mechanism ought to be able to forecast—since we are assuming that the machine is completely clocklike and has no cloudlike "slippage" in its gears—all its dispositions. However, I believe there is an important sense in which even this is not quite true, unless stated very carefully and with all the necessary qualifications. The "results" of performing certain "operations" with the machine, definable in terms of what sorts of physical sign patterns pop up in the "answer" (cumulative bank) register are, after all, among the dispositional properties of the machine. And some of the strict uniformities (and statistical generalizations) in these "results" cannot, oddly enough, be predicted by a knower who has not made the cognitive identification of certain functional consequences of the machine's micro-structural features with the abstract concepts of arithmetic.

Consider the following example: A set of instructions is provided for carrying out division operations, and for recording their results, such that one obtains a sequence of "outcomes" (semantically uninterpreted by the Martian) that in fact constitute successive answers to the question "Is this integer prime?" We also have the Martian concurrently performing the task of keeping track of the proportion of such outcomes that have cumulated up to the nth integer, although of course he doesn't know that is what he is doing. Finally, we assume that the Martian knows logarithms (or at least that we can instruct him in the sheer mechanics of entering a logarithm table). Then it can be asked whether the proportion of outcomes of the "prime type" accumulated up to any point in this sequence of operations exceeds the reciprocal of the natural logarithm (i.e., the Martian is, so to speak, "empirically" examining Gauss's law of the density of distribution of primes). We now ask the Martian to predict, from his complete causal understanding of the machinery, how far along in the sequence he will have to go before he can be certain that the cumulated proportion of outcomes of the "prime" type will at some point have exceeded the Gauss approximation, instead of falling on the low side as it will at first. Now this number, Skewe's number, is

$$10^{10^{10^{34}}}$$

which is believed to be larger than the cardinal number of all of the atoms in the universe. So of course the Martian will never reach this value, before which—we don't know how much before—the prime proportion "flips over" so that Gauss's asymptotic formula errs on the high side of the actual value. We Earthlings, who *have* made the coordination—or for that matter an Earthling who *knows nothing about the machinery but only knows that the calculator "does arithmetic"*—can correctly state a lower bound for the number of such consecutive operations necessary to achieve this proportion of outcomes; whereas the Martian, or anyone else who has not made the coordination between the machine's structural properties and the axioms of arithmetic, could not make such a prediction. It seems obvious that this constitutes in some sense a genuine "cognitive edge," and that it is therefore false, or at least very misleading, to say that one who has described the internal mechanical structure, but has not made the explicit identification of the machine's states and operations with arithmetical concepts, would have "said everything that can be said." He would have said something which, given the appropriate explicit definitions and interpretations, *suffices to derive everything* that can be said, given the further assumption that he is an omniscient mathematician who is able to derive all the theorems that validly flow from the arithmetical postulates embodied in the structure of the machine. But if you don't say something that can be said, it is misleading to characterize your description as having said everything that can be said, even if what you have said is capable of entailing everything which can be said.

Nevertheless, recognizing this fact does not force us to postulate a "something more" going on *causally* in the machine, i.e., we do not infer from this that there is some sort of an additional arithmetic spook at work which sees to it that the calculator "obeys the postulates of arithmetic." Whether or not any such additional causal entity needs to be invoked depends upon whether our analysis of the situation amounts to a projection or a reduction, in Reichenbach's sense.[21] A desk calculator, an electronic

[21] H. Reichenbach, *Experience and Prediction* (Chicago: University of Chicago Press, 1938), pp. 110–114.

computer, or—if physicalism be true—a human brain is a reductive complex of its elements. Nevertheless we have to insist that even in the case of a reductive complex, there is an important sense in which one may not have said everything that can be said about the complex, even though he may have said everything about the elements, and have included certain ways of stating everything about their relations, such that everything about the reductive complex follows of necessity from the statements which he has formulated. Thus if I recognize that a wall is a reductive complex of the bricks, and then I give the bricks numbers 1, 2, 3, and so forth, and state that Brick #1 is adjacent to Brick #2 and Brick #3 is located immediately above the first two and symmetrical with respect to them, and so forth, the vast conjunction of all true statements of this kind entails a "molar" statement about the wall. (They are not equivalent, since, as Reichenbach points out, the molar statement about the wall, while entailed by this conjunction of statements about the bricks, does not entail this conjunction; because the same molar statement about the wall is also a consequence of alternative conjunctions about the bricks.) What I have said about Bricks #1, #2, #3, and so forth may entail the "molar" statement that the wall is 50 feet high; but if I do not make this latter statement, it is misleading to say, at least for certain purposes and in certain contexts, that I have "said everything that can be said." And it is at least theoretically possible for an individual to have a "complete understanding" of each of the statements about the elements and, depending upon the complexities of the structure, not to be (psychologically) able to derive a molar consequence that validly flows from these statements.

This is not an appropriate place, even if I had the technical competence, to enter upon a detailed consideration of the formal or structural relationships that must obtain between a physical system and a specified molar means-end process in order for the system to be capable of performing the specified process. And I certainly do not mean to suggest that Professor Popper is unfamiliar with the problems in this area. I rather imagine he knows more about them than I do. Nevertheless, I must point out that his paper reads as if he believed a proposition which I am confident that he does not believe, to wit, "If a predicate 'P' designating a property P does not appear in a language adequate to describe a sequence of events related by causality, those events being considered at a certain level of analysis, it follows that the property P cannot, without inconsistency, be predicated of the system as a whole, or at another level of analysis." I do not myself know of any compelling reason for holding such a meta-proposition; and it is pretty clear that adopting it would generate some difficult (and, as I think, needless) puzzles. Example: Suppose we are talking political science, it would be a major lacuna in any characterization of Dwight Eisenhower to omit the statement that Eisenhower was a Republican. But even the most consistent identity theorist would consider it a category mistake to predicate of one of Mr. Eisenhower's cerebral neurons that the neuron was Republican. Must we say that since none of Eisenhower's neurons was Republican, therefore Eisenhower could not be such? Or, at the molar level, since none of Eisenhower's letter-forming actones[22] (e.g., engraphing the mark 'a') may be meaningfully characterized as Republican, hence he wrote no Republican-oriented manuscripts? If the activities of the living human brain were—as I do not assert—completely "clocklike"; or if they were largely clocklike but with a certain irreducible element of "cloudiness"; or if they were extremely "cloudy," with only a small "clocklike" element present; in any case, no description of the cerebral processes at the micro-level, formulated in neurophysiological language, will include the predicate 'Republican.' It does not seem to me that this point about the appropriateness of certain predicates being confined to what one may loosely call "the whole person and his molar acts" has any relation to the question where the human nervous system is located on the cloud-to-clock continuum.

To stay away from the technical complexities of modern computer theory, consider an ordinary Hollerith punch-card machine. We have a batch of cards in a military personnel unit which are encoded in a certain way, i.e., the row and column positions have been, by some physical procedure, set into correspondence with properties, whether simple physical ones or extremely complicated social ones, of the military personnel whom the cards "represent." Professor Popper need not fear that I am surreptitiously avoiding the problem by shifting it backward to the encoding process. On the one hand, certain aspects of the encoding process do not involve "intentional mental acts" on the part of any encoder; but of course some do, and the ones that do will present a problem of microanalysis in relation to molar analysis of the same kind we are here considering. I am not, I trust, arguing circularly that there is no difference between the human brain and a Hollerith machine, a view I would vigorously repudiate.

[22] H. A. Murray, *Explorations in Personality* (New York: Wiley, 1938), pp. 55–59, 96–101.

I am only saying that a Hollerith machine encodes information by a correspondence rule relating one set of properties, sometimes very complex ones, to another set, namely, a hole or non-hole at a specified locus on the card. That a human operator rationally intends the encoding process for his conscious purposes is true but irrelevant for my purpose at this point. Suppose that at induction a soldier fills in a response-box set opposite to a named occupation on a checklist. We cannot presume that in filling in the box the soldier *has* to call up thoughts or images of his occupational activity, although of course he *might* do so. The more usual situation would be that a man who in civilian life has acquired the skills involved in making bread, and who has a pre-induction history of being paid to do this, will also be an individual who has acquired the verbal disposition to token 'baker' in response to the question 'What is your occupation?' Notice that there is no necessary overlap in the physical subsystems of the soldier's brain involved in these vocational activities and in his self-descriptive tokenings, nor does the self-descriptive tokening of 'baker' necessarily involve any concurrent or antecedent tokenings designating the activities of a baker. These dispositions are correlated in the English-speaking population by virtue of what R. B. Cattell calls "an environmental mould," that is to say, a cluster of topographically dissimilar dispositions which go together in a given culture or subculture by virtue of the fact that any human organism that learns the one will also have learned the other.[23] The correlation is similar to that which exists between a person's motoric skills in making an incision into living flesh and his disposition to respond verbally to the question "What is a Billroth II?" There is negligible overlap between these dispositions, either on their stimulus side or on their response side; and there is nothing about either one of them which suggests any appreciable overlap in the functioning of the cerebral machinery. The fact remains that they would be very highly correlated, because anyone who possesses certain incision-making skills at a given level of proficiency is certain to be a surgeon, and surgeons in the course of their training also learn to state verbally what a Billroth II is.

To return to our inductee-baker, he fills in a square box on an occupational checklist which the machine further encodes by punching a hole in a certain position on the card representing this soldier. Ditto for his height,

[23] R. B. Cattell, *Description and Measurement of Personality* (Yonkers-on-Hudson: World, 1946), pp. 64–66, 74, 496; *Personality* (New York: McGraw-Hill, 1950), pp. 33–36.

weight, and eye color. (These can be coded mechanically, if desired.) Now suppose we want, for some strange reason, to select from a regiment all the enlisted men who are over six feet tall but weigh less than 180 pounds, who have blue eyes, and who were bakers in civilian life. The machine's board is wired accordingly, and the cards are run through the sorter ending up with a stack of cards in which are punched the serial numbers of all soldiers having this particular combination of properties. The functioning of this machine is very far toward the clocklike rather than the cloudlike end of Professor Popper's continuum. With a little care we can render it as clocklike as desired. Now suppose concerning any particular card which emerges from such a sorting process, we say, "Give me a detailed causal account of how this particular card happened to drop out during the sorting." This question can be answered in physical language describing the structures and processes of the machine without any reference to the vocational activities of the bakers, or to concepts of height, weight, and eye color. Nothing is left out of this causal account. If we start with the initial conditions on how the machine is wired, and how a batch of cards representing the entire regiment is punched, nothing *need* be said which cannot be completely expressed in terms of such concepts as brushes, electrical contacts, punched holes, the geometry of the coordinate positions at which holes are punched, and the like.

Now this causal account of the card-sorting operation, which *leaves nothing out* (in one perfectly legitimate sense of the phrase 'leaves nothing out'), does not preclude a sentence of the following kind being literally true: "The sorter is picking out the cards of blue-eyed bakers over six feet tall and weighing less than 180 pounds." This is a perfectly good account of what the sorter is "doing." This sentence employs concepts which did not appear at the lower level of analysis of the machine's inner workings. Furthermore, it employs concepts which are *not translatable into the minimum vocabulary adequate to give an account of the machine's mechanical and electrical workings.* Is there any puzzle here? If so, it is resolved by recognizing the role of the previous encoding process, in which certain complex properties possessed by the soldier were set into a certain correspondence with loci punched on the cards. And corresponding to the fact that the cards can be repeatedly sorted is the logical particle 'and' joining the predicates 'blue-eyed,' 'baker,' 'weight less than 180 pounds,' and 'height over six feet.'

In order for me to be capable of making rational inferences or having intentions, it is necessary that my brain have a certain kind of structure. There are alternative physical arrangements that are equally capable of providing this necessary structure, the human brain being one of them. As long as the brain is capable of some kind of consistent encoding procedure, it can "represent" external facts, such as someone's being a baker, by nervous connections which, *when examined at their own level of analysis,* do not partake, however faintly, of "bakerhood." And even if the sequence of activation of individual nerve-cell dispositions were completely clocklike, this would not show, or even tend to show, that our beliefs, intentions, or volitions find no place in the world or that they have no causal efficacy.

It may be illuminating at this point to reexamine a famous puzzle about intentionality propounded by Sir Arthur Eddington.[24] Suppose a man from Mars arrives on the earth mysteriously possessed of such a Utopian knowledge of Earthling neurophysiology that he is able, by a combination of behavioral and microtechniques (such as single-unit stimulation and the like), to give a complete causal account, *in neurophysiological terms,* of all the activities and dispositions of any given member of *Homo sapiens.* In particular, he observes (and was able to predict) that on November 11, 1918, great numbers of people in many cities of the world stand about in the public square waving their arms and shouting. Now, says Eddington, there is apparently "nothing left out" of this causal account; and yet the Martian would not know the most important thing there is to know about this social occurrence—namely, that these people are celebrating the armistice. This is true, if the Martian confines his attention to the momentary activities, but it is false if he allows himself to consider dispositions as well. There is surely no reason for saying, given Eddington's own assumption of a Utopian state of Martian neurophysiology, that the Martian is forbidden to include the dispositions of nerve cells in his description of the state of affairs. If these micro-dispositions are included, I contend that Eddington is incorrect in saying that the Martian would not understand the "meaning" of the celebration. The reason is very simple: *Given a complete micro-account of the neural dispositions, one possesses all of the information necessary to construct a descriptive semantics.* He would, for example, know *that* (and he will also know *why*) persons waving their arms and

<hr>

[24] A. S. Eddington, *Science and the Unseen World* (New York: Macmillan, 1929). The book is inaccessible to me at this writing so that while the basic puzzle is due to Eddington, the formulation of conditions is mine and may not accord precisely with his original setup.

shouting in the public square would be disposed to reply, if asked why they are carrying on in this crazy way, "The war is over." And, of course, his complete catalogue of neuronal dispositions would locate the word "war" in the descriptive semantic space, i.e., he would know what the word 'war' *means* to English-speaking human beings. Since we know that one can learn a language by recording the molar dispositions of its speakers (as, for example, an explorer or missionary *must* be able to do when he is the first visitor to a tribe of aborigines), a fortiori one would know the language if he knows all the micro-dispositions. Because, of course, the micro-dispositions entail the molar dispositions, but not conversely; and not all molar dispositions are realized in any finite behavior sample. Now it is perfectly true that the Martian is not *forced* to carry out any such descriptive-semantic research. He may, if he is only interested in neurophysiology, confine his explanations, predictions, and concepts to the micro-level. Whether such a confining to the micro-level "leaves something out" (in the causal sense) depends upon how far back in the causal chain it is desired to analyze the celebration. The immediate causal ancestor of a man's standing in Times Square and shouting would be, say, his looking at a newspaper headline "War is over." But the remoter causal ancestor is a complex of behavioral events at Compiègne, which the Martian would describe by a phrase in his language that is approximately synonymous with the English expression 'agreement to a cessation of hostilities.'

Can anything philosophically important about the mind-body problem, or the cloud-clock problem, be inferred from the fact that *if* physicalism is assumed true, the Martian *need not* pursue the causal chain that far backward but, on the other hand, that he *can* do so; or from the fact that he can predict "armistice behavior" successfully without tracing the causal chain back, confining himself to the momentary brain-cell dispositions; or from the fact that he could even infer the "meaning" of the celebration? I think not. These considerations do help to illuminate matters somewhat. Thus, for example, we are thereby reminded of the distinction between a statistical regularity of descriptive semantics (inferable by the Martian from nerve-cell dispositions) and the *nonpsychological* concept of a *semantic rule*, which is not a behavioral regularity but a prescription that the Martian formulates in his own Martian metalanguage.[25] A Utopian knowledge of the nerve-cell dispositions would be a Utopian knowledge of de-

---

[25] See Sellars, publications cited in footnote 20 *supra*.

scriptive semantics, and a Utopian knowledge of descriptive semantics, together with a Utopian knowledge of the tokening dispositions of the celebrators, would obviously inform the Martian—*assuming he himself has the level of abstraction equipment which Eddington must presuppose in order for him to carry out the hypothesized microanalysis*—what the "content" of the celebration is all "about." All of which aids in dissipating the paradoxical flavor of the situation, but cannot help us to decide whether (a) physicalism and (b) determinism are true doctrines about mind.

Professor Popper is disturbed by the notion that a "clockwork" view of the human mind implies that the behavior scientist would be able to "write" the symphonies of Beethoven through his knowledge of Beethoven's physical states, even though the scientist himself were completely ignorant of musical theory.[26] Why does this distress him? He says "all this is absurd." But is it really absurd? I will go him one better (partly to test the limits of my own convictions in the matter!). Consider the following example. Suppose a Utopian neurophysiologist studies the brain of a mathematician who is currently working on Fermat's Last Theorem. We will assume that this neurophysiologist knows the kind and amount of applied mathematics he needs to carry on ordinary calculations upon physiological measures, but that he is completely ignorant of pure mathematics, including number theory. Thus we assume that he has never even heard the phrase 'Fermat's Last Theorem,' let alone understands what it designates. Let us further suppose (with Professor Popper) that the cerebral mechanism has certain clocklike features but others that are cloudlike. In particular, let us suppose that there occur occasions on which the strengths of the neuronal activity in two systems of cell assemblies "competing" for command of the output channel are so close that a difference in only a few critically located "trigger" neurons firing or not will show up as a molar output difference. It is irrelevant for our present purposes whether at another level of analysis—say, by the physical chemist—this cloudlike feature arises from the quasi-random character of distributions of initial conditions of intra-neuron particles whose individual chemical and physical transitions are, nevertheless, completely clocklike; or whether it arises from a fundamentally indeterministic feature of nerve-cell action quantum-theoretical in nature, as has been postulated by some physicists and neurophysiologists.[27] In either case, the point is that a randomizing component

[26] Popper, "Of Clouds and Clocks," p. 11.
[27] E.g., N. Bohr, *Atomic Theory and the Description of Nature* (New York: Mac-

is built into the functioning of our mathematician's cerebral system, super-imposed upon the clocklike features that are involved in his being thoroughly trained in mathematical manipulation (so that he always treats an exponent differently from a base, and the like). Suppose our Utopian neurophysiologist, ignorant of number theory, is able to show *at the micro-level* that there exists a set of alternative tokening dispositions, each of which is itself a chain of subdispositions to perform particular mathematical operations. That is, our neurophysiologist sees that the mathematician is "capable of" (= has non-zero probability of emitting) several alternative work-product sequences on a given day. Within each of these alternative chains, there are points at which the cerebral machinery functions clocklike, and there are other points at which it functions cloudlike. (And even if it functioned clocklike at all points in the chain, the cloudlike selection of the *initial member* of a chain is unpredictable by the neurophysiologist.) So he doesn't know *which* of the chains will actually take place, but he can list all the physically possible alternatives. And if his psychophysiology is truly Utopian he can associate probabilities with each of

millan, 1934); J. C. Eccles, "Hypotheses Relating to the Brain-Mind Problem," *Nature*, 68 (1951), 53–57; J. C. Eccles, *The Neurophysiological Basis of Mind* (Oxford: Oxford University Press, 1953), pp. 271–286; A. S. Eddington, *The Philosophy of Physical Science* (Cambridge: Cambridge University Press, 1939), pp. 179–184; *The Nature of the Physical World* (New York: Macmillan, 1929), pp. 310–315; *New Pathways in Science* (New York: Macmillan, 1935), pp. 86–91; P. Jordan, *Science and the Course of History* (New Haven, Conn.: Yale University Press, 1955), pp. 108–113; I. D. London, "Quantum Biology and Psychology," *Journal of General Psychology*, 46 (1952), 123–149; P. E. Meehl, "Determinism and Related Problems," chapter VIII, especially pp. 190–191, and footnotes 30, 31, pp. 213–215, and Appendix E, "Indeterminacy and Teleological Constraints," pp. 328–338, in P. E. Meehl et al., *What, Then, Is Man?* (St. Louis: Concordia, 1958; while I no longer hold the theological position presupposed in that discussion, the treatment of determinism and speculative brain processes still appears to me as essentially defensible); Meehl, "The Compleat Autocerebroscopist," pp. 122–124; M. H. Pirenne and F. H. C. Marriott, "The Quantum Theory of Light and the Psychophysiology of Vision," in S. Koch, ed., *Psychology, a Study of a Science*, vol. 1: *Sensory, Perceptual and Physiological Formulations* (New York: McGraw-Hill, 1959), pp. 288–361; F. Ratliff, "Some Interrelations among Physics, Physiology, and Psychology in the Study of Vision," especially pp. 442–445, in S. Koch, ed., *Psychology, a Study of a Science*, vol. 4: *Biologically Oriented Fields* (New York: McGraw-Hill, 1962). For criticism of the notion that quantum-indeterminacy at the single-unit micro-level could be relevant to psychological determinism at the level of molar behavior or experience, see E. Schroedinger, *Science and Humanism* (Cambridge: Cambridge University Press, 1951), pp. 58–64; A. Grünbaum, "Causality and the Science of Human Behavior," in H. Feigl and M. Brodbeck, eds., *Readings in the Philosophy of Science* (New York: Appleton-Century-Crofts, 1953); Lizzie S. Stebbing, "Causality and Human Freedom," in her *Philosophy and the Physicists* (London: Methuen, 1937), pp. 141–242; and section X, pp. 13–14, of Professor Popper's lecture.

these alternatives. (It is perhaps better to assume that the Utopian neurophysiologist is a considerably evolved species as respects his brain, studying a mathematician of *Homo sapiens*. Otherwise there may be information-theoretical difficulties involved in $Brain_1$ carrying out the requisite microanalysis of $Brain_2$.[28] These can presumably be avoided by setting no time limit on the neurophysiologist's derivation, so that he may continue work for months or years after the mathematician has quit. Or we may assume breakthroughs in computer engineering permitting superhuman computer brains. Or we may substitute "quasi-Omniscient Jones," who represses number theory, for the physiologist.)

Now, his microanalysis of each chain will obviously enable him to characterize the motor output—that is, the effector movements of the muscles of the mathematician's hand; and, consequently, from skeletal structure and biomechanics he knows what each virtual sequence of sign designs will be, i.e., *he knows what mathematical expressions the mathematician would write down*, if he carried out a given ("possible") cerebral sequence. Viewed thus, as the mere graphical residues of a molar class of finger movements (Neurath's "mounds of ink"), the potential work product might be devoid of meaning to our physiologist, yet its potential occurrence would be derivable by him from the Utopian microanalysis. So our Utopian neurophysiologist is able to list a set of mutually exclusive and exhaustive "behavior outcomes," namely, all the mathematician's potential work products for the day, although he does not know which one will actually take place but has only the probabilities associated with each. Finally, let us suppose that one of these "possible work products" is a valid proof of Fermat's Last Theorem. But, regrettably, it is a sequence having (for this particular mathematician) an extremely low probability; and it is not the sequence which in fact eventuates on the given day. ("The potential proof remains unactualized.") Having worked on the problem for several weeks, our unlucky mathematician becomes discouraged, and thereafter pursues other interests.

Now this mathematician was in some sense "capable of" a proof of Fermat's Last Theorem (assuming for the moment that a valid proof of this theorem does exist) but he in fact never discovers it. However, the neurophysiologist has now before him a list of alternative potential work products, only one of which ever came into being, and that actualized one is

---

[28] J. R. Platt, *The Step to Man* (New York: Wiley, 1966), pp. 147–149. The point has been made by several writers.

not a valid proof. The psychophysiologist takes the whole stack of hypothetical work products (each of which is directly derivable as a consequence of the effector movements terminating a chain of CNS events) to the Department of Mathematics. I remind you that the neurophysiologist doesn't know anything about number theory. He doesn't "understand what the mathematician is working on." Yet, the low-probability valid proof, which was never actually carried out by the mathematician, would be recognized as a valid proof by the Department of Mathematics. Thus the neurophysiologist in some sense could "discover" a valid proof of Fermat's Last Theorem without understanding mathematics, by studying the brain of a mathematician who, while in some sense potentially capable of developing such a proof, never in fact does so. I readily agree that this sounds counterintuitive. But I do not see anything contradictory about it. And I think the reader will agree with me that it is interesting.

I have argued above that statements about human behavior or experience which attribute causal efficacy to reasons have a meaning which should be acceptable both to a philosopher-logician and to a determinist psychologist, but that such statements are elliptical so that unless carefully unpacked they are likely to be misleading. Thus I have said that, strictly speaking, a valid argument, considered as a certain formal structure (an abstract universal) is not an event "in the world," at least in any ordinary sense; and therefore it cannot function as a causal agent with respect to an event, e.g., a human locomotion, manipulation, or phonation. The ontology of universals, the reality of abstract entities, and the more technical aspects of the traditional nominalism-realism debate are not—it is hoped—relevant, because a serious discussion of them is, regrettably, beyond my competence. My colleague Professor Maxwell thinks I am wrong, or at least terminologically ill-advised, to say that logical relations (such as deducibility) are not "in the world." As he—rather compellingly—puts it, "*Everything* has got to be 'in the world'; where else *could* it be?" Professor Popper even writes recently of a "third world," whose denizens are abstract ideas.[29] I confess I do not understand Sir Karl here; but perhaps Professor Maxwell's demand is met by my agreement with Carnap and Sellars on linguistic structures; see footnotes 16–20 and associated text *supra*. In any event, by saying that there are Platonic universals but they

[29] K. R. Popper, "Epistemology without a Knowing Subject," in B. Van Rootselaar and J. F. Staal, eds., *Logic, Methodology and Philosophy of Science*, vol. III (Amsterdam: North-Holland, 1968), pp. 333–373.

are not in the world, I have made my position more difficult, and Professor Popper's easier, to maintain. I have argued that when a bit of rational behavior is being fitted into the causal framework, the question whether certain logical categories (such as the category Illicit Distribution) are required in formulating the behavioral laws or quasi-laws depends upon the level at which the behavior analysis is being conducted. If we are attempting to formulate psychological laws either in mentalistic language or in molar behaviorese we will find such formal categories indispensable, because we will be unable to characterize a stimulus class which functions as a discriminative stimulus for such verbal responses as 'Illicit Major' on the part of a logician-subject *unless that stimulus class is characterized by reference to its syntactical structure.* Whether the mentalistic or molar-behavioristic psychologist actually employs the logician's *terminology* or not is irrelevant, inasmuch as he will be driven, in his account of the subject's behavior or experience, to introduce a specification of the syntactical features of the stimulus input, which specification will in fact *be* the logician's definition of 'Illicit Major.' But I have also maintained that the same is not necessarily the case, given a complete Utopian micro-description, although the complete Utopian micro-description will *entail* (within the nomological network) the same syntactical statements at the molar level which would have to be invoked by the molar behaviorist in predicting or explaining rational behavior. This is because the whole organism and its molar activities are reductive complexes of the micro-structures and micro-events, and hence the statements about the whole person follow from the statements about his component parts and part processes, analogously to the way in which statements about a wall follow from conjunctions of statements about the bricks. But we have also seen that there is an important sense in which, unless one *asserts* these molar statements which are consequences of the statements about the elements, he has not literally "said everything that can be said" about such a reductive complex.

The anti-determinist or, perhaps more strongly, the ontological dualist may object to this analysis with the following: "You say that your refurbished behaviorism, including as it does a physical$_2$ microanalysis, and a recognition of the molar-indispensability of certain logical categories such as *valid form,* does justice to the logician's legitimate claims, while still maintaining physicalism and determinism in the domain of mental life. In this you attempt to please both parties; you want to have your cake and eat it too. I do not know whether the hard-nosed behaviorist will buy this, but I, as a firm believer in the genuine efficacy of reasons, cannot buy it. Because, while you tell me that the micro-account *entails* those statements at the molar level which are characterizations of stimulus inputs as logical forms, the fact remains that you also maintain the dispensability of concepts like *valid reason* in a complete causal analysis. Because while you admit that one who fails to assert some of the consequences of those statements which he does assert has failed (in a sense) 'to say everything that can be said,' nevertheless it remains true that you hold it possible to present a complete causal account of human actions without reasons entering the causal chain *as reasons*. That is, you maintain that it would in principle be possible, within a Utopian neurophysiology, to detail the processes in a person's brain confining oneself to physiological descriptions at the level of 'neuron-language,' such that the resulting molar output, e.g., punching somebody in the nose, could be predicted and completely understood causally at this level of analysis; and it is obvious that no reference to the good reasons he may have had for punching somebody in the nose would occur in such a micro-causal analysis. This is what I mean by insisting that you are depriving rules and reasons and validity of all genuine efficacy in human affairs. If you can give a complete causal account of what a person does and why he does it without at any point mentioning the reasons *for which* he does it, then it seems to me that you have, in effect, eliminated the reasons from any significant role. You throw a sop to me and my friends the indeterminists, emergentists, Cartesians, etc., by telling us that certain conjunctions of micro-statements entail molar-level statements which—given suitable metalanguage definitions of logical notions like 'implies' and 'negates'—in turn entail statements about a person's *reasons*, in our full sense of 'reasons.' But you also insist that one *need not do this* in giving the complete micro-causal account. You, so to speak, 'permit' us to mention reasons; but you insist that you yourself are not *compelled* to mention them. But, surely, if they need not be mentioned, they are dispensable. And this we cannot admit."

Since I myself admit—nay, I insist, as against a certain kind of behaviorist—that reasons are psychologically efficacious, i.e., that the hearing of reasons and the thinking of reasons and the tokening of valid arguments play a role, and for rational men may play the crucial or determinative role, in the guiding of their actions and utterances, it is the more obligatory upon me to answer this objection. It seems to me that the core issue here can, without prejudice, be formulated thus: Has one "dispensed with" the causal efficacy of a configural property of a physical state or system as playing a role in a causal explanation *whenever he avoids explicitly characterizing that configural property*, confining himself to description at a lower level of analysis ("lower" in the sense of a reductive complex), provided that (a) what he *does* assert in his lower level description can be shown to entail nomologically the configural statements and (b) if asked, he concedes—as he must in consistency—that these configural consequences are entailed by his lower level statements? Is there not a considerable element of conventional or stipulative usage involved here, about which it is pointless and fruitless to argue? One who describes a physical system omitting dispositional statements about it which flow as necessary consequences of the statements he has made, might be said, on one convention, to have "left something out," because he did not say everything that could, and (strictly speaking) everything that *must*, if the question is raised, be said. But so long as he is not inconsistent, so long as he is quite willing to *admit* the necessary consequences of what he has said, and those consequences of course *include* the entailed configural properties of the system, the locution "He thinks these configural properties are irrelevant to an adequate account" is surely misleading. In what sense can I be said to think that any feature of a physical system is "irrelevant," if I concede that this feature is a necessary consequence of features to which I have attributed relevance, and further, that if the system *lacked* these (entailed) configural features, its "output" characteristics (e.g., tokening an implied conclusion when one has tokened the premises) would be different from what they in fact are?

In assessing the conventional element in whether we would think it convenient and clarifying to say that such a scientist "leaves something out" or "considers something causally inefficacious," one consideration might be whether the physical system includes a subsystem which functions as a kind of controller, guider, evaluator, or selector, with respect to another subsystem, such that, among the intermediate or molar-level theorems that flow from the axioms of which the system is a model, there is a statement which says, roughly speaking, that the monitor or selector system will "accept" or "reject" a certain product or message from the monitored or controlled subsystem, depending upon whether that product or message possesses or fails to possess such-and-such formal properties. Thus, for example, when we program an electronic computer to perform certain computational checks upon its own work and to report to us the presence of inconsistencies; or when we program it to inform us that our own program is itself defectively written—in such cases it seems very natural, and not just a computer engineer's whimsy, to use the connective 'because' in sentences like the following: "The computer rejects these data because they include entries in a correlation matrix which exceed unity." It is true that in this kind of case it is also possible to describe and explain the operation of the monitoring or evaluating subsystem in micro-terms. But it remains true that we can identify two such functional subsystems, and we can correctly (and literally) say that the monitoring subsystem *classifies* the states and outputs of the monitored subsystem with regard to their possession or nonpossession of certain formal properties. It seems to me that one can arrange a continuously graded series of physical systems, each of which is a physical embodiment of certain formal rules (i.e., each of which is, in the logician's sense of the word, a "model" or provides an "admissible interpretation" of a formal calculus) from one extreme at which it would be a very marked departure from ordinary usage to employ the connective 'because' followed by a characterization in terms of validity or logical structure, to another extreme at which a failure to include this intermediate or molar-level characterization would be looked upon as some sort of prejudice or inadvertence. Take, for instance, the case of a beam balance, which we do not ordinarily think of as performing logical or mathematical operations. We place three one-gram weights in one pan, and we place two one-gram weights in the other pan, and we observe as a causal consequence of these physical operations that the beam becomes and remains nonhorizontal. It would not ordinarily occur to anyone to describe this state of affairs by saying, "The balance tips *because of a truth of arithmetic*, namely that $3 > 2$." But there is a perfectly legitimate sense in which such a statement would be literally correct. If there is a Platonic sense in which the truths of arithmetic are not "in the world," there is another sense in which they are, namely, that since these theorems are analytic, all physical objects do in fact exemplify them. (Cf. Wittgenstein, "It used to be said that God could create everything, except what was contrary to the laws of logic. The truth is, we could not say of an 'unlogical' world how it would look. To present in language anything which 'contradicts logic' is as impossible as in geometry to present by its coordinates a figure which contradicts the laws of space; or to give the coordinates of a point which does not exist. We could present spatially an atomic fact which contradicted the laws of physics, but not one which contradicted the laws of geometry."[30]) If we move along this continuum of "rule representation" from the beam balance (which "exemplifies," "instantiates," "physically embodies" the axioms of arithmetic, as well as the formalism expressing the laws of mechanics—the former necessarily, the latter contingently) to the ordinary desk calculator, it still seems somewhat inappropriate, but much less so, to characterize its operations by using the connective 'because' followed by an arithmetical truth. I note that even here we are more likely to do so in an extreme or "special" instance, such as an inadvertent division by zero, where we say, "The machine keeps running and doesn't pop up with an answer, because you divided by zero." (We here correlate the mechanical fact that it would "run forever" if the gears didn't wear out with the arithmetical notion that

$$\frac{N}{X} \to \infty \text{ as } x \to 0,$$

or roughly put, that if division by zero were allowed the answer would be "infinity.") The fact that the gear wheels in the calculator are toothed in isomorphism with the decimal system, and that they are arranged from left to right in isomorphism with the way in which we place numerals in the decimal system to represent the powers of 10, facilitates our intuitive appreciation of a more explicit embodiment of the "rules of arithmetic" in the machine's structure than we readily feel in the beam balance case. Just how natural it seems to employ the locution 'because' followed by an arithmetical truth seems, in the case of a desk calculator, to depend partly upon the complexity of the operation involved, a psychological aspect which does not reflect any fundamental physical or logical difference. For example, suppose I am given a printed instruction the rationale of which I do not understand, as follows: I first set a number into the upper (cumulator) register; then I proceed to subtract the consecutive odd numbers, 1, 3, 5, . . . until I get all zeros in the register; then I record the number which appears in the lower (counting) register. If I now clear the machine and operate upon this recorded result with itself through the multiplication key, the machine presents an "answer" in the upper (cumulator) register, and that answer is the number that I started out with. Suppose I am baffled by this, and I ask the question "What is the explanation of this

[30] L. Wittgenstein, *Tractatus Logico-Philosophicus* (London: K. Paul, Trench, Trubner, 1922), Propositions 3.031–3.0321; see also K. R. Popper, "Why Are the Calculi of Logic and Arithmetic Applicable to Reality?" in his *Conjectures and Refutations* (New York and London: Basic Books, 1962), pp. 201–214.

remarkable mechanical phenomenon?" I would probably be satisfied, unless I were specifically interested qua mechanician in the internal workings of a desk calculator, by someone's saying, "Oh, that happens because it is a theorem of arithmetic that the sum of the first k odd numbers is equal to the square of k." It seems to me that whether one views this use of 'because' as literal or figurative is a matter of adopting a semantic convention, rather than a psychological, ontological, or epistemological issue about which there can be a genuine cognitive disagreement. If, for example, one were to require, in stipulating what constitutes a legitimate use of the word 'because' followed by a theorem of some formal science (logic, set theory, arithmetic, differential equations) that the physical system should in some suitable sense be *tokening the theorem* (rather than merely exemplifying it), then he would say that the use of 'because' in the present instance would be incorrect usage (or, at best, metaphorical). But there seems to be no compelling reason for adopting such a stringent stipulation regarding the word 'because.' We employ logic and mathematics to describe the world, whether in its inanimate or animate features. A configural feature of a physical system, whether animate or inanimate, is literally characteristic of it. A formal theorem, whether of logic or mathematics or set theory or whatever, that is exemplified by the system's states and lawful transitions is, I submit, literally attributable to factual (contingent) structure-cum-events of the physical order. Unless some strong counter-consideration were advanced, such as the danger of confusion or of anthropomorphic projection (e.g., of feeling states or experienced motives) into an inanimate system, it is hard to see why such locutions should be conventionally forbidden.

If it is now objected that 'because' cannot be stipulated as allowable usage without doing great violence to both ordinary language and technical conventions, on the ground that the rule exemplified is not in the causal chain, I am at a loss how to reply beyond repeating what I have already said, to wit, that while the *rule* is not in the physical order, an *embodiment* (model, satisfier) of the rule *is* in the physical order. I would say further that the legitimate element of "necessity" which most logicians today would be willing to concede (in spite of Hume) is clearly present in this type of situation. That is to say, if we reconstruct a post-Humean notion of causal necessity as a combination of (a) logical necessity, (b) analysis of reductive complexes, and (c) the distinction between fundamental nomologicals and derivative nomologicals, then we properly assert that the calculator gives the answers which are *arithmetically necessary*, and that it does so "necessarily," given the presupposition that the laws of physics hold and that the machine is not broken, worn out, or the like.

When we move to the modern electronic computer, an additional element enters which makes it still more natural to refer to logical and mathematical theorems in explaining the machine's behavior, namely, that the machine contains a physically identifiable subsystem which stores "instructions" of a nature less generic than the ever-binding laws of logic and arithmetic. And as these instructions become more and more complicated, as when we instruct a computer to examine a certain result with respect to some property (such as whether it is odd or even, or whether it is greater than a certain value) and, depending upon the result of this examination, to operate upon this result in one or another way, then we feel quite at home with such explanations phrased in terms of rules of logic and arithmetic and the word 'because.'

There is, however, still something lacking in the computer which we might wish to require before we employ the word 'because' followed by a reference to a logical rule, namely, a physical subsystem which corresponds to a psychological *motive*. If we can distinguish motives from nonmotivated intentional states, one can say that a computer has intentional states, i.e., applies rules, but does not have motives, i.e., it does not desire things. I myself can discern no division point (other than the phenomenal or consciousness criterion) on the complexity-of-goal-seeking dimension which is other than arbitrary. When Samuels's checker-playing computer[31] (which *learned* to defeat him, the programmer!) examines a set of possible eight-move sequences, and selects the initial move of that sequence which optimizes certain features of the resulting position, I would insist that this is an unquestionably intentional, goal-directed, criterion-applying process, except for the "sentience" component. The same can be said for the Logic Theorist computer program, which cooked up a shorter and more elegant proof of the *Principia Mathematica* Theorem *2.85, done in three steps rather than Russell and Whitehead's nine, relying on fewer axioms, and rendering a certain lemma superfluous.[32] (This kind of thing weakens Professor Popper's argument from "novelty" or "creativity," I think—even

---

[31] A. Samuel, "Some Studies in Machine Learning Using the Game of Checkers," *IBM Journal of Research and Development*, 3 (1959), 210–229.

[32] A. Newell, J. C. Shaw, and H. A. Simon, "Empirical Explorations with the Logic Theory Machine," *Proceedings of the Western Joint Computer Conference*, 11 (1957), 218–230; "Report on a General Problem Solving Program," *Proceedings of the*

Paul E. Meehl

though the example deals with the propositional calculus where neither Church's Theorem nor Gödel's troubles us.)

Here again one has a problem of stipulation. I myself would include, as a necessary ingredient of "desire," the subjective, phenomenal, experiential component, which I presume to be lacking in an electronic computer (regardless of the logical complexity of its "intentional" features). As to "goal," I am quite neutral. As to "selection," I say the computer *selects*. At this point the sapience aspect of the mind-body problem borders on the sentience aspect, which is not the subject matter of this paper. It is, I think, arguable that adopting a convention requiring a raw-feel experiential aspect as a necessary component of the construct *motive* would be very inconvenient. It might preclude the animal psychologist or ethologist from attributing motives to animals at certain levels, where the phylogenetic continuity in many goal-directed aspects of behavior does not seem to be interrupted, but where it becomes increasingly dubious whether any such subjective or raw-feel component is present. But more importantly, while one may entertain (as I do) grave doubts about the validity of considerable portions of the Freudian picture of the mind, I am prepared to argue that one of its core characteristics, the same basic idea of the controlling influence of motive-like variables or states which are not reportable by the subject as having an inner-phenomenological aspect, seems rather well corroborated. It would be very inconvenient, both theoretically and in clinical practice, if we were forbidden to refer to an individual's motives except in those cases in which he is able to give an introspective report of them. I suspect that most psychologists would, like myself, put greater emphasis here upon theoretical generality than upon vulgar speech. Hence the preference to use 'motive' after the manner of Freud or Tolman, the conscious/unconscious distinction being made not by noun choice but by an adjective (conscious motives are contrasted with unconscious motives). But of course if someone wants a noun (e.g., 'desire') that always means *conscious motive*, well and good. These semantic preferences are stipulative, and it is silly to hassle over them. What is *not* stipulative is the

International Conference on Information Processing (Paris: UNESCO, 1959), pp. 256–264. See generally E. Hunt, "Computer Simulation: Artificial Intelligence Studies and Their Relevance to Psychology," in P. R. Farnsworth, M. R. Rosenzweig, J. T. Polefka, eds., Annual Review of Psychology, 19 (1968), 135–168. A. Newell, J. C. Shaw, and H. A. Simon, "Note: Improvement in the Proof of a Theorem in the Elementary Propositional Calculus," C.I.P. Working Paper no. 8, January 13, 1958 (in ditto form, available from Dr. Simon).

empirical finding that much of human behavior is controlled by internal state variables or events that (a) have most of the usual causal properties of reportable motives but yet (b) are not reportable as having a subjective-experience aspect. It is not easy to improve on Freud and Tolman[33] in spelling this out.

If a human brain, like an electronic computer, has a structure which makes it susceptible of storing instructions concerning the allowability of certain kinds of transitions, it would seem appropriate to say that a person accepts a particular argument "because it is valid" and rejects another one "because it is invalid." That one could carry out a complete causal analysis without referring to these logical meta-categories (because he might do it instead at the micro-level) does not invalidate the literal truth of this statement, although it is commonly thought to do so. One way of seeing this is to put the question "Would the argument be accepted by the individual if it were not formally valid?" The answer, of course, is that if the argument were not formally valid, then it follows (from the fact that the micro-laws entail the macro-laws and the macro-state is a reductive complex of the micro-states that—literally—compose it) that the cerebral mechanism would reject the argument. Hence, if the critic says, "According to your view the validity of the input argument makes no difference in what happens," we would have to reply that the critic is simply mistaken in saying this. Because it can be shown from the microanalysis that if the input argument were invalid, the macro-behavior, i.e., the logician's tokening response, would be to say, "This argument is an Illicit Major." Any configural property of an input which "makes a difference," in the sense that if it were lacking, the individual would say "No," but if it is present the individual will say "Yes," surely must be said to "make a difference" in the most stringent use of that expression.

While the complexities of the reconstruction vary widely, and while the presence of a state or event which constitutes a "guiding motive" (properly so-called) makes a great difference, the fundamental point I am making seems to be exemplified both in inanimate and animate contexts; and, within animate systems, is exemplified both in the "psychological" and the "purely physiological" domains. In physics there are problems such that the state to which a system moves can be derived by alternative meth-

[33] S. Freud, "The Unconscious," in J. Strachey, ed., *Standard Edition of the Complete Psychological Works of Sigmund Freud*, vol. 14 (London: Macmillan, 1957), pp. 159–216; Tolman, *Purposive Behavior in Animals and Men*.

ods which are not contradictory but which do represent analyses at different levels of description. Thus, we may invoke highly general principles of a quasi-teleological sort (such as conservation principles or least-action principles), but we may sometimes achieve the same result (less easily and elegantly) without invoking these principles, by proceeding at the micro-level of causal analysis. Or, again, a certain theoretical concept may be one which has a summary function, or which characterizes a complex configuration by reference to certain summarizing quantities, so that the attribution of a certain value of the summarizing quantity follows necessarily from what would be a huge conjunction (or disjunction of conjunctions) of statements about the components. In such situations, there is a non-controversial sense in which one who omits mention of the summarizing quantity may be said to have "left something out of his description," because he did not say everything that might correctly be said. Putting it more strongly, he omitted saying something that is necessarily true on the basis of those things he has in fact said. But I think we should not say that he has given a *defective* causal account because of this omission, inasmuch as the omitted statement concerning the summarizing quantity is virtually present (in the sense of logical entailment) given the nomologicals, and/or explicit definitions, in the statements he has made. Example: I place a block of ice in the center of a room which is being kept warm by a roaring fire in the fireplace. An omniscient physicist provides me with the monstrous conjunction of micro-statements regarding the collisions of individual air molecules with the molecules at the icecake's surface and a blow-by-blow quantum-mechanical account of the manner in which the intra-molecular forces holding each particular molecule in its position in the ice crystal are counteracted, so that this molecule becomes free of the crystal, i.e., becomes part of the fluid. At no point does he employ the terms 'crystal,' 'melt,' 'fluid,' and even his references to the internal geometry of the ice crystal are clumsily formulated by a complicated conjunction that avoids reference to planes, lattices, and the like. Now the true statements involving all of these macro- and intermediate-level concepts are, given the explicit and contextual definitions of these terms in physical theory, to be found among the sentences that follow nomologically from this vast conjunction of sentences characterizing the micro-states and micro-events which he does utter. I do not believe that anyone could object to such an account, other than on esthetic grounds or because of its cumbersomeness. That is, one could not object by saying that the causal account has been rendered somehow incomplete by the failure of the physicist to include mention of all the sentences, and therefore the words that would normally occur in such sentences, at a more molar level of causal understanding. No one would object to this account by saying, "That's all very well, but it won't do as a complete causal explanation of what took place, because you have described the situation as if the difference in temperature between the cake of ice and its surroundings was irrelevant, i.e., that such a temperature differential had no efficacy, that the fact that the ice was colder than the fire-heated air made no difference." To say that this micro-account is defective because it suggests that the temperature difference between the ice and the surrounding air "had no effect," "was irrelevant," "made no difference," or that it was some kind of a "supernumerary," "mere parallel," or "epiphenomenon," would be misleading. The complete characterization of the situation concerning the surrounding air, and the causal explanation of that situation by reference to the chemical changes taking place in the fireplace, obviously do "make a difference," in the literal sense that if those circumstances had been other than they in fact were, the micro-events described in the huge conjunction of statements about the freeing of the individual molecules of water from their crystalline state would not have been true. The conjunction of a vast set of sentences about the molecular motions of the air together with the conjunction of statements about the molecules in the ice, entail a statement that the summarizing quantity known as "temperature" will be higher in the one than in the other.

Consider a nonpsychological case in the animate domain. A biochemist describes the processes which go on in a man's blood chemistry over a period of years at the biochemical level. A physical chemist explains the micro-details of how these various values of blood concentrations bring about a deposition of lipids in the intima of the coronary artery. A physicist gives us a detailed micro-account (in terms of Euler's equations, etc.) of the hydrodynamic situation at this site, resulting from the narrowing of the arterial lumen. A physiologist provides an explanation of what happens in the individual cells of the cardiac muscle itself as a consequence of the reduced oxygen supply and deficient rate of removal of metabolic products, including a micro-characterization of processes which a cytologist, if asked, would recognize as constituting "death of the individual cell." Now one might legitimately complain of this account (especially if he were in the life-insurance business, or a close friend of the family's, or the attending physician) that it fails to state something that was literally true, something that should appropriately be said at another level of analysis, namely, that the patient suffered a myocardial infarction as a result of a coronary occlusion. But it would be misleading for this objector to say that the team (physical chemist + biochemist + physicist + physiologist + cytologist) had "left something out of the account," if by that is meant that there was some further event, entity, or process at work influencing the chain of causality and that this something had been omitted from the description. It would be very wrong of the critic to say, "You have described this as if it made no difference in causing the man's death that he had a coronary occlusion which produced a myocardial infarction." Nothing of the sort. The conjunction of lower level statements made by our five-man basic science team, when taken together with the explicit definitions of the science of pathology, *is* an assertion that the patient suffered a coronary occlusion leading to a myocardial infarction.

Again, suppose we consider two flywheels having the same mass but different physical dimensions, whereby one has a much larger radius of gyration than the other. We inquire about the torque necessary to accelerate these flywheels to a specified rate of revolution, but we do this by considering the transmission of the applied force through the material particle by particle, and by literally summing these billions of components, rather than performing the usual integration analytically from the geometry of the two flywheels. Of course it turns out that in spite of their equal masses, a greater torque is required to achieve a stated angular acceleration in the case of the flywheel whose mass can be considered as concentrated at a greater distance from the center. But we do not explicitly mention this in our causal explanation. Would anyone object to our causal account, saying it was "incomplete" because it "treated the radius of gyration as a something lacking in relevance or effect"?

Whether or not one would "normally" or "naturally" employ locutions of an intentional or quasi-purposive type in vulgar speech is, of course, largely lacking in scientific or philosophical interest. So long as we understand the causal and logical categories and their relationships to one another in the various contexts, whether we opt for one or another label is uninterestingly stipulative. My intuitions about what the usage of vulgar speech would be in a given setting are, like everyone else's, armchair speculations based on anecdotal impressions and lacking in such scientific support as might be obtained through a properly designed sampling procedure with the best available methods of psycholinguistic investigation. What sentences the alleged "plain man" would be willing (or reluctant) to token in regard to the behavior and "mental processes" of a digital computer which beats him at a game of checkers is one of the dullest topics imaginable, and cuts no philosophical or scientific ice so far as I am concerned. My own references in this paper to what (in my armchair opinion) we would "ordinarily" or "naturally" say are intended pedagogically and psychotherapeutically, and nothing really hinges upon these educated guesses of mine. I am not interested in armchair psycholinguistics, whether Oxbridge or Minneapolis style; and I should not dream of deciding a scientific or philosophical question on the basis of what I guessed would be the verbal behavior of a (hypothetical) uninformed layman were he asked to think about difficult and obscure matters which he does not in fact think about, and lacks adequate conceptual equipment to think about.

It is, nevertheless, instructive from the standpoint of curing one's own intuitive resistances to pinpoint their source, considering a variety of examples with an eye to analyzing carefully those in which one experiences considerable ambivalence with respect to the use of quasi-mentalistic or quasi-purposive labels as applied to an inanimate system. Speaking for myself (and I invite the reader to consider whether this may be true of him also), it is my impression that when the element of sentience either is excludable in high probability or is irrelevant (because it is not supposed to be causally efficacious, or because it is clearly present in *both* systems under comparison), the human/ subhuman or even animate/ inanimate dichotomy sometimes receives less subjective weight in our readiness to employ purposive or intentional locutions than does the question whether the process under study has in it some features of "matching" or "comparison" of the actual with the ideal. So that when an inanimate physical system has a structure which enables it to function as a kind of "judge" or "comparator," which "applies criteria," we readily employ mentalistic verbs such as 'sort,' 'classify,' or 'test'; whereas we are reluctant to employ such language, *even with respect to an organic system,* if this element of judging, of comparing, of determining whether something satisfies a condition is utterly lacking.

It is true that in such cases we are aware of the fact that a human mind constructed the inanimate mechanism for a certain purpose, and it is sometimes argued that *this* understood origin and (human) purpose is what justifies the use of such mentalistic language in speaking of a calculating machine. There is no doubt considerable truth to this, and I have no wish to play it down. However, I wish to maintain that such an empirically based comprehension of the human designer's purpose in building the inorganic machine is, while typically present, not a necessary condition for properly applying some (not all) of these "criteria"-flavored words. For example, there are machines the function of which, *part by part*, a layman with a rudimentary understanding of mechanics and mathematics could be brought to understand, but the overarching industrial or scientific purpose to which they were put might baffle him. He might not have anything like an adequate comprehension of the desired "end product" envisaged by the engineer who built the machine to satisfy certain human motives, but he might nevertheless be capable of *characterizing the properties of* the end product satisfactorily. Or, to take a science-fiction example, suppose we find, on geological excavation in the pre-Cambrian rock strata, several complicated apparatuses on which no written instructions are provided in a language we understand, but there is a small plate showing a picture of what appears to be a large bird fastened in the machine. We get ourselves an adult male ostrich and find that he "fits into the machine" with a little adjusting here and there, and when we start the thing running it turns out that what it does is to pluck prime-numbered feathers in a line running down the ostrich's back. This would be pretty spooky, and we might have a very difficult time understanding what a pre-Cambrian somebody was up to, but after experimenting with several ostriches on several of these "extinct machines," we would be entitled to say that, odd though it is, there is good evidence that the "purpose" or "function" of the machine was to pluck out the prime-numbered feathers from ostriches. Point: One does not need to understand the overarching "why," the *ultimate* "end in view," of the maker of the machine in order to infer something about the characteristic end product, as *proximate* "end in view," which results from the machine's operation. In fact, we do not always presume a designer (or plan or prevision) when asking quasi-teleological questions; witness the atheist zoologist or anatomist who undertakes research to answer the question "What is this organ *for*?"—a question which is, given careful formulation, surely sensible aside from one's views on natural theology. The most incisive and illuminating discussion of this problem that I know of is by Nagel.[34]

[34] E. Nagel, *The Structure of Science* (New York: Harcourt, Brace and World, 1961), chapter 12. See also C. G. Hempel, "The Logic of Functional Analysis," in L.

At the risk of boring the reader, let me consider one last example, to highlight the problem of causal analysis in relation to abstract universals, when the latter are allowed to go unmentioned in a (purportedly complete) causal account. We have a simple industrial testing machine, a plate with an elliptical-shaped hole in it, the hole being made slightly larger than any of a batch of elliptical-shaped tiles which the machine is to "test." The machine has "arms" that place each tile in position above the hole, and then proceed to rotate the tile slightly in both directions from its initial position, the rotation being smooth (or at least by steps of very small angles) so that an optimal placement will not be inadvertently "missed." The relationship between the distribution of sizes of the ellipses and the amount of "play" in the hole is such that an elliptical tile whose major and minor axes are in any ratio between 6:8 and 7:8 will be capable of falling through the hole and dropping into a collecting box below. Otherwise the machine throws the tile aside. Now suppose someone provides a detailed account of exactly what happens in the sequence of operations involved in testing 100 consecutive tiles on this machine. He describes the form of the testing slot by stating the coordinates of a very large number of points on its edge (located, say, 1 millimeter apart). Thus he does not write any mathematical function in the familiar algebraic form, although he does in effect write a function for it in the sense of providing a finite set of ordered pairs of numbers. After having thus "tested" 100 tiles, ending up with 97 of them in the box (these having passed through the test slot successfully) and 3 "rejects," and having given a detailed account of the sequence of events as each tile was being tested, our nonmathematical mechanic offers this as a complete causal account of what happened. But now a critic advances the following: "You are assuming, in your alleged complete causal account, that the elliptical shape is of no relevance, that the relation of the major to minor axes in the tiles makes no difference." He makes this objection on the ground that we have not made any reference to the word 'ellipse,' or said anything about major and minor axes. Would this criticism be valid? We would admit that critic is now pointing out something which is literally and physically true about the sequence of events, namely, that *whether or not a particular tile ends up in the box or as a reject depends upon whether it meets or fails to meet*

Gross, ed., *Symposium on Sociological Theory* (New York: Harper, 1959); reprinted with alterations in C. G. Hempel, *Aspects of Scientific Explanation* (New York: Free Press, 1965), pp. 297–330.

a certain geometrical specification. This specification, that of 'being an ellipse' (with a certain range of tolerance permitted in the ratio of the axes), has been "left out" of our account. So we have not said everything that could be said. But would one infer from this that mechanism or physical determinism must be false as an ontology of testing machines, or that a reified platonic universal (some sort of ideal elliptical tile laid up in heaven) must get into the act somewhere to see to it that things go properly?

An ellipse is an abstraction, a universal belonging to the domain of a formal science, which certain material objects may "model" (in the usual sense that a model is a set of entities that satisfies a calculus). What we have presented, in our allegedly complete physicalistic description of the slot, when we specified the coordinates of the points running along the edge 1 millimeter apart, is a set of statements which collectively entail that a tile which passes through is of elliptical shape, with axes in the range 6:8 to 7:8.

Keeping in mind my exclusion of the two considerations beyond the scope of this paper (the subjective-experiential aspect of the mental and the motivational or purposive), let us imagine a critic who adopts Popper's view about the causal efficacy of universals to advance the following: Your account leaves out any reference to the abstract universal *ellipticity*; you describe what goes on as if ellipticity was irrelevant, as if it made no difference to what happens. But the truth of the matter is that ellipticity is the core of the whole process of sorting by this machine; it is precisely the fact and amount of ellipticity of a particular tile that makes the difference between the two grossly different outcomes of a test—to be 'accepted' or to be 'rejected' by the machine. Evaluating ellipticity is what the machine does, and you have left that out entirely. How can your account be complete, when it treats the abstract universal *ellipticity* as an irrelevancy, as something one can mention or not as he pleases, since it makes no difference to what happens?

In order to decide how much truth there is in this objection, we must first explicate what it means to say that the fact of ellipticity is treated by the microanalyst as "making no difference to what happens." The most straightforward explication of the phrase 'such-and-such a factor *makes no difference*' would seem to be that the outcome, given the factor's presence, is indistinguishable in all respects from the outcome that would have eventuated assuming the factor to have been absent. So we unpack the criticism "You say that ellipticity makes no difference" as being an assertion by the critic to the following effect: "The nonmathematical mechanician, by putting forth (as allegedly complete) a causal analysis which makes no mention of ellipticity, thereby implies that if the shape of the testing matrix had been other than elliptical, the results would have been the same as they in fact were. Or, in terms of an individual tile (axes 7.5:8) which was accepted, if that tile had been circular, or had axes in the ratio 5:8, it would nevertheless have been accepted." But of course this is false, and is not being asserted by the mechanician. Not only does the mechanician avoid asserting this false counterfactual; he does not assert anything which implies it. On the contrary, what he says *does* imply a counterfactual contradictory to the one imputed, namely, "If the shape of the sorting machine slot had been other than elliptical, the outcome would have been different, i.e., the tiles which dropped through and the tiles which were cast aside ('rejected') would have been different from the tiles that were in fact passed and cast aside." It is one thing to point out, quite rightly, that the mechanical account of the machine's operations fails to mention something which is true, and which follows necessarily from what was mentioned. In this sense the critic is correct in saying "Something has been left out of the account." But this something which has been left out is not a something which involves any new ontological commitments about the furniture of the world, nor is it something which gets us into trouble with the thesis of mechanical determinism. What *would* involve some sort of additional ontological commitments, and would presumably mean that the causal account of the micro-mechanical determinist was defective (i.e., literally left something out, failed to mention a causally significant property), would be an asserted or implied counterfactual, "The elliptical form as a universal instantialized by this particular machine is irrelevant to what the machine does." But that counterfactual is neither asserted nor implied; on the contrary, its contradictory is implied by the mechanical account, even though that account does not use the word "ellipse" or any short synonym thereof. What the account does contain is the large conjunction of sentences giving coordinates of points on the edge, and these coordinates satisfy the equation of an ellipse. Putting aside the necessary refinements of tolerance, physical discontinuity, etc., the mechanician's clumsy conjunction of statements entails (given the definition of an ellipse) that "ellipticity makes a difference in what happens." What more does the critic want?

It is difficult to summarize the argument presented in the preceding pages. There are five related lines of thought. First, I warn against the temptation to identify "determining factors" in human belief with "nonrational" (e.g., Freudian, Marxian) factors, emphasizing that we have all learned logic as well as other things and that "to think logically" is part of our cerebral computer's programming. Second, I hold that *logical* categories are unavoidable for the psychologist who wishes to deal with human behavior. Third, I argue that the Platonic universals of logic (e.g., Rule of Detachment, *modus tollens*, *dictum de omni et nullo*) are physically modeled by certain subsystems of the human brain, so that—absent countervailing nonrational forces of Freudian or Marxian type—it tends (statistically) to "think rationally." Fourth, I hold that when a physical system models a calculus, a knower K_c who understands the calculus and knows how to derive theorems within it will have a genuine cognitive edge over a less well-informed knower K_m who knows *everything* K_c knows about the machine's parts + arrangement + laws of mechanics, but who lacks K_c's expertise with the calculus. This genuine cognitive edge is literally *physical* in its content, inasmuch as K_c can actually make correct predictions about the future movements of the machine which K_m cannot make. Fifth, I argue that even if a complete physicalistic micro-causal account might be given of human, rational decision-making—an account that contains no explicit reference to *reasons*—the truth of such a complete micro-causal account would be compatible with a "molar" account truly asserting that *reasons decisively influenced the choice.*

As I stated at the beginning, none of these arguments is intended to show that complete psychological determinism obtains, a thesis which I consider open on present evidence, and unnecessary for the current conducting of psychological research.

Nuisance Variables and the Ex Post Facto Design

In a recent important contribution Kahneman[1] has pointed out a psychometric difficulty in the use of matched groups, analysis of covariance, and partial correlation as methods of holding constant the influence of a variable which we cannot control experimentally. Anyone acquainted with psychological and sociological literature will surely agree with Kahneman's initial sentence, "Spurious correlations and confounding variables present a characteristic and recurrent problem to the social scientist." The particular aspect of this many-faceted problem with which Kahneman deals is the fact of statistical "undercorrection" which arises from imperfect reliability in measuring the variable to be controlled. The literature abounds in examples of failure to recognize this difficulty, and hardly any faculty member goes through an academic year without sitting on several doctoral orals in which the candidate—not to mention his adviser—is blissfully unaware of the magnitude of the error that may be thereby introduced, sometimes of vitiating proportions. The present paper is in no sense to be viewed as a criticism of Kahneman's contribution. However, I am afraid I shall make matters worse by pointing out the co-presence of a source of error which is at times equally serious as the one to which Kahneman addresses himself but which usually works in the opposite direction. Furthermore, I have no constructive suggestion to offer, and I am unaware that anybody has presented one. It is my opinion that the high prior probability of a

AUTHOR'S NOTE: This work was supported in part by the National Institute of Mental Health (Research Grant M-4465) and in part, through my summer appointment as professor in the Minnesota Center for Philosophy of Science, by the Carnegie Corporation of New York. A shorter version has appeared as PR-69-4 of Reports from the Research Laboratories of the Department of Psychiatry, University of Minnesota (1969). A paper with less philosophical and more psychometric emphasis is in preparation for submission to a psychological journal.

[1] D. Kahneman, "Control of Spurious Association and the Reliability of the Controlled Variable," *Psychological Bulletin*, 64 (1965), 326–329.

joint (and typically countervailing) influence of the source of error pointed out by Kahneman and the source of error I shall emphasize brings about the circumstance that many traditionally acceptable designs in psychology and sociology are methodologically unsound. To put it most extremely, the so-called ex post facto "experiment"[2] is fundamentally defective for many, perhaps most, of the theoretically significant purposes to which it has been put. It is perhaps no exaggeration to say that the net influence of Kahneman's criticisms and my own, if valid, is to make a scientifically sound ex post facto design well-nigh impossible with presently available methods.

Frequently research in biological and social science presents the problem of "spurious association" (a concept which in itself deserves a more thorough philosophical analysis than it has, to my knowledge, been given by either statisticians or social scientists). Typically these are research problems in which the organisms under study are in some way "self-selected"[3] with respect to an experience, setting, or property which is one of the variables of research interest, or a variable known to be correlated with the latter. That is to say, we have to deal with situations unlike the laboratory experiment in which a randomizing procedure is *externally applied* to a sample of organisms in such fashion that the sources of uncontrolled variance can be said in advance to distribute themselves randomly over experimental treatments. We have to deal with the case in which we, as investigators, do not select what part of the city a child lives in or what college he goes to or what religion his parents profess, but instead must take the "experiment of nature" (as Adolf Meyer would have called it) as it comes. Such investigations, lying somewhere between anecdotal, clinical, or "naturalistic" impressions and laboratory experiments, attempt to combine the necessity of taking the organisms as they come with such scientific procedures as accurate observation, quantitative assessment of variables, and mathematical analysis of the data. (I do not wish to convey the impression that the *only* reason we proceed thus is the fact that we are physically or ethically unable to manipulate and randomize all of the variables, since a case can also be made, and would be made by many clini-

---

[2] F. S. Chapin, *Experimental Designs in Sociological Research* (New York: Harper, 1955); E. Greenwood, *Experimental Sociology: A Study in Method* (New York: King's Crown Press, 1945).

[3] E. Greenwood, *Experimental Sociology: A Study in Method* (New York: King's Crown Press, 1945), pp. 126–129.

cians, social scientists, and ethologists, that observing the phenomena in their "natural setting" may also have distinct qualitative advantages over the artificial situation presented by the laboratory. This, of course, is not to say that the one is any more or less "real" than the other. Anything which happens is real. We merely recognize that a tiger in the laboratory, or a tiger in the zoo, does not live in the same kind of stimulus field, and hence does not maintain the same kind of long-term psychological economy, as one in the Bengal jungle.) Example: If we investigate schizophrenia, with an eye to either its genetic or its environmental determiners, we have to take the schizophrenics as they come. This is because neither our scientific information nor our ethics permits us to produce schizophrenia experimentally, or to predetermine who is a potential schizophrenic and assign such persons randomly to non-schizophrenogenic family environments. Example: If we are interested in economic behavior (say, incentive-pay problems), we cannot have any assurance that a short-term laboratory microcosm involving learning nonsense syllables and "payment" in extra grade points represents an adequate experimental analogue, let alone an identical kind of psychological situation (only reduced in temporal scale), to the question with which we started.

I make these observations of familiar truths to avoid any possibility of being misunderstood as saying that only laboratory experiments, in which control and randomization can be effectively imposed by the investigator, are intrinsically appropriate or scientific. Such a view is far from my philosophical position. There are good reasons, some practical and some methodological, for studying behavioral phenomena "in the state of nature." These reasons are sometimes so good that even the ex post facto design may be preferable to the laboratory method, and will in many cases be better than leaving an important problem completely unresearched.[4] The criticisms I shall advance are aimed at forestalling fallacious inferences of the kind commonly made from such designs, but more importantly, are

[4] D. T. Campbell, "Reforms as Experiments," *American Psychologist*, 24 (1969), 409–429. Note also that in that paper, although Campbell is dealing with more informative situations where time changes are available, he lists "Selection" (= differential recruitment of comparison groups) as a source of bias. As I see it, this rubric would cover two of the three difficulties I am raising for the static case. See also D. T. Campbell and J. C. Stanley, "Experimental and Quasi-Experimental Designs for Research on Teaching," in N. L. Gage, ed., *Handbook of Research on Teaching* (Chicago: Rand McNally, 1963), where the Chapin-Greenwood ex post facto design is totally rejected, but on the basis of "regression artifacts," a source of bias more akin to Kahneman's problem than to mine.

made with the hope of inducing the mathematically competent and statistically creative among us to work on a problem whose importance is, I am persuaded, greatly underestimated by most social scientists.

There are three distinguishable aspects of what I take to be one core difficulty with the method of statistical matching in non-laboratory designs. Their precise logical relationship is not clear, but they are prima facie distinguishable, so I shall discuss them separately. I do not thereby prejudge, nor will the sequel premise, that the three are fundamentally different.

For convenience of exposition, and without, I hope, being prejudicial to any issues, I assume in what follows that Kahneman's problem does not exist. That is, I presuppose (counterfactually) that we possess infallible (perfectly reliable and valid) measures of the "nuisance variable" which we intend to "control" by matching, analysis of covariance, or partial correlation. I do not see that it makes any fundamental difference what kind of statistical control we employ. I should imagine that any novel method of control which was "after-the-fact statistical" in character, i.e., which relied upon some kind of generation of equivalent samples, or some kind of statistical correction for an alleged nuisance variable's influence, would suffer from the same methodological taint.

The first problem which arises is what I shall label *systematic unmatching*. This is most clearly exhibited by the method of matched pairs, in which we artificially constitute a nonrandom sample of the original population by selecting pairs of subjects who are pairwise equated on the nuisance variable. In such cases we are usually interested in the causal influence of an "input" variable X on an "output" or "consequence" variable Y, and we do not have experimental control of X, so that the organisms are somehow, directly or indirectly, *self-assigned* to "treatments" (levels of X). Here the usual reason why we match or partial out some third variable Z is our methodological suspicion that Z may exert a significant causal influence upon both X and Y and, consequently, that the prima facie association between X and Y (with Z left to vary freely, neither controlled experimentally nor partialed out by some statistical device) would reflect an output difference which is "spurious." My first thesis, in a nutshell, is the following: If one is a psychological determinist, or even a quasi-determinist, he must assume that for any but the most trivial and "unpsychological" examples of input variable X, the naturalistic self-selection of the organisms for treatments or levels of X must itself be determined. Hence, the result of holding constant an identified nuisance variable Z will, in general, be to systematically *unmatch* pair members with respect to some fourth (unidentified) nuisance variable W.

Stated in the abstract this thesis seems pretty hard to avoid, but it may sound like a hairsplitting academic point. So let me concretize it to show how serious a problem it presents for the researcher. Let us suppose we are interested in the "influence" of amount of schooling upon subsequent income. We cannot control who stays in school and who drops out before graduating from high school. Even if we could ethically and politically control it, by stopping some students and continuing others, we would be thereby defining a new type of population psychologically, whose statistics would hardly be generalizable to the "natural population" of our original problem.

We enter the files of students in a certain city school system and we divide them into those who did and those who did not complete the twelfth grade. We find that the high school graduates are earning markedly higher salaries twenty years later, i.e., at the time of the investigation. We are not so naive as to take this finding at face value, because we recognize that there might be certain individual-differences variables, located "within the organisms themselves," that would be relatively stable over time and that would, on the one hand, influence income and, on the other hand, also influence the individual's self-selection for values of the input variable, i.e., school level attained. An obvious example of such a nuisance variable is intelligence. We realize that the differences in income might be due (partly) to the fact that the high school graduates were as a group more intelligent than the dropouts, and that this difference in IQ would be (partly) causative of continuance versus noncontinuance of education. So we enter the files for IQ and perform a statistical correction, either by a method such as analysis of covariance which utilizes the total N, or by defining subsamples of the original sample in which individuals are matched pairwise for IQ. If such matched groups differ in income, we conclude (fallaciously) that the difference is "not attributable to intelligence."

I say "fallaciously" because, of course, Kahneman's point applies here. That is, the unknown true intelligence level of an individual lies somewhere between the best estimate we could get by knowing how far he persisted in school, clearly one of the several fallible indicators of brains, and the IQ we find in the files, another fallible indicator of brains. His true intelligence lies somewhere between these, at a position which is sometimes estimable but more often not. But we are passing Kahneman's objection here and assuming counterfactually that the files contain infallible measures of intelligence.

Now if there is in fact a correlation between brains and persistence in school, matching dropouts with completers for infallible IQ surely results in the samples we generate being unmatched for some *other* determining factor or factors capable of influencing the probability of school continuance. And, on the average, the members of a pair will presumably be more badly unmatched on these other factors (having been matched on IQ) than they would have been if we had let the chips fall where they may. Example: A stupid adolescent who continues through high school may do so because his parents put a very high emphasis upon educational achievement, and a bright one may drop out because his parents do not value such performance. The introjection of parental values is surely one of the major variables reflected in almost any kind of achievement, educational or vocational. This introjection would presumably function as a nuisance variable W which is left uncontrolled by matching on Z $(= IQ)$. More importantly, the matching of groups on variable Z tends, on the average, to increase systematic unmatching on W. Thus, a dropout matched at 125 IQ with a continuer will be an extreme (low) deviate on, say n Ach;[5] whereas a continuer matched at 90 IQ with a dropout will be an extreme (high) deviate on n Ach. Within each pair, large systematic differences in n Ach (or any other unmeasured nuisance variable influencing self-selection for school continuance) will be practically guaranteed by the matching procedure. Or again, individual differences in "sociopathic-like" (low-anxiety, defective impulse-control, acting-out tendencies) will surely affect the dropout incidence.[6] If two boys are equal on an infallible IQ measure but one has graduated from high school and the other one has dropped out, there is a good chance that they differ on this component, which is not one which our file data normally enable us to assess. I hope these examples show that, rather than being a minor blemish on the ex post facto design, the likeli-

[5] D. C. McClelland, J. W. Atkinson, R. A. Clark, and E. L. Lowell, *The Achievement Motive* (New York: Appleton-Century-Crofts, 1953); D. C. McClelland, *The Achieving Society* (Princeton, N.J.: Van Nostrand Reinhold, 1961).

[6] S. R. Hathaway and E. D. Monachesi, *Adolescent Personality and Behavior: MMPI Profiles of Normal, Delinquent, Dropout, and Other Outcomes* (Minneapolis: University of Minnesota Press, 1963).

hood of systematic unmatching represents a major methodological weakness which is likely to corrode the entire investigative enterprise.

A second difficulty, which I shall call the *unrepresentative subpopulation problem*, is the first one as seen from the population-sampling point of view. If we match pairwise for a nuisance variable, such as a demographic factor that is known or supposed to be sizably correlated with each variable of interest, what we do (willy-nilly) by the matching procedure is to identify samples from subpopulations that differ systematically from the entire population of interest. If the nuisance correlations are small, the "improvement" achieved statistically will be negligible, i.e., the matching was relatively pointless. If it is large, the systematic departure of the resulting subpopulations from the original population in certain parameters will be correspondingly increased. In the extreme case we may be working with samples from a subpopulation which differs very markedly from the population of original interest. This means that our statistical generalization must be carefully confined to the unrepresentative subpopulations specified by the matching operation, and while that can of course be done, it will frequently leave us without an answer to the main question which aroused our research interest in the first place. Example: Suppose we have evidence to indicate that there is a relationship between the incidence of schizophrenia and socioeconomic class. We want to study the properties of a certain psychometric device, such as the Rorschach or MMPI, or some kind of cognitive performance such as abstraction ability or visual perception, in schizophrenics versus manic-depressives. I daresay that almost any competent Ph.D. candidate would take it completely for granted that his design would require a matching for socioeconomic and educational level. He finds the expected sizable difference between his manic-depressives and schizophrenics with regard to socioeconomic level on some suitable measure (e.g., the Hollingshead Index), and in order to "control" for its "spurious influence" (I put these phrases in quotes not ironically but to indicate that one does not have a clear notion precisely how the *statistical* control is related to the control of *causal influence*, discussed below as a third difficulty) he does not sample randomly from the entire hospital population of the two diagnostic groups but instead he constitutes a matched sample in which each schizophrenic is paired with a manic-depressive having the same social class index. Depending upon how he goes about this matching, our investigator may or may not be able to specify a statistically definable subpopulation, but let us assume that he can. He then samples randomly from these subpopulations to get the actual group of patients he studies on the output variable (abstraction ability or perceptual speed or Rorschach F+ or whatever it may be). Now it is obvious that this subpopulation is an atypical one, because the matching procedure will practically guarantee that on the average his schizophrenics are of somewhat higher socioeconomic class than the schizophrenic hospital population generally; and, similarly, the manic-depressive subpopulation from which he samples is now a biased subpopulation from the universe of manic-depressives. That is, the schizophrenic group sampled is pulled upward from their population social class value and the manic-depressive group is pulled downward from their population social class value; otherwise, of course, successful matching would not have been achieved. The expected result of such a procedure is a marked reduction in variance, which is the usual empirical finding.[7] One cannot avoid the consequence that either this degree of departure is large enough to be worth worrying about or it isn't. If it is not large enough to worry about, there was no merit to engaging in the matching operation; if it is large enough to be worth worrying about, then one has a new problem by virtue of the fact that he is now studying unrepresentative (higher class) schizophrenics and unrepresentative (lower class) manic-depressives. And of course psychologically this is a very serious difficulty. Presumably some schizophrenics, as well as semi-compensated or compensated schizotypes,[8] either remain in or gravitate to a lower social and educational class because of the general social incompetence associated with schizotypy,[9] the obvious exceptions being individuals possessed of rare gifts that society rewards in special domains, e.g., esthetic or intellectual talents. By contrast, manic-depressives are "clinically well" (and, except to the very skilled eye, not detectably different from normal persons) between their psychotic episodes; further,

---

[7] See Chapin, *Experimental Designs in Sociological Research*, chapters III–V, for several examples.

[8] P. E. Meehl, "Schizotaxia, Schizotype, Schizophrenia," *American Psychologist*, 17 (1962), 827–838; P. E. Meehl, *Manual for Use with Checklist of Schizotypic Signs* (Minneapolis: Psychiatric Research Unit, University of Minnesota Medical School, 1964); S. Rado, *Psychoanalysis of Behavior* (New York: Grune and Stratton, 1956); S. Rado, "Theory and Therapy: The Theory of Schizotypal Organization and Its Application to the Treatment of Decompensated Schizotypal Behavior," in S. C. Scher and H. R. Davis, eds., *The Outpatient Treatment of Schizophrenia* (New York: Grune and Stratton, 1960), pp. 87–101; S. Rado and G. Daniels, *Changing Concepts of Psychoanalytic Medicine* (New York: Grune and Stratton, 1956).

[9] W. H. Dunham, *Community and Schizophrenia* (Detroit: Wayne State University Press, 1965).

there are certain features of the manic-depressive inter-psychotic character structure which are highly rewarded economically in the American culture, such as the social extraversion, the competitive striving, and a special sort of narcissism which these persons possess in spite of their superficial affiliative tendencies. A schizophrenic who remains in the sample after the matching operation is likely to differ from his more typical schizophrenic brethren in dimensions such as achievement motive, ego strength, energy level, frustration tolerance, social skills, perseverance, and goodness knows what all, variables likely to be significant influences with respect to the psychometric or experimental output measure under study. And the same is true, but in the other direction, for the manic-depressives. Additional biasing effects, almost inevitable given the relatively poor reliability of psychiatric diagnosis, will be a heightened proportion of misdiagnosed cases in both directions, and an inflated proportion of so-called "schizo-affective psychoses," who are atypical of either a manic-depressive or a schizophrenic population. I do not see how it is possible to make any valid correction for this kind of influence, since we are here talking about numerous unknown nuisance variables that become jointly definitive of unspecifiably deviant subpopulations. But what we can say, if we are psychological determinists, is that the two groups of patients under investigation are both unrepresentative of their respective diagnostic categories.

The third component of this problem is so obvious that one would be embarrassed to dilate upon it, except for the fact that a remarkable number of social scientists seem almost oblivious of the point. I shall call it causal-arrow ambiguity. While every sophomore learns that a statistical correlation does not inform us about the nature of the causality at work (although, except for sampling errors, it does presumably show some kind of causal relation latent to the covariation observed), there has arisen a widespread misconception that we can somehow, in advance, sort nuisance variables into a class which occurs only on the input side of the causal arrow and another class which occurs only on the output side.[10] This is, of course, almost never the case. The usual tendency, found widely among sociologists and quite frequently among psychologists (particularly among those of strong environmentalist persuasion), is to assume sub silentio

[10] For a brief but very clear analysis of the possible ways in which correlations among three variables may arise see O. Kempthorne, An Introduction to Genetic Statistics (New York: Wiley, 1957), pp. 283–286. In discussing "adjustments" (for nuisance variables) Kempthorne warns (p. 284) that "the adjustment of data should be based on knowledge of how the factor which is being adjusted for actually produces its effect.

381

that there is a set of demographic-type variables, such as social class, domicile, education, that always operate as nuisance variables to obscure true relationships or generate "spurious" ones, functioning primarily or exclusively on the *input* side from the standpoint of causal analysis. This automatic assumption is often quite unjustified. Example: We study the relationship between some biological or social input variable, such as ethnic or religious background, upon a psychological output variable, such as IQ or *n Achievement*. We find that Protestants differ from Catholics or that whites differ from blacks. But we find further that the ethnic or religious groups differ in socioeconomic class. We conclude, as an immediate inference and almost as a matter of course, that we have to "control" for the socioeconomic class variable, in order to find out what is the "true" relationship between the ethnic or religious variable and the psychological output variable. But of course no such immediate inference is defensible, since on certain alternative hypotheses, such as a heavily genetic view of the determiners of social class, the result of such a "control" is to bring about a spurious reduction of unknown magnitude in what is actually a valid difference.

Another example is the objection to the use of certain kinds of test items on measures of intelligence, when that objection is put *solely* in terms of the statistical fact that social class differences exist on these items. I cannot enter here into the substantive merits of that controversy, which is extraordinarily complex, and to which no adequate general solution seems to exist at present. No one would deny that if a certain kind of cognitive performance involves a content to which lower class children have inadequate environmental exposure (a notion which would have a high prima facie plausibility even without any research), such an item is not a "good item," assuming we are interested in the assessment of basic capacity variables. But what I do wish to query is the usual assumption among many psychologists and sociologists that *of course* whenever we find that a given kind of test item discriminates social class, it follows rather directly that it is an inappropriate item, such that measures compounded out of such items are to that extent "biased" or "invalid." This immediate inference is fallacious.

That it is fallacious can easily be discerned by considering the statistical

An arbitrarily chosen adjustment formula may produce bias rather than remove the systematic difference. It is this fact which tends to vitiate the uses of the analysis of covariance recommended in most books on the analysis of experiments."

consequences of a counter-hypothesis, and noting that they are indistinguishable from those of the conventional one. Suppose, to take the extreme case, that socioeconomic level were *completely* determined by abstract-conceptual intelligence, and that abstract-conceptual intelligence were *completely* determined by the genes; then it would follow as a consequence that high-valid items would discriminate social class perfectly. Analogy: We make a file study of the incidence of positive tuberculin tests in a random sample of patients seen in an outpatient clinic, and discover that test positives occur more frequently among the *lumpenproletariat* than they do among Cadillac drivers. We do not conclude forthwith that the tuberculin test is "invalid" because it is "biased" against the poor! Why not? The reason nobody concludes this is, of course, that we all *already know* how the direction of the causality runs. Similarly, in agricultural experiments we know that an analysis of covariance in which the nuisance variable statistically controlled is, say, a soil characteristic will give us the "right answer," because our well-corroborated causal model *tells us in advance* the direction of the causal arrow. Nobody in his right mind supposes that the yield of corn in August causally determines random table entry or certain properties of the soil present during the preceding spring and summer; therefore we are confident that an analysis of covariance will give us the causal answer in which we are interested as agricultural experimenters. The same is rarely the case when the behavior scientist partials out or matches with respect to a nuisance variable, because the latter may itself be (and, in general, *will be*) a *dependent* variable with respect to a variety of nuisance factors which we can perhaps say something plausible about, but which we do not know how to measure or control. There is no general justification for the routine assumption that demographic and allied variables such as religion, size of community, educational level, ethnic and religious background, and social class should be taken as always functioning solely on the input side and, therefore, as always appropriately "controlled" by a matching operation or by some similar type of statistical correction.

I would go further than this and suggest that it is not only incorrect to insist that groups must routinely be matched on such demographic or other nuisance variables, but that, for all we know, in some unknown proportion of designs the net effect of such matching is not to improve the validity of the inferences made but is actually to introduce systematic error. I do not wish to maintain that matching makes matters worse more often that it makes them better, but I consider it an open question on the present evidence. If I were advising a doctoral candidate who asked me whether he should control for educational and social class in comparing schizophrenics and manic-depressives with respect to the presence of psychometric thought disorder, I would honestly not know what to tell him. I suspect I would have to tell him that if he didn't match, he would be in danger of flunking his doctoral oral, because most of the members of the committee would be operating on the traditional assumption that he should have done so; but that so far as I myself was concerned, the unrepresentative character of the resulting matched samples would be such that I wouldn't know what he would be entitled to conclude if he got a difference, and even less if he failed to get a difference, on the output variable having followed such a matching procedure.

This line of argument does not conflict with what we teach students in courses on experimental design with regard to the purely *statistical* influence of matching procedures upon design sensitivity or what Fisher calls "precision." It is, of course, true that a matching procedure will (if successful) have the effect of reducing the error term which appears in the denominator of a significance test, and in that sense will give us higher power. But that statistical truism is in no way incompatible with the claim I am making here, namely, that we are thereby defining different subpopulations and consequently that the parameters we are estimating may not be the parameters we were originally interested in.

Perhaps the most succinct (but still general) way of formulating the problem of controlling nuisance variables statistically in a nonexperimental context would be "How are we entitled to interpret the associated counterfactual conditional!" I set aside the super-positivistic approach that purports to eschew any such counterfactual, claiming to confine itself to the observations plus the formalism—a sort of "psychologist's Copenhagen interpretation"—since I have not found any theoretically interesting cases in which this "minimum interpretation" is *consistently* adhered to. And this is hardly surprising, since if one genuinely intends to utilize the statistical formalism *solely* for predictive purposes, there is no rational basis for introducing such statistical "control." That is, it makes no sense to speak of a correlation as "spurious" or "in need of correction" *unless* a possible error in causal-theoretical interpretation is envisaged. Thus the correlation between years of schooling and subsequent salary—I of course neglect the separate problem of ordinary sampling errors—stands on its own feet, and if you want to forecast income from schooling, the "influence" of IQ as a shared statistical component can be neglected (= allowed to operate) at the purely descriptive level. In every instance that I have come across in which the investigator felt it necessary to employ partial correlation, analysis of covariance, or artificially concocted matched samples to "avoid the influence" of an alleged nuisance variable, the rationale of such a procedure lay in his wish to conclude with a causal-theoretical inference or, at least, a counterfactual conditional of some kind.

When a social scientist of methodological bent tries to get clear about the meaning, proof, and truth of those counterfactuals that interpret statistical formalisms purporting to "control the influence" of nuisance variables, he is disappointed to discover that the logicians are still in disagreement about just how to analyze counterfactuals.[11] It appears that the logi-

[11] While the problem of interpreting conditionals is an ancient one (see, for example, M. Hurst, "Implication in the Fourth Century B.C.," *Mind*, 44 (1935), 484–495, and B. Mates, "Diodorean Implication," *Philosophical Review*, 58 (1949), 234–242), and the issues of present controversy were adumbrated by W. E. Johnson in the 1920's (see W. E. Johnson, *Logic, Part I*, Cambridge: Cambridge University Press, 1921, chapter III, and *Logic, Part III, The Logical Foundations of Science*, Cambridge: Cambridge University Press, 1924, chapter I), the current concern over the logical analysis of counterfactuals and their relation to natural laws was precipitated by the papers of R. Chisholm, "The Contrary-to-Fact Conditional," *Mind*, 55 (1946), 289–307 (reprinted in H. Feigl and W. Sellars, eds., *Readings in Philosophical Analysis*, New York: Appleton-Century-Crofts, 1949, pp. 484–497), and N. Goodman, "The Problem of Counterfactual Conditionals," *Journal of Philosophy*, 44 (1947), 113–128. An extensive literature on the subject followed these seminal contributions. See, perhaps best read chronologically, F. L. Will, "The Contrary-to-Fact Conditional," *Mind*, 56 (1947), 236–249; S. Hampshire, "Subjunctive Conditionals," *Analysis*, 9 (1948), 9–14 (reprinted in Margaret Macdonald, ed., *Philosophy and Analysis*, New York: Philosophical Library, 1954); E. L. Beardsley, " 'Non-Accidental' and Counterfactual Sentences," *Mind*, 46 (1949), 573–591; H. Hiz, "On the Inferential Sense of Contrary-to-Fact Conditionals," *Journal of Philosophy*, 48 (1949), 586–587; K. R. Popper, "A Note on Natural Laws and So-Called 'Contrary-to-Fact Conditionals,' " *Mind*, 58 (1949), 62–66; W. Kneale, "Natural Laws and Contrary-to-Fact Conditionals," *Analysis*, 10 (1950), 121–125 (reprinted in Macdonald, ed., *Philosophy and Analysis*); D. Pears, "Hypotheticals," *Analysis*, 10 (1950), 49–63 (reprinted in Macdonald, ed., *Philosophy and Analysis*); A. R. Anderson, "A Note on Subjunctive and Counterfactual Conditionals," *Analysis*, 12 (1951), 35–38; D. J. O'Connor, "The Analysis of Conditional Sentences," *Mind*, 60 (1951), 351–362; J. Weinberg, "Contrary-to-Fact Conditionals," *Journal of Philosophy*, 48 (1951), 517–528; T. Storer, "On Defining 'Soluble,' " *Analysis*, 11 (1951), 134–137; G. Bergmann, "Comments on Storer's Definition of 'Soluble,' " *Analysis*, 12 (1952), 44–48; J. Anderson, "Hypotheticals," *Australasian Journal of Philosophy*, 30 (1952), 1–16; R. Brown and J. Watling, "Counterfactual Conditionals," *Mind*, 61 (1952), 222–233; R. Brown and J. Watling, "Hypothetical Statements and Phenomenalism," *Synthese*, 8 (1950–52), 355–366; B. J. Diggs, "Counterfactual Conditionals," *Mind*, 61 (1952), 513–527; Erna F. Schneider, "Recent Discussion of Subjunctive Conditionals," *Review of Metaphysics*, 6 (1953), 623–649; J. L. Watling, "Propositions Asserting Causal Connection," *Analysis*, 14 (1953), 31–37; A. R. Anderson,

Paul E. Meehl

cal (and epistemological) analysis of counterfactuals is a task involving some of the deepest and oldest of philosophical problems (e.g., the modalities, extensional logic's adequacy, substance and property, character of natural laws, identity, the kinds of "contingency" and "necessity," the meaning of 'accidental' in a determinist framework, the theory of proper names and definite descriptions). I had intended to include something

reviews of articles by Schneider, Diggs, Storer, Bergmann, and Brown and Watling, *Journal of Symbolic Logic*, 19 (1954), 68–71; R. Chisholm, "Law Statements and Counterfactual Inference," *Analysis*, 15 (1955), 97–105; J. C. Cooley, "Professor Goodman's 'Fact, Fiction, and Forecast,' " *Journal of Philosophy*, 54 (1957), 293–311; W. Sellars, "Counterfactuals, Dispositions, and the Causal Modalities," in H. Feigl, M. Scriven, G. Maxwell, eds., *Minnesota Studies in Philosophy of Science*, vol. II (Minneapolis: University of Minnesota Press, 1957), pp. 225–308; J. L. Watling, "The Problem of Contrary-to-Fact Conditionals," *Analysis*, 17 (1957), 73–80; P. B. Downing, "Subjunctive Conditionals, Time Order, and Causation," *Proceedings of the Aristotelian Society*, 59 (1959), 129–140; K. R. Popper, "On Subjunctive Conditionals with Impossible Antecedents," *Mind*, 68 (1959), 518–520; W. Kneale, "Universality and Necessity," *British Journal for the Philosophy of Science*, 12 (1961), 89–102; N. Rescher, "Belief-Contravening Suppositions," *Philosophical Review*, 70 (1961), 176–196; R. W. Walters, "The Problem of Counterfactuals," *Australasian Journal of Philosophy*, 39 (1961), 30–46; J. L. Mackie, "Counterfactuals and Causal Laws," in R. J. Butler, ed., *Analytical Philosophy* (New York: Barnes and Noble, 1962), pp. 66–80; H. A. Simon and N. Rescher, "Cause and Counterfactual," *Philosophy of Science*, 33 (1966), 323–340; G. C. Nerlich and W. A. Suchting, "Popper on Law and Natural Necessity," *British Journal for the Philosophy of Science*, 18 (1967), 233–235; K. R. Popper, "A Revised Definition of Natural Necessity," *British Journal for the Philosophy of Science*, 18 (1967), 316–321; G. Molnar, "Kneale's Argument Revisited," *Philosophical Review*, 78 (1969), 79–89. Much of this linguistic analysis, while inherently interesting, has little or no value for the social scientist seeking methodological clarification on his scientific use of counterfactuals. I found the papers by Hiz, Popper, Kneale, Sellars, Mackie, Simon and Rescher, and Molnar most illuminating. See also brief or related discussions in R. B. Braithwaite, *Scientific Explanation* (New York and London: Cambridge University Press, 1953), pp. 295–300; A. W. Burks and I. M. Copi, "Lewis Carroll's Barber Shop Paradox," *Mind*, 59 (1950), 219–222; A. W. Burks, "Laws of Nature and Reasonableness of Regret," *Mind*, 55 (1946), 1–3; A. W. Burks, "The Logic of Causal Propositions," *Mind*, 60 (1951), 363–382; A. W. Burks, "Dispositional Statements," *Philosophy of Science*, 22 (1955), 175–193; R. Carnap, "Testability and Meaning," *Philosophy of Science*, 3 (1936), 419–471, and 4 (1937), 1–40; R. Carnap, *Philosophical Foundations of Physics* (New York: Basic Books, 1966), pp. 196–215; C. G. Hempel and P. Oppenheim, "Studies in the Logic of Explanation," *Philosophy of Science*, 15 (1948), 135–175 (reprinted in C. G. Hempel, *Aspects of Scientific Explanation*, New York: Free Press, 1965); C. G. Hempel, "Problems and Changes in the Empiricist Criterion of Meaning," *Revue Internationale de Philosophie*, 11 (1950), 41–63; C. G. Hempel, *Philosophy of Natural Science* (Englewood Cliffs, N.J.: Prentice-Hall, 1966), pp. 56–58; W. Kneale, *Probability and Induction* (Oxford: Oxford University Press, 1949), pp. 70–78; C. I. Lewis, *An Analysis of Knowledge and Valuation* (LaSalle, Ill.: Open Court, 1946), pp. 211–233; E. Nagel, *The Structure of Science* (New York: Harcourt, Brace and World, 1961), pp. 68–73; A. Pap, *An Introduction to the Philosophy of Science* (Glencoe, Ill.: Free Press, 1962), chapters 15 and 16, pp. 273–306; A. Pap, "Dispositional Concepts and Extensional Logic," in Feigl, Scriven, Maxwell, eds., *Minnesota Studies in Philosophy of Science*, that I hoped would be new and constructive at this point of my discussion, but deadline obligations and my status as a philosophical amateur have combined to make me more realistically modest in aims. I hope, however, that what I have to say at present about counterfactuals does not depend on precisely how the logicians ultimately agree to "fix them up." I am encouraged in this hope by the fact that agreement does exist about the important role of the explicandum, and—to a considerable extent—about criteria for a satisfactory explication. One main area of agreement—of direct relevance to the social scientist's problem—is the intimate connection between a counterfactual's legitimacy and the natural-law/ accident-universal distinction. One way (the main way, some hold) in which a natural law differs from an accidental universal is that the former legitimates a counterfactual while the latter does not. "If Kosygin had not learned Russian, he would be unable to speak it" is presumably a sound social-science counterfactual, relying on the laws of psycholinguistics. But we cannot rely on the accidental universal "All persons who discuss politics with Meehl speak English" to legitimate a counterfactual "If Kosygin were to discuss politics with Meehl, he would speak English."

As I read the record, there are some counterfactuals we wish to exclude because we doubt that they are meaningful, but we want to assure that criteria adequate to exclude them will not inadvertently forbid other similar-appearing counterfactuals which do seem intuitively meaningful, and of great importance in the discourse of science and common life. Take, for example, what may be labeled (nonprejudicially) as 'counter-identicals,' i.e., counterfactual statements concerning a named or definitely described individual, where the protasis falsifies one of his properties. In spite of Leibniz, the scientist, lawyer, physician, and ordinary man will—I think correctly—insist that many such counter-identicals are meaningful and useful. We surely do not wish to adopt a semantic convention which denies the status of wff to, say, "If defendant had driven his car with ordinary care, plaintiff would not have sustained injury," or "It was fortunate

vol. II, pp. 196–224; K. R. Popper, *The Logic of Scientific Discovery* (New York: Basic Books, 1959), pp. 420–441; W. V. Quine, *Methods of Logic* (New York: Holt, 1959), pp. 13–17; H. Reichenbach, *Elements of Symbolic Logic* (New York: Free Press, 1947), chapter VIII, pp. 355–404; W. Sellars, "Concepts as Involving Laws and Inconceivable without Them," *Philosophy of Science*, 15 (1948), 287–315; C. D. Broad, "The 'Nature' of a Continuant," in his *Examination of McTaggart's Philosophy*, vol. I (Cambridge: Cambridge University Press, 1933), pp. 264–278 (reprinted in Feigl and Sellars, eds., *Readings in Philosophical Analysis*, pp. 472–481).

for me that I had a flu shot, since everyone else in the family fell dreadfully ill with flu." Contrast these counter-identicals with this one (example courtesy of Dean Kenneth E. Clark): "If Meehl and I had lived in the sixth century, he would have been an archbishop, and I would have been Merlin's research assistant." Is this counterfactual legitimate? Hard to say, but if so, it will take some doing to unpack satisfactorily. Worse is "If Caesar had been born in 1900, he would have been a fascist." Still worse is "If my maiden aunt were a tram car, she would have wheels."

If one conceives of an individual as a bundle of properties, there is a difficult problem in unpacking all such counter-identicals. I believe the best way to do it is to begin with a distinction between the actual world and other imagined (hypothetical) worlds belonging to the same world family, where 'world family' designates the infinite set of conceivable worlds sharing nomologicals but differing in particulars.[12] Assuming that this can be done satisfactorily (and no one has, to my knowledge, offered a criticism of Sir Karl Popper's 1967 paper attempting to rigorize it), I think we could then offer a translation of counter-identicals in terms of world lines in some unrealized world of our world family, sharing coordinates with the named or described individual's actual world line up to the critical event (e.g., failure to obtain his flu shot as planned), and diverging thereafter. His properties and most of his relations would be identical with those of the actual individual up to that space-time point, but would diverge—perhaps increasingly—thereafter. In stipulating semantic rules for the well-formedness of a counter-identical, there would doubtless be a certain arbitrariness about which of an individual's properties are, so to say, "privileged properties," such that a counterfactual denying them is forbidden. Intuitively one feels that it is essential to the person called 'Caesar' that he be an ancient Roman, but it is not essential to Meehl that he receive a flu shot. Of course a rule excluding "If Caesar had been born in 1900 . . ." is laid down in the interest of avoiding strange and counterintuitive discourse and preventing unprofitable puzzles, and we do not wish to forbid too much. Thus it makes sense to begin a counterfactual with "If an American child born in 1900 had Caesar's complement of genes" (wildly improbable but not, I submit, counter-nomological) but this admissible case need not be forbidden by a rule adequate to forbid the

---

[12] Sellars, "Concepts as Involving Laws and Inconceivable without Them"; Popper, *The Logic of Scientific Discovery*, p. 430; Nerlich and Suchting, "Popper on Law and Natural Necessity," pp. 233–235; Popper, "A Revised Definition of Natural Necessity," pp. 316–321.

counter-identical beginning with "If Caesar [proper name, denoting an individual who satisfies a certain definite description] had been born in 1900 . . . " My hunch is that a sufficiently tolerant set of exclusion rules could be rigged up, keeping in mind that an adequate logician's translation of legitimate proper-name or definite-description counter-identicals need not—I think will not—show the individual's name recurring on the right-hand side of the equation; just as in Russell's theory of definite descriptions itself, we have learned to accept the fact that an unpacking adequate to avoid paradoxical metaphysics leaves us without 'the present King of France' as a single semantic element on the right-hand side. But the development of these suggestions must wait for another occasion.

Accepting provisionally the world-family concept and the associated distinction between nomologicals and accidental universals, we see that the interesting cases for social-science methodology would remain problematic even after the cute counter-identical puzzles of logic seminars had been liquidated. This is because the social-science cases of interest are not (by and large) in danger of counter-definitional meaninglessness but, instead, may suffer from counter-nomological falsity or contradictoriness. That is, the problematic counterfactuals of psychology and sociology do not typically find us wondering "What does it *mean*, does it make any *sense*?" but rather, "Is it consistent and true? *Could* [nomologically] the counterfactual hypothesis be satisfied, given the nomologicals presupposed? And, *if* it could, does the counterfactual conclusion *follow* within that nomological system?"[13] Since our warrant for asserting counterfactuals consists of the nomologicals of our world family, plugging in counterfactual particulars so as to yield a different world of the family, we must avoid unwittingly contradicting ourselves in the antecedent. Consider statements like: "Imagine that these organisms, which in fact have properties P_1, P_2, . . . , P_k, Q_1, Q_2, . . . , Q_m, had instead possessed properties P_1, P_2, . . . , P_k, Q_1', Q_2', . . . , Q_m'; then . . ." (Note that this way of talking is ubiquitous in biological and social science—we cannot even understand the notion of a *control* group without admitting such formulations!) To get to the counterfactual conclusion following '. . . then . . .' we rely on natural laws. But what if the natural laws relied on forbid the counterfactual antecedent P_1, P_2, . . . , P_k, Q_1', Q_2', . . . , Q_m'? How do we know that these are compossibles, i.e., that the counterfactual conjunction is not nomologically forbidden?

[13] H. Hiz, "On the Inferential Sense of Contrary-to-Fact Conditionals," pp. 586–587.

I am not talking about what might be a logician's technical problem, i.e., the nonexistence of a general algorithm for stepwise deciding whether this conjunction would instantialize a counter-theorem. No, the problem is not so esoteric as that. The problem lies in the incompleteness of the social scientist's nomological network. Underlying (derivationally and causally) the known laws of social science are the unknown ones—the "true reasons why" the known laws are the way they are. Furthermore, very odd but true, *some of the laws are, from a philosopher's viewpoint, not nomologicals but accidental universals.* This is because many "laws" of biological and social science are structure-dependent and history-dependent in a special way, so that while their logical form (taken singly) is that of laws of nature, they are not derivable from the fundamental nomologicals (laws of physics). Many "taxonomic" laws are pseudo-nomological, which is one reason why examples like "All crows are black" are unsuitable for most philosophy-of-science discussions. Unfortunately it is not always easy to ascertain when a biological or social-science generalization (taken as true and well evidenced) is really akin to "All silver melts at 960.5° C"—a nomological—and when it is akin to "All the coins in my pocket are silver," an accidental universal. It may be objected that the melting point of silver is also structure-dependent, but this, while true, does not prevent the generalization's being a true nomological, because we can (theoretically) include a characterization of the micro-structure in our "theoretical" *definition* of the technical term 'silver,' in which case the structure dependence is fully represented in the antecedent. I.e., we have "If a substance is silver [= has such-and-such micro-structure], it melts at 960.5° C," a proposition presumably entailed within (complete) physical theory as a consequence of the fundamental nomologicals. Viewed this way, the generalization is a theorem within a formalized physical theory (and, note carefully, would be nontrivially true for all worlds in our world family even if no silver existed in some of them). In biology, the statement "A mammal dies if deprived of oxygen" is of this sort, since its structure dependence can analogously be represented in an adequate theoretical (anatomical + physiological) definition of 'mammal.' By contrast, the taxonomic generalization "All mammals have paired gill-slits at some stage of their development" is an accidental universal, as is "If a species of animal has a heart, it has kidneys." These taxonomic property correlations are—like Meehl's friends' English-speaking and his silver coins—"historical accidents," reflecting the course of evolution which could have been different given the same fundamental nomologicals but differing initial conditions of the earth.[14]

Most of the statistical "laws" (correlations) investigated in disciplines such as differential psychology, personology, clinical psychology, and soociology are more akin to the accidental universals of taxonomy than to genuine derived nomologicals. The social scientist who works in these fields studies covariations between selected *dispositions* manifested by individuals ("traits," "capacities," "temperamental or cognitive parameters") and also the correlations of these with a variety of *status* variables and *life-history* antecedents. The nomological network and initial conditions that gave rise to these statistical associations are horrendous in number and complexity. They involve factors ranging in kind from genetic drift in the remote past when a certain ethnic group was forming to the child's internalization of religious and political ideologies.

It is hardly necessary to give examples, which abound on every side, but I will provide one extreme case to convey the flavor. Suppose a clinical psychologist working in neurology finds (as he would if he bothered) that the normal siblings of children with Tay-Sachs's disease (infantile form of amaurotic family idiocy) are somewhat less prone to physical aggression than random "control" children, and that they show a pattern of superior verbal and inferior spatial abilities on standardized tests. He might be misled into some pretty fruitless genetic, neurological, or social speculations if he were somehow ignorant of the religio-ethnic category *Jewish*. As it happens, we can provide a plausible explanation of these strange correlations, but we have to rely on several different sorts of information from very different disciplines. The fact that Tay-Sachs's disease is almost (not quite) confined to Jews presumably arises from some ancient accident of genetic drift under migration (this mutation can hardly have any reproductive advantage), combined with the cultural fact of a zealous religiously based avoidance of miscegenation. The lesser physical aggression of Jewish children is cultural, partly based upon traditional contempt for violence ("The goyim use their fists as a substitute for brains," as one of my Jewish patients put it) and the Jews' centuries-old persecuted minority status which renders physical counteraggression a poor tactic. There are data

[14] I can still recall vividly my astonishment, during the first conversation I ever had with Professor Carnap in the middle 1950's, when—in response to my objection to his tentative definition of 'derived nomological' that it would render many "laws" of biological and social science as accidental universals—he replied calmly, "But of course they are; it is, however, quite harmless to call them laws, for most purposes."

showing a rather pronounced verbal/spatial disparity among Jews[15] so that the "Jewish factor" also underlies the association between this trait relation and Tay-Sachs's disease in a sibling. The differential ability pattern for Jews itself remains to be explained, however. Easy cultural explanations are available (e.g., Talmudic value of words) but one cannot entirely exclude a genetic contribution as partially responsible. In any case, our present-day trait correlations are the end result of the confluence of factors ranging from random genetic mutations and drift to the "historical accident" that a Middle East tribe of gifted nomads invented ethical monotheism five or six thousand years ago!

The puzzling Tay-Sachs correlations are rendered easily explicable by the clear-cut character of the clinical entity (pathognomic signs, early appearance, regular course) and its simple mode of inheritance (Mendelian recessive of complete penetrance). When we deal with nonpathological traits or trait clusters involving only moderate correlations among continuous variables ("loose-knit syndromes") the causal unscrambling job is much harder. Consider, for example, the association between socioeconomic level, child-rearing practices, and impulse control (inhibition of overt aggression, ability to postpone gratification, frustration tolerance). Social learning doubtless plays the major role in producing these correlations, but it would require environmentalist dogmatism to rule out the possibility of some contribution of polygenic "temperament" factors. There may be inherited dispositions that act through several distinct causal chains, converging upon the same correlational result. Basic CNS parameters affecting one's capacity to inhibit, one's rage readiness, anxiety proneness, delay tolerance, social dominance, etc., could contribute by concurrently influencing (1) the educational and vocational level attained by the parents, (2) the social models they provide for the child, (3) the child's genetic disposition to respond to social controls, (4) the parental reactions to the child's modes of responding, (5) the over-all gratification/ frustration level in the home, and so forth.

We now know that such "temperamental" traits as aggressiveness, social dominance, anxiety susceptibility, liking for alcohol, exploratory tendency, rate of recovery of sex drive after copulation, and general activity level are partially gene-determined in the mouse; that the Basenji dog

[15] G. S. Lesser and S. Stodolsky, "Learning Patterns in the Disadvantaged," *Harvard Educational Review*, 37 (1967), 546–593; G. S. Lesser, G. Fifer, and D. H. Clark, "Mental Abilities of Children from Different Social-Class and Cultural Groups," *Monographs of the Society for Research in Child Development*, 30 (1965), 1–115.

breed differs markedly from the beagle hound in its capacity to develop a canine "conscience" through affectionate socializing experiences with humans; and that in the human species, a sizable genetic component of variation ("heritability") obtains for several personality traits, including general intelligence, several "special abilities" (e.g., dexterity, mechanical, spatial, verbal), pattern of vocational interests, self-control, anxiety proneness, impatience, social introversion, the phenomenology of emotional experience, and the needs for autonomy, affiliation, aggression, and self-exhibition. (I have recently seen a manuscript reporting unexceptionable research findings to the effect that Chinese neonates are more "placid" than Caucasians when tested under standard conditions during their first 72 hours after delivery!) The weight of presently available evidence and the rapid rate at which more of the same is accumulating is such that any rational social scientist should view *as a wide-open research problem* the role of genetic variations in determining inter-trait, trait-history, and trait-status correlations.[16]

I stress the genetic factors partly, in all frankness, to combat the environ-

[16] D. G. Freedman, "Constitutional and Environmental Interactions in Rearing of Four Breeds of Dogs," *Science*, 127 (1958), 585–586; K. Lagerspetz, "Studies on the Aggressive Behavior of Mice," *Annales Academiae Scientarum Fennicae*, 131 (1964), 1–131; E. Slater and J. Shields, "Genetical Aspects of Anxiety," in M. H. Loder, ed., *Studies in Anxiety*, Special Publication no. 3, *British Journal of Psychiatry* (Ashford, Kent, 1969), pp. 62–71; Sandra Scarr, "Genetic Factors in Activity Motivation," *Child Development*, 37 (1966), 663–673; T. E. McGill and W. C. Blight, "Effects of Genotype on the Recovery of Sex Drive in the Male Mouse," *Journal of Comparative and Physiological Psychology*, 56 (1963), 887–888; G. Lindzey, H. Winston, and M. Manosevitz, "Social Dominance in Inbred Mouse Strains," *Nature*, 191 (1961), 474–476; I. Gottesman, "Heritability of Personality: A Demonstration," *Psychological Monographs*, 77, no. 9 (1963); Sandra Scarr, "Social Introversion-Extraversion as a Heritable Response," *Child Development*, 40 (1969), 823–832; Sandra Scarr, "Environmental Bias in Twin Studies," *Eugenics Quarterly*, 15 (1968), 34–40; A.-M. A. Bloch, "Remembrance of Feelings Past: A Study of Phenomenological Genetics," *Journal of Abnormal Psychology*, 74 (1969), 340–347; J. Shields, *Monozygotic Twins Brought up Apart and Brought up Together* (London: Oxford University Press, 1962). On behavior genetics generally, see J. L. Fuller and W. R. Thompson, *Behavior Genetics* (New York: Wiley, 1960); G. E. McClearn and W. Meredith, "Behavioral Genetics," in P. R. Farnsworth, Olga McNemar, and Q. McNemar, eds., *Annual Review of Psychology*, 17 (1966), 515–550; G. E. McClearn, "The Inheritance of Behavior," in L. Postman, ed., *Psychology in the Making* (New York: Knopf, 1962), pp. 144–252; J. Hirsch, "Individual Differences in Behavior and Their Genetic Basis," in E. Bliss, ed., *Roots of Behavior* (New York: Hafner, 1962), pp. 3–23; J. Hirsch, ed., *Behavior-Genetic Analysis* (New York: McGraw-Hill, 1967); B. K. Eckland, "Genetics and Sociology: A Reconsideration," *American Sociological Review*, 32 (1967), 173–194; D. C. Glass, ed., *Biology and Behavior: Genetics* (New York: Rockefeller University Press, 1968); M. Manosevitz, G. Lindzey, and D. D. Thiessen, eds., *Behavioral Genetics: Method and Theory* (New York: Appleton-Century-Crofts, 1969).

mentalist brainwashing which most of my philosopher readers will have received from their undergraduate social-science classes; but mainly because the commonest error in handling nuisance variables of the "status" sort (e.g., income, education, locale, marriage) is the error of suppressing statistically components of variance that, being genetic, ought not to be thus arbitrarily relegated to the "spurious influence" category.[17]

[17] See B. Burks and T. L. Kelley, "Statistical hazards in nature-nurture investigations," *Twenty-Seventh Yearbook of the National Society for the Study of Education, Nature and Nurture, Part I: Their Influence upon Intelligence* (Bloomington: University of Indiana Press, 1928), pp. 9–38. Professor Jane Loevinger, upon reading a draft of the present paper, called my attention to this 42-year-old contribution, which I confess never to have read. My sole justification for retaining those portions of my paper that essentially repeat the old Burks-Kelley arguments is that the social-science literature shows that many of my brethren must never have read them either. I have sometimes wondered whether it is only in the inexact sciences that rather simple methodological truths have to be noticed afresh after the passage of an "academic generation" or two. Does this strange phenomenon occur also in physics and chemistry? In psychology one is uncomfortably aware of the truth of Gide's remark, "It has all been said before, but you must say it again, since nobody listens." For an often-ignored job fifteen years after Burks and Kelley, see J. Loevinger, "On the Proportional Contributions in Nature and in Nurture to Differences in Intelligence," *Psychological Bulletin*, 40 (1943), 725–756. An excellent methodological discussion of genetic factors in relation to social class —the nuisance variable most often "controlled for" in social-science research—is I. Gottesman, "Biogenetics of Race and Class," in M. Deutsch, I. Katz, A. R. Jensen, eds., *Social Class, Race and Psychological Development* (New York: Holt, Rinehart and Winston, 1968), pp. 11–51. I have found remarkably little explicit discussion of the causal-arrow ambiguity problem in writings by professional statisticians, the most helpful exception being Kempthorne, *An Introduction to Genetic Statistics*. Presumably this is because they (a) take it as perfectly obvious, (b) think in terms of agricultural research, where background knowledge usually excludes one of two causal directions, and (c) deal with experimental manipulations rather than "passive observation" of cross-sectional relations presented by experiments of nature, such as we perforce study in differential psychology and sociology. When the causal-arrow ambiguity problem is briefly considered in connection with analysis of covariance by statisticians, their concern is over the possible influence of "treatments" upon the "concomitant [= nuisance] variable," such that a regression-based adjustment of output means would lead to an underestimate of treatment effects. This is not quite the same as our present problem, although closely related. See, for example, M. S. Bartlett, "A Note on the Analysis of Covariance," *Journal of Agricultural Science*, 26 (1936), 488–491; W. G. Cochran, "Analysis of Covariance: Its Nature and Uses," *Biometrics*, 13 (1957), 261–281; H. Scheffé, *The Analysis of Variance* (New York: Wiley, 1959), pp. 198–199; B. Ostle, *Statistics in Research*, 2nd ed. revised (Ames: Iowa State University Press, 1963), pp. 456–457. What is more surprising is the lack of explicit discussion in expositions of analysis of covariance written by psychologist-statisticians for social-science readership. For a brief, clear, and persuasive refutation of the still-prevalent notion that statistical weights in multivariate prediction systems somehow quantify "[causal] influence," see L. Guttman, "An Outline of the Statistical Theory of Prediction," in P. Horst, ed., *The Prediction of Personal Adjustment*, Social Science Research Council Bulletin, 48 (1941), 286–292. This 29-year-old SSRC bulletin is insufficiently known and still very much worth study.

Since socio-psychological correlations are the outcome of so complex a causal situation, the formulation of legitimate counterfactuals is extraordinarily difficult. It should be noted that this complexity obtains not merely because of the sheer *number* of relevant factors so commonly mentioned, but also because in the life histories of a group of subjects there are numerous possibilities of *correlated initial and boundary conditions* (e.g., an upper class subject has heard better grammar and may also possess family-name leverage at college admission), *subject-selected learning experience* (e.g., if you never give studying a try you can't discover that getting A's can be fun), *social feedback loops* (e.g., aggressive personal style elicits counteraggression by social objects, which may further increase the subject's own aggression), *autocatalytic processes* (e.g., poor performance yields situational anxiety as a by-product, which further accelerates performance decline), and *critical junctures in "divergent" causality* (e.g., atypical carbohydrate breakfast → mid-morning hypoglycemia → temper outburst at boss → failure to get expected promotion → last straw for ambitious wife → divorce scandal → alcoholism → suicide).[18]

[18] Roughly, in divergent causal chains, small initial-condition fluctuations determine very different remote outcomes; in convergent situations, small fluctuations "average out" so that whether any one individual initial event is E or ~E has a negligible effect on the system's direction of movement. See I. Langmuir, "Science, Common Sense and Decency," *Science*, 97 (1943), 1–7; I. D. London, "Some Consequences for History and Psychology of Langmuir's Concept of Convergence and Divergence of Phenomena," *Psychological Review*, 53 (1946), 170–188; P. E. Meehl, *Clinical versus Statistical Prediction: A Theoretical Analysis and a Review of the Evidence* (Minneapolis: University of Minnesota Press, 1954), pp. 37–67. Langmuir's distinction is of course implicit in numerous historical and fictional treatments of the theme "small causes, great effects." A familiar example is speculation about whether World War I would have broken out if the obstetrician who delivered Wilhelm II had been more skillful, as a result of which the Kaiser would have been spared his withered arm, hence would have felt less need to overcompensate, etc. "For want of a nail the shoe was lost; for want of a shoe the horse was lost; for want of a horse the rider was lost; for want of a rider the battle was lost; for want of a victory the kingdom was lost." Machiavelli (*The Prince*) points out that all of Cesare Borgia's careful planning went for nothing because he could not have foreseen that he would be lying desperately ill at the very moment the Papacy was vacated by the death of his father (Alexander VI). For fictional emphasis on the critical role of minor, quasi-random fluctuations see L. Sterne, *Tristram Shandy*; L. Tolstoy, *War and Peace*; J. O'Hara, *Appointment in Samarra*; J. H. Wallis, *Once off Guard*. See also I. D. London, "Quantum Biology and Psychology," *Journal of General Psychology*, 46 (1952), 123–149; P. Jordan, *Science and the Course of History* (New Haven, Conn.: Yale University Press, 1955), pp. 108–113; J. R. Platt, *The Step to Man* (New York: Wiley, 1966), pp. 174–177; M. H. Pirenne and F. H. C. Marriott, "The Quantum Theory of Light and the Psychophysiology of Vision," in S. Koch, ed., *Psychology, a Study of a Science*, vol. 1: *Sensory, Perceptual, and Physiological Formulations* (New York: McGraw-Hill, 1959), pp. 288–361; F. Ratliff, "Some Interrelations among Physics, Physiology, and Psychology in the Study of Vision,"

Paul E. Meehl

The correlational statistics relating trait, status, and history variables within a defined social group depend causally upon the "accidental universals" (more precisely, the "accidental joint frequency distributions") that happen to prevail in that society, given its gene pool, geographic setting, economic system, class structure, political institutions, legal forms, and so on. In attempting to formulate quantitative counterfactuals on the basis of these statistics, we implicitly assume that imagined alterations in selected particulars would be nomologically possible without an entailed disturbance in the statistical structure (the numerical claims of the counterfactual being based upon that structure's parameters). As of this writing it remains unclear to me when, if ever, this assumption is warranted, although it does seem that some situations make it more plausible than others. The trouble is that, while I cannot produce any clear criteria, I have the impression that the "safest" cases are those in which well-confirmed theoretical knowledge already exists. (In agricultural experiments we can be confident about the causal status of soil heterogeneity as a nuisance variable; hence calculating what Fisher labels "adjusted yields" in an analysis of covariance leads fairly directly to a legitimate counterfactual concerning the output averages.) If I am essentially correct in this impression, the social scientist's position is discouraging because he wants typically to rely upon his quantitative counterfactuals as a basis for causal theorizing rather than the other way round.

To concretize the discussion, consider again the example of treating a student's IQ as a nuisance variable in a research study which aims to ascertain the relationship between educational level attained and subsequent adult income. Since the textual interpretation of the counterfactual corresponding either to an analysis of covariance or to the now largely abandoned partial correlation presents an identical problem, I shall use partial correlation because the statistics of the situation is easier to discuss. The working formula for a partial correlation, being expressed in terms of algebraic operations (taking products and differences) upon the three zero-order correlations, obscures what really underlies the process of "partialing

especially pp. 442–445, in S. Koch, ed., *Psychology, a Study of a Science*, vol. 4: *Biologically Oriented Fields* (New York: McGraw-Hill, 1962). As an extreme, dramatic, but perfectly possible example of Jordan's Verstärkung or Langmuir's divergence, suppose Adolf Hitler to have been lost on a dark night while serving as message-runner in World War I. A quantum-indeterminate event in his retinal receptors is amplified neurally as a result of which he turns his head toward a faint light source just as an enemy sniper fires slightly off target. Six million Jews would have escaped liquidation thereby!

out" a nuisance variable such as IQ. In deriving the partial correlation formula, what do we do? Let x = educational level attained, y = adult income, and z = IQ. In the algebra underlying the final partial correlation formula, which purports to tell us "what the correlation between income and schooling *would be*, except for the influence of IQ [as a nuisance variable]," designated by the partial correlation coefficient $r_{xy \cdot z}$, what we do algebraically in the derivation is to construct a set of residuals constituting a difference variable u, obtained by regressing the first variable of interest x upon the nuisance variable z; we then consider the set of residuals constituting a constructed variable v obtained by regressing the other variable of interest, y, upon the nuisance variable z; and then we correlate these residuals. The resulting coefficient of correlation r_{uv} is called the partial correlation between x and y with z held constant $(= r_{xy \cdot z})$. Since it turns out in the algebra that the magnitude of this new coefficient is computable directly from the zero-order correlations without actually going through the steps of computing all of these residuals u_i and v_i on the individual subjects, the cookbook user of partial correlation is not, so to speak, forced by the working formalism (unless he refreshes himself on the derivation) to look the counterfactual problem squarely in the face when asking himself how this final derived number is to be textually interpreted.

Let us examine one of those residuals as it appertains to an individual subject of our research investigation. Plugging in the value of his IQ in the best fitting x-on-z regression equation (I assume linearity as a condition for the Pearson r to be an adequate descriptive statistic), we "estimate" how far he should go in school. Similarly, plugging his IQ into the regression equation of y-on-z, we estimate how much money he should be earning at age 35. We then find that he didn't go precisely as far in school as our regression equation would "predict," nor does he earn exactly as much money at age 35 as the other regression equation would "predict." That is, there is a discrepancy u_i between what we would expect him to earn and what he actually earns, and a discrepancy v_i between how far we would expect him to go in school and how far he actually went in school. It is these two discrepancy values u_i, v_i which are correlated over the entire group of individuals. The question of interest is, how is each of these to be interpreted as applied to him? Can we say, for example, "If this subject had had a higher IQ by so-and-so many points, then he would have proceeded farther in school, by such-and-such many grades"? Does the regression line of schooling upon IQ legitimate such a counterfactual? I do not assert dogmatically that it does not; but it seems to me evident that there is considerable doubt about whether it does. Do we mean, for example, "If everything else that happened to him was exactly as it in fact was, but his IQ had been so-and-so many points higher, then he would have gone such-and-such many more grades in school"? Is that the intended translation? If it is, is it a valid counterfactual legitimated by the regression equation? I for one do not know, and I doubt that anybody else knows either. It might very well be a counter-nomological, since it might require a violation of some laws of social psychology for his parents, teachers, and peers to treat him exactly as they in fact did, given that his IQ was significantly changed from what it in fact was. It does not, for example, require any far-out speculating to be fairly certain that a child with an IQ = 140 living in a somewhat anti-intellectual proletarian family would be reacted to rather differently by his siblings, and by his high-school dropout father, than would a child, similar genetically in all other respects, but with an IQ of 95! It won't do to solve this by main force, simply saying, "Well, we are going to insist upon translating the counterfactual so as to ensure that everything else happens to him exactly as it did, given his actual IQ." An easy way to exclude this heavy-handed approach is to point out that there is a necessary quantitative interdependence between such factors as social reinforcement and the behavior of the individual under study. To say that we are going to assume that everything else is just as it was in this type of situation is rather like saying that we are going to assume that everything about a pigeon's reinforcement schedule in a Skinner box could be "just as it was," while concurrently assuming that the pigeon responded at twice as high a rate. Under such circumstances, either you have to decide that the pigeon will end up receiving a larger number of total reinforcements, or—if it is insisted that the total pellets delivered are to be held constant in the counterfactual—then there must be an alteration of the reinforcement schedule. You can't have it both ways. Furthermore, in the human case matters become very complicated because of the fact that humans can talk to themselves about the schedules they're being put on by their social environment. If we insist, say, that the proportion of times a school teacher says "right" versus "wrong" in the child's second-grade school experiences be held constant, then giving him another 30 points in IQ will require that the teacher say "wrong" on quite a few occasions when the child "knows that he is in fact right." Obviously this will have a profound effect on his attitudes regarding work, achievement, payoff, elders as representatives of the larger society, and so on.

I do not of course mean to argue that there cannot be any counterfactuals involving "corrections for nuisance variables" that are (a) meaningful and (b) true. My point is that it is frequently—I incline to say typically —difficult to decide about their meaningfulness, and even more difficult to decide about their truth. One can rarely interpret counterfactually a residual about a regression line or plane with confidence that he knows what the counterfactual means and that it is a valid consequence of the relevant nomologicals.

Part of the trouble here is as discussed above, that the statistical system under study is a resultant of the influence—interactive and frequently mutual, i.e., involving feedback—of a large number of variables, known and unknown, and we happen to have selected three of them for study, none of them having been experimentally manipulated by us. From the standpoint of the statistician aiming at a safe (minimum) interpretation, a partial correlation coefficient between variables x and y with z held constant is nothing but the zero-order correlation obtained when we regress x upon y within a narrow z-slice, provided that relationship is invariant over z-slices (rarely tested!). That is, we define a plane located in the three-variable box which is parallel to the xy-plane and located z units out on the z-axis. The locations of the person points in this box are the end result of a multitude of causal factors, varying all the way from a single mutated gene that renders particular individuals mentally deficient to the interpretative vagueness of certain legal language in the Civil Rights Act. There is nothing about the formalism for characterizing the distribution of person points confined to a given z-slice—a process which is of course unobjectionable when given the statistician's minimum interpretation—that enables us to formulate a counterfactual without having to worry about what the whole box would look like if the world were different in certain important ways, biologically and socially, from the way it in fact is. It is easy to see this by considering what a very strong counterfactual, textually interpreting a partial correlation, would read like. We often speak of the partial correlation as telling us what the "true correlation would be if the nuisance variable were held constant." Suppose we attempt the counterfactual "If there were no IQ differences in the population, then the correlation between years of schooling and subsequent income would be $= r_{xy \cdot z}$." This strong counterfactual is clearly impermissible on two counts.

399

First, the antecedent is (effectively) counter-nomological in genetics, given the probabilistic mechanisms of gene assortment. (If this objection were to be rebutted by pointing out that the laws of genetics are themselves—strictly speaking—"accidental universals," structure-dependent outcomes of our world's cosmic history, one rejoinder would be that for the social scientist, operating at *his* level of explanation, the laws of biology can be taken as nomologicals.) Second, even if we allow the antecedent, we surely cannot assume that the statistical structure would be as it is if human beings all had the same g-factor. (For expository simplicity I have treated IQ as g-factor, which is of course a gross distortion. The IQ is a fallible measure of g-factor, and g-factor is itself the result of polygenic hereditary components interacting with life-history parameters. Needless to say, this oversimplification only weakens my argument.) In fact such a supposition would almost certainly be erroneous. The whole educational system would probably have evolved very differently. Teachers' attitudes and beliefs about students would be radically different from what they are. Employers' interpretations of the school record at job entry would be quite unlike what they are in our world. It would be pointless for me to compile a long list of "social-facts-that-would-be-otherwise" in documenting something so obvious as the theses: If a major source of achievement-related individual differences were removed, society would be considerably changed; and the statistical structure relating trait, history, and status variables would be so materially different that quantitative counterfactuals based upon the received structure's parameters are all invalid.

This paper was criticized by two sociologist reviewers on the plausible but specious grounds that the matched-case method has been replaced in sociological research by the use of multivariate designs. Aside from the fact that current social-science generalizations and theory rely in part upon earlier investigations employing matching, and the fact that matching has by no means been completely replaced by multivariate analysis in social-science research, I must emphasize that these critics do not see the main point I am making. The core difficulty is *not* eliminated when we substitute multivariate analysis for case-matching, as should be obvious to anyone who understands the mathematics underlying the derivation of multivariate estimates. Thus, for example, in the analysis of covariance, the "influence" of a nuisance variable is sought to be removed algebraically, by calculating an F-test on the means of the output variable of interest upon residuals obtained when this output variable of interest has first been regressed upon the nuisance variable and the output means "adjusted" accordingly. As in the older partial correlation formula, what we are actually doing in the analysis of covariance may be obscured (to the "cookbook user" of statistical formulas) by the fact that computational method bypasses the actual calculation of these individual case residuals about the nuisance variables' regression line. It cannot be overemphasized in the present context that analysis of covariance as a method of control by statistics rather than by experimental manipulation *suffers from precisely the same inherent methodological vice in the social sciences as does the method of matched groups.* In the matched-group method, the investigator physically constitutes a nonrepresentative "artificial" subpopulation for study. In multivariate analysis, he concocts statistically, by the making of certain algebraic "corrections," a virtual or idealized sample, the members of which are fictional persons assigned fictional scores, to wit, the scores the investigator algebraically infers they *would* have had on the output variable of interest if the alleged causal influence of the nuisance variable were removed. The empirical meaning of this "virtual," fictional, idealized, inferred-score population is totally dependent upon our giving a correct interpretation to the presupposed causal counterfactual.[19] One might even maintain—although I do not wish to press the point—that modern multivariate analysis is *farther* removed from physical reality than the old matched-group procedure, because the latter at least deals with an actual physical subpopulation, a set of real scores obtained by existent individuals, atypical though they may be; whereas the multivariate method, by its very nature, deals with a fictional or "virtual" score distribution whose elements were generated computationally by the investigator.

As I said above, it is not clear what exactly is the relationship between the three aspects of the problem which I have christened "systematic unmatching," "unrepresentative subpopulations," and "causal-arrow ambiguity." But it seems to me that taken together, and combined with the problem (operating in the other direction) discussed by Kahneman, they force us to the conclusion that a large portion of current research in the behavioral sciences, while meeting the conventionally accepted standards

[19] H. A. Simon and N. Rescher, "Cause and Counterfactual," *Philosophy of Science*, 33 (1966), 323–340. This is a very illuminating article, the best I have seen on the subject, and includes a formal proof that no statistical manipulations performed on static data can resolve the causal-arrow-ambiguity problem.

of adequate design, must be viewed as methodologically unsound; and, more specifically, I suggest that the ex post facto design is in most instances so radically defective in its logical structure that it is in principle incapable of answering the kinds of theoretical questions which typically give rise to its use.

Some Methodological Reflections on the Difficulties of Psychoanalytic Research

Being here in the somewhat ill-defined role of a "methodologist" with psychoanalytic experience, I shall first make a few general comments reflecting my own views on philosophy of science. Since it is impossible to develop or defend them in a brief presentation, let me simply say that these views accord generally with the consensus of those who claim expertise—if such exists, as I believe—in that field. That they are not widely accepted in psychology reflects a failure to keep up with developments, many psychologists espousing a philosophical position that is some thirty years out-of-date.

Whatever the verisimilitude[1] of Freud's theories, it will surely be a matter of comment by future historians of science that a system of ideas which has exerted such a powerful and pervasive influence upon both professional practitioners and contemporary culture should, two-thirds of a century after the promulgation of its fundamental concepts, still remain a matter of controversy. That fact in itself should lead us to suspect that there is something methodologically peculiar about the relation of psychoanalytic concepts to their evidential base.

Let me begin by saying that I reject what has come to be called "opera-

AUTHOR'S NOTE: Revised and expanded version of a paper read at a joint session of the Division of Experimental Psychology and the Society for Projective Techniques, seventy-fourth annual convention of the American Psychological Association at New York City, September 6, 1966 (see *American Psychologist*, 21 (1966), 701). The present version is appearing as part of a symposium in *Psychological Issues* and is printed here by the kind permission of Dr. George S. Klein, editor.

This work was largely done during my summer appointments as professor in the Minnesota Center for Philosophy of Science, under support from the Carnegie Corporation of New York. I should also like to express my indebtedness to colleagues Herbert Feigl, Carl P. Malmquist, and Grover Maxwell for the stimulation, criticism, and clarification provided, as always, at Center discussions.

[1] K. R. Popper, *The Logic of Scientific Discovery* (New York: Basic Books, 1959); *Conjectures and Refutations* (New York: Basic Books, 1962).

Paul E. Meehl

tionism" as a logical reconstruction of scientific theories. Practically all empiricist philosophers (e.g., Carnap, Feigl, Feyerabend, Hempel, Popper, Sellars), thinkers who cannot by any stretch of the imagination be considered muddleheaded, obscurantist, or antiscientific in their sympathies, have for many years recognized that strict operationism (in anything like the form originally propounded by Bridgman) is philosophically indefensible.[2] In saying this, they do not, however, prejudge those issues of scientific research *strategy* that arise between a quasi-operationist like Skinner and a psychoanalytic theorist. And it is commendable that Skinner and his followers (unlike some psychologists) have been careful to avoid invoking "philosophy of science" in their advocacy of either substantive views or research strategy.

Associated with my rejection of operationism is the recognition that biological and social sciences are forced to make use of what have come to be known[3] as "open concepts," the "openness" of these concepts having two or three distinguishable aspects which space does not permit me to develop here. One important consequence of this openness is that we must reject Freud's monolithic claim about the necessity to accept or reject psychoanalysis as a whole. This is simply false as a matter of formal logic, even in explicitly formalized and clearly interpreted theoretical systems, and such a systematic "holism" is a fortiori untenable when we are dealing with what is admittedly a loose, incomplete, and unformalized conceptual system like psychoanalysis. It is well known that proper subsets of postulates in physics, chemistry, astronomy, and genetics are continually being changed without "changing everything else" willy-nilly, and it is absurd to suppose psychoanalytic theory, unlike these advanced sciences, is a corpus of propositions so tightly interknit that they have to be taken "as a whole."

I would also reject any requirement that there should be a *present mapping* of psychoanalytic concepts against constructs at another level of analysis, such as neurophysiology or learning theory. All that one can legitimately require is that psychoanalytic concepts ought not to be incom-

---

[2] But see F. Wilson, "Is Operationism Unjust to Temperature?" *Synthese*, 18 (1968), 394–422; "Definition and Discovery: I, II" *British Journal for the Philosophy of Science*, 18 (1967), 287–303, and 19 (1967), 43–56.

[3] Following the late, great philosopher Arthur Pap, "Reduction Sentences and Open Concepts," *Methodos*, 5 (1953), 3–30; *Semantics and Necessary Truth* (New Haven, Conn.: Yale University Press, 1958). See also L. J. Cronbach and P. E. Meehl, "Construct Validity in Psychological Tests," *Psychological Bulletin*, 52 (1955), 281–302.

SOME METHODOLOGICAL REFLECTIONS

patible with well-corroborated theories of the learning process or nervous system function. But the situation in these two fields is itself so controversial that this negative requirement imposes only a very weak limitation upon psychoanalytic theorizing.

I would also combat the tendency (found in some psychonomes) to treat the terms "experimental" and "empirical" as synonymous. An enterprise can be empirical (in the sense of taking publicly observable data as its epistemic base) *without* being experimental (in the sense of laboratory manipulation of the variables). Such respectable sciences as astronomy, geography, ecology, paleontology, and human genetics are obvious examples. We should not conflate different dimensions such as the following: experimental-naturalistic; quantitative-qualitative; objective-subjective; documentary-behavioral. It is obvious, for example, that one can carry out objective and quantitative analysis upon a nonexperimental document (e.g., diary, personal correspondence, jury protocol).

I should make clear that while I am not an "orthodox Popperian," I find myself more in sympathy with the logic and methodology of science expounded by Sir Karl Popper[4] than with that of any other single contemporary thinker. While I share with my Minnesota colleagues Feigl and Maxwell reservations about Sir Karl's complete rejection of what he calls "inductivism," I agree with Popper in emphasizing the extent to which theoretical concepts (often implicit) pervade even the so-called "observation language" of science and of common life; and I incline to accept refutability (falsifiability) as the best criterion to demarcate science from other kinds of cognitive enterprises such as metaphysics.

There is a certain tension between these views. What I have said about operationism, open concepts, and the scientific status of nonexperimental investigation makes life easier for the psychoanalytic theorist; but the Popperian emphasis upon falsifiability tends in the opposite direction.

As a personal note, I may say that, as is true of most psychologists seriously interested in psychoanalysis, I have found my own experience on the couch, and my clinical experience listening to the free associations of patients, far more persuasive than any published research purporting to test psychoanalytic theory. I do not assert that this is a good or a bad thing, but

[4] Popper, *The Logic of Scientific Discovery* and *Conjectures and Refutations;* and see M. Bunge, ed., *The Critical Approach: Essays in Honor of Karl R. Popper* (New York: Free Press, 1964).

405

Paul E. Meehl

I want to have it down in the record. In the "context of discovery"[5] this very characteristic attitude is worth keeping in mind.

The inventor of psychoanalysis took the same view, and it might be good research strategy to concentrate attention upon the verbal behavior of the analytic session itself. If there is any strong empirical evidence in support of Freud's ideas, this is perhaps the best place to look, since this is where he hit upon them in the first place! We have today the advantage which he regrets not having, that recording an analysand's verbal behavior is a simple and inexpensive process. Skinner points out that what makes the science of behavior difficult is *not*—contrary to the usual view in psychoanalytic writing—problems of *observation*, because (compared with the phenomena of most other sciences) behavior is relatively macroscopic and slow. The difficult problems arise in slicing the pie, that is, in classifying intervals of the behavior flux and in subjecting them to powerful conceptual analysis and appropriate statistical treatment. Whatever one may think of Popper's view that theory subtly infects even so-called observation statements in physics, this is pretty obviously true in psychology because of the trivial fact that an interval of the behavior flux can be sliced up or categorized in different ways. Even in the animal case the problems of response class and stimulus equivalence arise, although less acutely. A patient in an analytic session says, "I suppose you are thinking that this is really about my father, but you're mistaken, because it's not." We can readily conceive of a variety of rubrics under which this chunk of verbal behavior could be plausibly subsumed. We might classify it syntactically, as a complex-compound sentence, or as a negative sentence; or as resistance, since it rejects a possible interpretation; or as negative transference, because it is an attribution of error to the analyst; or, in case the analyst hasn't been having any such associations as he listens, we can classify it as an instance of projection; or as an instance of "father theme"; or we might classify it as self-referential, because its subject matter is the patient's thoughts rather than the thoughts or actions of some third party; and so on and on. The problem here is not mainly one of "reliability" in categorizing, although goodness knows that's a tough one too. Thorough training to achieve perfect interjudge scoring agreement *per rubric* would still leave us with the problem I am raising.

[5] H. Reichenbach, *Experience and Prediction* (Chicago: University of Chicago Press, 1938), but see I. Lakatos, "Criticism and the Methodology of Scientific Research Programmes," *Proceedings of the Aristotelian Society*, 69 (1968), 149–186.

There are two opposite mistakes which may be made in methodological discussion on the evidential value of verbal output in a psychoanalytic hour. One mistake is to demand that there should be a straightforwardly computable numerical probability attached to each substantive idiographic hypothesis, of the sort which we can usually compute with regard to the option of rejecting a statistical hypothesis. This mistake arises from identifying "rationality in inductive inference" with "statistical hypothesis testing." One need merely make this identification explicit to realize that it is a methodological mistake. It would, for instance, condemn as "non-rational" all assessment of substantive scientific theories, or the process of inference in courts of law, or evaluation of theories in such disciplines as history or paleontology. No logician has succeeded in constructing any such automatic numerical "evidence-quantifying" rules, and many logicians and statisticians are doubtful whether such a thing could be done, even in principle. It is obvious, for instance, that a jury can be put in possession of a pattern of evidence which makes it highly rational to infer beyond a reasonable doubt that the defendant is guilty; but no one (with the exception of Poisson in a famous ill-fated effort) has tried to *quantify* this evidential support in terms of the probability calculus. Whether a distinction can be made between quantifying the corroboration of nomethetic theories[6] and quantifying the probability of particularistic (= idiographic) hypotheses is difficult to say, although we should pursue that line of thought tenaciously. Ideally, I suggest, a Bayes Rule calculation on the idiographic constructions of psychoanalysis should be possible.

The opposite error is the failure to realize that Freud's "jigsaw-puzzle" analogy does not really fit the psychoanalytic hour, because it is simply not true (as he admits elsewhere) that all of the pieces fit together, or that the criteria of "fitting" are sufficiently tight to make it analogous even to a clear-cut criminal trial. Two points, opposite in emphasis but compatible: Anyone who has experienced analysis, practiced it, or listened to taped sessions, if he is halfway fair-minded, will agree that (1) there are sessions where the material "fits together" so beautifully that one is sure almost any skeptic would be convinced, and (2) there are sessions where the "fit" is very loose and underdetermined (fewer equations than unknowns, so to speak), this latter kind of session (unfortunately) predominating.

---

[6] An algorithm for which, says Lakatos, p. 324, is precluded by Church's theorem. I. Lakatos, "Changes in the Problem of Inductive Logic," in I. Lakatos, ed., *The Problem of Inductive Logic* (Amsterdam: North-Holland, 1968), pp. 315–417.

The number of theoretical variables available, and the fact that the theory itself makes provision for their countervailing one another and reversing qualities (e.g., the dream work's sometime expression of content by opposites), lead to the ever-present possibility that the ingenuity of the human mind will permit the therapist to impose a construction which, while it has a certain ad hoc plausibility, nevertheless has low verisimilitude. What we would like to have is a predictive criterion, but the trouble is that the theory does not claim to make, in most cases, highly specific content predictions. Thus, as Freud himself pointed out, while we can sometimes make a plausible case for the occurrence of certain latent dream thoughts which were transformed through the dream work into the manifest content of a dream, the same set of dream thoughts *could* have been responsible for a manifest content completely different. Similarly, in paleontology, the fossil data may be rationally taken to lend support to the theory of evolution, but there is nothing in the theory of evolution that enables us to predict that such an organism as the rhinoceros will have been evolved, or that we should find fossil trilobites. Or, again, the facts may strongly support the hypothesis that the accused had a motive and the opportunity, so he murdered the deceased; but these assumptions would be equally compatible with his having murdered him at a different time and place, and by the use of a knife rather than a revolver. I do not myself have any good solution to this difficulty. The best I can come up with is that, lacking a rigorous mathematical model for the dream work, and lacking any adequate way of estimating the strengths of the various initial conditions that constitute parameters in the system, we should at least be able to apply crude counting statistics, such as theme frequencies, to the verbal output that occurs during the later portions of the hour when these are predicted (by psychoanalytically skilled persons) from the output at the beginning of the hour. I look in this direction because of my clinical impression that one's ability to forecast the *general theme* of the associative material from the manifest content of the dream plus the initial associations to it, while far from perfect, is nevertheless often good enough to constitute the kind of clinical evidence that carries the heaviest weight with those who open-mindedly but skeptically embark upon psychoanalytic work. Let me give a concrete example of this (one on which I myself would be willing to lay odds of $90 to $10, and on more than a mere "significant difference" but on an almost complete predictability within the limits of the reliability of thematic classification). If a male patient dreams about fire and water, or dreams about one and quickly associates to the other (and here the protocol scoring would be a straightforward, objective, almost purely clerical job approaching perfect interscorer reliability), the dominant theme in the remainder of the session will involve *ambition* as a motive and *shame* (or triumph) as an affect. In 25 years as a psychotherapist I have not found so much as a single exception to this generalization. This kind of temporal covariation was the essential evidential base with which Freud started, and I suggest that if sufficient protocols were available for study, it is the kind of thing which can be subjected to simple statistical test. Since there is no obvious phenotypic overlap in the content, a successful prediction along these lines would strongly corroborate one component of psychoanalytic theory, namely that involving the urethral cluster. Now I believe that there are many such clusterings which could in principle be subjected to statistical test, and my expectation is that, if performed, they would provide a rather dramatic support for many of Freud's first-level inferences, and a pretty clear refutation of others.

Whether or not one is a convinced "Bayesian" is largely irrelevant here, provided we can set *some* safe empirical bounds on the priors, which we can presumably do for the "expectedness" (of our test-observation) in the denominator of Bayes's Formula, relying on statistics from a large batch of unselected interviews. Even expectedness values $\cong$ ½ can become a basis for fairly strong corroborators if there are several all "going the right direction." And if we are real, feisty, honest-to-goodness "Bayesian personalists" about probability, it might be plausibly argued that a fair basis for assigning the priors would be guesstimates by academic psychologists largely ignorant of Freud. This basis of prior-probability assignments permits us to go outside the analytic session into those diverse contexts (for which explicit statistics are lacking) of daily life, history, biography, mythology, news media, personal documents, etc.—data sources which collectively played a major role in convincing the nontherapist intelligentsia that Freud "must have something." Example: No philosophically educated Freudian would have trouble guessing which of these four philosophers wrote a little-known treatise on *wind*: Kant? Locke? Hume? Santayana? A Freudian would call to mind Kant's definition of a moral act as one done *solely* from a sense of duty (rather than, say, a spontaneous loving impulse or a desire to give pleasure); the pedantic punctuality of his daily walk, by which the Königsberg housewives allegedly set their clocks; his remarkable statement that "there can be nothing more dreadful than that the actions of a man should be subject to the will of another"; and his stubborn refusal over many years to speak with a sister following a minor quarrel. But I doubt that a panel of (otherwise knowledgeable) psychologists, ignorant of Freudian theory, would tend to correctly identify Kant as having a scholarly interest in wind—even if we helped them out by adding the fact of Kant's excessive concern with constipation in his later years. The same is no doubt true of my rash prediction (upon first descending the stairs inside the Washington monument) that the wall plaques would show more financial contributions by fire departments than by police departments. (They do.) Point: The very "absurdity" or "farfetched" character of many psychoanalytic *connections* can be turned to research advantage, because the prior probabilities of such-and-such correlations among observables are so very differently estimated by one thinking outside the Freudian frame.

As must be apparent from even these brief and (unavoidably) dogmatic remarks, I locate the methodological difficulties of testing psychoanalytic theory differently from many—perhaps most?—who have discussed it, whether as protagonists or critics. For example, I do not waste time defending[7] the introduction of unobservable theoretical entities, knowing as I do that the behaviorist dogma "Science deals only with observables" is historically incorrect and philosophically ludicrous. The proper form of the "behavioristic" objection is, as always in sophisticated circles, to the *kind* of theoretical entity being invoked (read: its role in the postulated nomological network, including linkages to data statements). Methodological insight quickly shifts our attention away from such philosophical issues to examination "of the merits," as the lawyers would say. Let me emphasize that I do not rely tendentiously upon philosophy-of-science considerations as a *defense* of psychoanalytic theory either. To rebut a dumb objection is merely to rebut a dumb objection; it does not make a scientific case. Those of us who are betting on a respectable verisimilitude in the Freudian corpus must beware of taking substantive comfort in this indirect way, as some "Chomskyites" are currently taking comfort from (easy) refutations of unsound philosophical positions employed by certain of their S-R-reinforcement opponents. We must try to be honest with ourselves even though we are (as always in science) "betting on a horse race." It simply won't do to get relaxed about the dubious methodological

[7] Else Frenkel-Brunswik, "Psychoanalysis and the Unity of Science," *Proceedings of the American Academy of Arts and Sciences*, 80 (1954), 271–350.

status of, say, a postulated "bargain between ego and superego" as explaining why Smith cuts himself shaving before visiting his mistress, on the ground that the superego is a theoretical construct, and that's peachy, since physicists can't see the neutrino either!

Having mentioned the neutrino, I am led to a comment on falsifiability in the inexact sciences. You will recall that when Pauli cooked up the neutrino idea in 1931—solely to preserve the conservation laws ad hoc!—the theory itself showed that the neutrino hypothesis was probably not falsifiable, because the imagined new particle had zero charge and zero restmass. It was not until 1956, twenty-five years later, that a very expensive, never-replicated experiment by Reines and Cowan successfully detected the neutrino (more) "directly." The auxiliary assumptions involved (e.g., would the cross section of cadmium nucleus be large enough?) were themselves so problematic that a negative experimental result could just as plausibly have counted against them as against the theory of interest. While Popper's stress on falsifiability (and the correlative idea that theories become well corroborated by passing stringent tests) is much needed by the psychologist, partly as an antidote to the current overreliance on mere null-hypothesis refutation as corroborating complex theories,[8] it has become increasingly clear that a too-strict-and-quick application of modus tollens would prevent even "good" theories (i.e., theories having high verisimilitude) from getting a foothold. "All theories are lies, but some are white lies, some gray, and some black." The most we can expect of psychodynamic theories in the foreseeable future is that some of them are gray lies. My own predilection is therefore for a neo-Popperian position, such as is represented by Feyerabend, Lakatos, and Maxwell.[9] But what precisely this methodological position means for the strategy of testing psychoanalytic theory is difficult to discern in the present state of the

[8] See W. Rozeboom, "The Fallacy of the Null-Hypothesis Significance Test," *Psychological Bulletin*, 67 (1960), 416–428; D. Bakan, "The Test of Significance in Psychological Research," *Psychological Bulletin*, 66 (1966), 423–437; P. E. Meehl, "Theory-Testing in Psychology and Physics: A Methodological Paradox," *Philosophy of Science*, 34 (1967), 103–115; D. T. Lykken, "Statistical Significance in Psychological Research," *Psychological Bulletin*, 70 (1968), 151–159.

[9] P. K. Feyerabend, "Attempt at a Realistic Interpretation of Experience," *Proceedings of the Aristotelian Society*, 58 (1958), 143–170; "On the Interpretation of Scientific Theories," *Proceedings of the Twelfth Congress of Philosophy* (Venice and Padua), 5 (1958), 151–159; "Explanation, Reduction, and Empiricism," in H. Feigl and G. Maxwell, eds., *Minnesota Studies in the Philosophy of Science*, vol. III (Minneapolis: University of Minnesota Press, 1962), pp. 28–97; "Problems of Microphysics," in R. G. Colodny, ed., *Frontiers of Science and Philosophy* (Pittsburgh: Uni-

Paul E. Meehl

philosophers' controversy. My own tentative predilection is for stronger theories,[10] such strong theories being subjected to more tolerant empirical tests than Popper or Platt seems to recommend. Discussion of this very complicated issue would take us too far afield, but suffice it to say that I now view the position presented in my 1967 paper as overly stringent, although its main point is still, I think, a valid one.

Perhaps the psychologist should first learn Popper's main lesson, including why Popper considers such doctrines as psychoanalysis and Marxism to be nonscientific theories like astrology (because all three are pseudo-"confirmable" but not refutable), and then proceed to soften the Popperian rules a bit. Whether these suggested "softenings" really conflict with a sophisticated falsificationism, or whether Popper himself would consider them objectionable, we need not discuss here.[11] Specifically, I advocate two "cushionings" of the Popperian falsifiability emphasis:

1. A theory is admissible not only if we know how to test it, but if we know *what else we would need to know* in order to test it.

2. A theory need not be abandoned following an adverse result, if there are fairly strong results corroborating it, since this combination of circum-

versity of Pittsburgh Press, 1962), pp. 189–283; "How to Be a Good Empiricist—A Plea for Tolerance in Matters Epistemological," in B. Baumrin, ed., *Philosophy of Science: The Delaware Seminar*, vol. 2 (New York: Wiley, 1963), pp. 3–39; "Realism and Instrumentalism: Comments on the Logic of Factual Support," in Bunge, ed., *The Critical Approach: Essays in Honor of Karl R. Popper*, pp. 280–308; "Problems of Empiricism," in R. G. Colodny, ed., *Beyond the Edge of Certainty* (Englewood Cliffs, N.J.: Prentice-Hall, 1965), pp. 145–260; "Reply to Criticism," in R. S. Cohen and M. W. Wartofsky, eds., *Boston Studies in the Philosophy of Science*, vol. II (New York: Humanities, 1965), pp. 223–261; "Review [of Nagel's *Structure of Science*]," *British Journal for the Philosophy of Science*, 17 (1966), 237–249; "On the Improvement of the Sciences and the Arts, and the Possible Identity of the Two," in R. S. Cohen and M. W. Wartofsky, eds., *Boston Studies in the Philosophy of Science*, vol. III (Dordrecht: Reidel, 1968), pp. 387–415; "Problems of Empiricism, II," in R. G. Colodny, ed., *The Nature and Function of Scientific Theory* (Pittsburgh: University of Pittsburgh Press, 1969); "Pro-Parmenides: A Defense of Parmenidean Apologies" (forthcoming); I. Lakatos, "Criticism and the Methodology of Scientific Research Programmes," *Proceedings of the Aristotelian Society*, 69 (1968), 149–186; "Changes in the Problem of Inductive Logic," in Lakatos, ed., *The Problem of Inductive Logic*, pp. 315–417; "Falsification and the Methodology of Scientific Research Programmes," in I. Lakatos and A. Musgrave, eds., *Criticism and the Growth of Knowledge* (Cambridge: Cambridge University Press, 1970); G. Maxwell, "Corroboration without Demarcation," in P. A. Schilpp, ed., *The Philosophy of Karl Popper* (LaSalle, Ill.: Open Court, forthcoming).

[10] J. R. Platt, "Strong Inference," *Science*, 146 (1964), 347–353; but see E. M. Hafner and S. Presswood, "Strong Inference and Weak Interactions," *Science*, 149 (1964), 503–510.

[11] But see the distinction between Popper$_0$, Popper$_1$, and Popper$_2$ in Lakatos, "Criticism and the Methodology of Scientific Research Programmes."

stances suggests that either (a) the auxiliary hypotheses and *ceteris paribus* clause of the adverse test were not satisfied, or (b) the theory is false as it stands but possesses respectable verisimilitude (i.e., is a gray lie), or both.

I think that these are sensible methodological recommendations that can be rationally defended within a "neo-Popperian" frame, and they do not appear to me to hinge upon resolution of the very technical issues now in controversy among logicians and historians of science. But I hasten to add that such "softening" of the pure, hard-line *modus tollens* rule must not be accompanied by a theoretical commitment such that we persist indefinitely in what Popper stigmatizes as "Parmenidean apologies," clinging to the cherished doctrine in spite of all adverse evidence.[12] When Parmenidean apologies are desirable, *which kinds* and *how long* to persist in them ("theoretical tenacity") are difficult questions.[13]

One big trouble with the application of neo-Popperian strategies to a theory such as Freud's is that the best case for either Parmenidean apologies or continuing use of a "gray-lie" theory in the face of strong and accepted falsifiers is the concurrent existence of strong corroborators, and this usually (not always) requires that the theory have made successful *point* predictions (i.e., predictions of antecedently improbable numerical values). The successful prediction of a mere directional difference is not of this kind, having too high a prior expectedness in Bayes's Formula absent the theory of interest. (If I am right, this atheoretical expectedness in the social and biological sciences approaches ½ as the power of our significance test increases, Meehl 1967.) Yet an attempt to formulate psychoanalytic theory so as to generate such high-risk numerical point predictions is hardly feasible at present. For one thing, the auxiliary hypotheses which are normally treated as (relatively) unproblematic in designing a test experiment are unavailable pending the development of powerful, well-corroborated non-psychodynamic theory (e.g., psycholinguistics). I must say that this state of affairs renders the prospects for cooking up strong tests rather gloomy.

From the standpoint of the experimental psychologist, for whom the experiment (in a fairly tough, restrictive usage of that term) is the ideal method of corroborating or discorroborating theories, the obvious draw-

[12] K. R. Popper, "Rationality and the Search for Invariants," address to International Colloquium on Philosophy of Science, July 11, 1965.

[13] To get the feel of them I recommend reading of P. K. Feyerabend, "Pro-Parmenides: A Defense of Parmenidean Apologies"; also "Realism and Instrumentalism: Comments on the Logic of Factual Support."

backs of the psychoanalytic hour as a data source are two, one on the "input" (= control) side and the other on the "output" (= observation) side. On the input side, unless the analyst's enforcement of the Fundamental Rule relies entirely upon the psychological pressure of a silence—a technical maneuver which is sometimes the method of choice but other times, I think, clearly not—we have the problem of the timing and content of the analyst's interventions as being themselves "biased" by his theoretical predilections. (It would be interesting to play around with the psychoanalytic analogue to a yoked-box situation in operant behavior research.) On the output side, the problem of "objectifying" the classification of the patient's verbal behavior is so complex that when you begin to think hard about it, the most natural response is to throw up your hands in despair. Tentatively I suggest two contrasting methods of such objectification, to wit: First, we rely upon some standard source such as *Roget's Thesaurus* or the Palermo-Jenkins tables or a (to-be-constructed) gigantic atlas of couch outputs emitted under "standard" conditions of Fundamental Rule + analyst silence, for determining whether certain words or phrases are thematically or formally linked to others. Such a "scoring system" bypasses the skilled clinical judge and hence avoids theoretical infection of the data basis. I need hardly point out its grave defect—so grave that a negative result would not be a strong falsifier—which is that the mainly idiographic theme indicators (those which make psychoanalytic therapy fun!) would be lost.

Alternatively, we permit the judgment of a skilled clinician to play a part in classifying the responses, but we systematically prevent his having access to other portions of the material (e.g., to the manifest content of the dream with which the patient commenced a session) so that he will not be "contaminated" by this material. Point: As much as any area of research in clinical psychology, the study of the psychoanalytic interview brings home the importance of solving, by ingenious methods, the perennial problem of "How do we get the advantages of having a skilled observer, who knows what to listen for and how to classify it, without having the methodological disadvantage that anyone who is skilled (in this sense) has been theoretically brainwashed in the course of his training?" In my view, this is *the* methodological problem in psychoanalytic research.

This brings me to my final point, which is in the nature of a warning prophecy more than a reaction to anything presently happening in psychoanalytic research. The philosophical and historical criticisms against classical positivism and naive operationism have (quite properly) included emphasis upon the role of theory in determining what, when, and how we observe. But most of the discussion of these matters has drawn its historical examples from astronomy, physics, and chemistry. In these examples, as I read the record, what the experimenter relied on in "making observations" was (relatively) nonproblematic and independently corroborated portions of the theoretical network for, say, constructing apparatus. The theory of interest was not "relied on" in that sense, although of course in another sense it was "relied on" in deciding what to do and what to look for. It seems to me important to distinguish these two sorts of reliance on theory, and if they are conflated under the broad statement "Theory determines what we observe," I think confusion results. Furthermore, it is misleading (for several reasons) to equate a mass spectrometer or a piece of litmus paper with a psychotherapist as an "instrument of observation." I seem to discern in some quarters of psychology a growing obscurantist tendency—partly anti-empirical but also even at times antirational—which relies upon the valuable and insightful writings of Kuhn and Polanyi[14] for what I can only characterize as nefarious purposes. It would be unfortunate indeed if efforts to objectify psychoanalytic evidence and inference were abandoned or watered down because of a comfortable reliance on such generalizations as "Scientists have commitments," "We often must stick to a theory for want of a better," "You have to know what you are looking for in order to observe fruitfully," "There is no such thing as a pure observational datum, utterly uninfluenced by one's frame of reference." These are all true and important statements, although the last one needs careful explication and limitation. I do not think general comments of this nature are very helpful in deciding how much an analyst subtly shapes the analysand's discourse by the timing of his interventions ("uncontrolled input"), or whether he classifies a bit of speech as "anal" in a theoretically dogmatic manner ("observer bias in recording output"). If the exciting developments in contemporary philosophy of science are tendentiously employed for obscurantist purposes, to avoid answering perfectly sensible and legitimate criticisms, it would be most unfortunate. The good old positivist questions "What do you mean," "How do you know" are still very much in order, and cannot be ruled out of order by

[14] T. S. Kuhn, *The Structure of Scientific Revolutions* (Chicago: University of Chicago Press, 1962); and M. Polanyi, *Personal Knowledge* (London: Routledge and Kegan Paul, 1958).

historical findings about where Einstein got his ideas! "Millikan relied upon a lot of physical theory, treated as unproblematic, when he 'observed' the charge on the electron" is a correct statement of the case. But such a statement is not, most emphatically *not*, on all fours with "Blauberman[15] is a qualified psychoanalyst, therefore we can rely upon his use of psychoanalytic theory when he classifies a patient's discourse as phallic-intrusive." What one *observes* in the psychoanalytic session is words, postures, gestures, intonation; everything else is inferred. I think the "lowest level" inferences should be the main object of study for the time being—we should be objectifying and quantifying "low-level theoretical" statements like "Patient is currently anxious, and the thematic content is hostile toward his therapist," rather than highly theoretical statements like "He has superego lacunae" or "His dammed-up libido is flowing back to anal channels." In the process of such objectifying-and-quantifying research, I can think of no better methodological prescription than the one with which Aristotle sets the standards of conceptual rigor as he begins his consideration of ethics, "It is the mark of an educated man to look for precision in each class of things just so far as the nature of the subject admits." No more—but no less, either.

[15] Lillian Ross, "The Ordeal of Doctor Blauberman," *New Yorker*, 37 (May 13, 1961), 39–48; reprinted in Lillian Ross, *Vertical and Horizontal* (New York: Simon and Schuster, 1963).

Popper and Laplace

This paper is designed as a consideration of one pattern of escape from the classic Laplacian determinism. The most convenient representation of the anti-Laplacianism that I am concerned with is Sir Karl Popper's 1965 lecture "Of Clouds and Clocks: An Approach to the Problem of Rationality and the Freedom of Man."[1] However, before embarking on an examination of Popper, I shall briefly review Laplace's standpoint.

Let us go back for a moment to the text of Laplace's A *Philosophical Essay on Probabilities.*[2] At the start of chapter II, Laplace asserts that all events, including the actions of the will, follow the laws of nature, and that it is only in ignorance of true causes that we deny this fact. The connection between events is guaranteed by the principle of sufficient reason. "We ought then," he says, "to regard the present state of the universe as the effect of its anterior state and as the cause of the one which is to follow."[3] Then follows the famous passage which I will paraphrase here. An intelligence which could satisfy the following three conditions would be able to correctly envisage both past and future: (1) To know all forces in nature. (2) To know the states of all things in nature. (3) To be able to analyze and calculate from these data. "The human mind offers, in the perfection which it has been able to give to astronomy, a feeble idea of this intelligence."[4] This last quotation reveals the source of Laplace's confidence in determinism. Laplace was responsible for major developments in celestial mechanics, where it seemed to him that planetary motion needed no cause external to the framework of the Newtonian program. The formation of the solar system was accounted for by Laplace's nebular hypothesis. Laplace's work toward a solution of the stability problem for the solar system

[1] The Arthur Holly Compton Memorial Lecture presented at Washington University, April 21, 1965 (St. Louis: Washington University, 1966).
[2] Trans. F. W. Truscott and F. L. Emory (New York: Dover, 1951).
[3] *Ibid.,* p. 4.
[4] *Ibid.,* p. 4.

convinced him that God is not required to keep things running as Newton had believed. Thus it was natural for a scientist deeply immersed in the most mathematically perfect of the sciences to say, "The curve described by a simple molecule of air or vapor is regulated in a manner just as certain as the planetary orbits; the only difference between them is that which comes from our ignorance."[5]

Let us now turn to Popper's account.

1. Clouds and Clocks

As the title of Popper's lecture indicates, he is primarily concerned with the implications of determinism-indeterminism for free will, the mind-body problem, and rationality. To characterize determinism and indeterminism succinctly, Popper utilizes the picture of a continuum of systems with clouds on the left representing systems of very random behavior, and clocks on the right representing systems of very orderly and predictable behavior. In this scheme, the solar system is a very good clock. The opposing views may now be formulated in terms of the cloud-clock picture. The "physical determinist" says that all clouds are really clocks, and that we are just more ignorant about the behavior of systems on the cloud side than we are about those on the clock side. This is the Laplacian view, which we will have occasion to examine later. The indeterminist, on the other hand, claims that all clocks are clouds to some extent. A better way of putting it is this. All physical systems are imperfect, and some are more imperfect than others. On the right, we have fine, precise systems like clocks, and on the left messy systems like clouds of gas molecules. Thus, according to the indeterminist, there is random behavior present to some degree in all systems. Chance is an objective feature of the universe, not merely a product of our ignorance. This position may be ascribed to Democritus and Charles Sanders Peirce, as well as to quantum physicists.

Popper declares that he is an indeterminist and that Peirce was correct in claiming that indeterminism is consistent with Newton's physics. Thus Popper is not just asserting that indeterminism is true because quantum physics now reigns. Hence, before we can proceed to the next stage of the analysis of the clouds and clocks lecture, we must take a detour through a reference he cites in the lecture, namely, "Indeterminism in Quantum Physics and in Classical Physics," published in the *British Journal for the Philosophy of Science* in 1950.

[5] *Ibid.*, p. 6.

2. Indeterminism in Classical Physics

Popper's indeterminism article is long and involved, and I will not analyze it in toto. However, we shall see whether it adequately backs up his
claim that indeterminism holds, in the sense that all clocks are clouds "to
some considerable degree."

In this paper, Popper takes indeterminism to be the view that not all
events are determined in every detail; determinism is the view that they
all are determined. 'Determined' is interpreted by Popper as 'predictable
in accordance with the methods of science.' This last phrase is itself susceptible of different interpretations. Does it mean 'predictable in actual
scientific practice' or 'predictable in some idealized version of science' or
'predictable in principle' or what? For the purpose of most of Popper's
arguments, 'predictable in some idealized version of science' will suffice
as an interpretation. However, it is important to keep in mind that many
traditional philosophical claims about determinism, if they are formulatable in terms of some sense of predictability, use the interpretation 'predictable in principle.' Less fashionably, determinism can be characterized
in terms of necessary connection between events. If all events in the world
are connected necessarily, then determinism holds. That is an ontological
approach; the epistemological account using the concept of prediction
approximates the necessary connection formulation in strictness only
when 'prediction' is taken as 'prediction in principle.' Laplace, of course, is
a determinist in the latter sense, with the demon doing the predicting-in-
principle.

Popper's strategy is this. Philosophical claims about determinism are
often not falsifiable. Popper substitutes a testable claim for the metaphysical statement that all events are completely determined (predictable in
principle). Then he challenges the testable formulation. The substitute
claim is that every finite prediction task, meaning a prediction of specified
precision on a closed isolated system, can be accomplished by some "predictor." A predictor is a machine which we may imagine to be constructed
according to the laws of classical physics for the predicting job at hand.
That is, Popper has taken the demon of Laplace and replaced it by a classically operating machine.

The predictor, unlike the demon, is part of the physical world and gets
its information by interacting with the system it is designed to measure.
Popper states that just as indeterminism is linked to the measurement
problem in quantum mechanics, there is an analogous problem of inter-

419

action in classical physics which leads to indeterminism in that case too. There are several arguments which Popper adduces to prove that there are situations in which predictors must fail. I will present very short sketches of arguments that approximate Popper's unnecessarily long and intricate versions.

The first argument which Popper gives is this. Relying on the claim that a predictor B amplifies the responses it picks up from a system A which it is trying to describe, Popper goes on to add another predictor C, which is describing the situation A + B. In order to measure A + B, C must interact weakly with it and amplify the signals. But then we assume that B is also observing C. This leads to a breakdown of the observer-observed relationship because of the strong two-way interaction. As far as I understand the argument, there is no need to assume that B strongly interacts with C. B has been constructed to observe A, not to interact with the rest of the world in the same way as it interacts with A. It is true that there could be physical systems like those Popper describes, but that isn't enough to establish Popper's claims. What is required to prove his claims is to show that it is impossible to construct a sequence of predictors leading to more and more precise predictions without limit. On succeeding pages, Popper tries to establish the impossibility of such a sequence, by arguing that if such a sequence converges, then after some predictor, say the nth, all the predictors would be so similar as not to be detectable by the nth predictor. Popper appears to be misled by the language of mathematicians about sequences. Given a small positive number epsilon, one can pick a term of the convergent sequence beyond which the terms stay within epsilon of the value to which the sequence converges (or within epsilon of each other). This fact about sequences in no way implies that the epsilon becomes physically insignificant or mathematically inconsequential at some stage. Altogether, Popper has not convinced me that his argument need be taken seriously.

The second argument to be considered is one in which Popper claims that there are unavoidable situations in which a predictor cannot actually come up with the prediction before the event to be predicted occurs. His arguments for this are again elaborate. I don't think elaborate arguments are needed here. Popper's point may be quickly granted. Consider a time interval so short that because the predictor itself operates on physical principles, it can't get the answer out fast enough.

Another argument purports to show that a given predictor cannot, at time t, have complete knowledge of its states up to and including time t. The act of compiling the knowledge of its past states will alter those states. Popper considers the possibility that the last statement on the machine record is self-coding by, say, Gödel numbering. This possibility is rejected by requiring that the machine must retain records of its calculations, too. Then the self-coding would presumably fail. I don't see the need to challenge this argument either.

The final argument I shall mention involves Gödel sentences to show that a predictor can be asked a question about one of its future states that cannot be answered before the state in question has in fact occurred. Again, the argument is devised to produce an exception to the generalization that every event can be predicted before the fact. I will substitute a similar argument for Popper's which is simpler to comprehend. Ask the machine whether, by time t, it will check that a specified formula F of the predicate calculus is a theorem of that system. This is a legitimate question to ask the machine, because the question is "really" about physics: namely, by time t, will the predictor have been in the physical state corresponding to a "yes" answer to that question. Now it is known that there is no effective procedure by means of which a predictor (computer) can decide whether a given formula of the predicate calculus is a theorem. For some formulas F, the machine will not be able to prove F before time t or refute F before time t, and thus cannot answer the stated question before t arrives.

After presenting a similar argument, Popper remarks in a footnote[6] that one can add model-theoretic methods to the calculator, i.e., enable it not only to prove but to check on the satisfiability of formulas. Then he counters that we could ask the machine whether it is consistent: that is, ask if it ever will produce a contradiction in its output. Of course it can't prove that it is consistent. Here Popper is clearly going too far. While he never questions the truth of classical physics, which is an assumption in all of these arguments, he suddenly asks the predictor to guarantee the consistency of, say, arithmetic. Why not, instead, ask the predictor whether it will ever make a wrong prediction. Clearly, the latter demand oversteps the requirements of this discussion, just as does Popper's question to the predictor about the consistency of arithmetic.

We have seen that Popper's arguments are in opposition to the claim

---

[6] K. R. Popper, "Indeterminism in Quantum Physics and in Classical Physics, Part II," *British Journal for the Philosophy of Science*, 1 (1950), 183.

that there are no situations which cannot be predicted. I have not tried to give a detailed evaluation of these arguments, but only to sketch the sort of lines of argument that Popper utilizes. For my purpose, no more detailed analysis is required because Popper states that "there does not appear any serious obstacle for a calculator to explain in any desired detail its own past states."[7] Thus Popper admits that none of his arguments block the after-the-fact explanation in any conceivable detail of any system of classical physics. Nevertheless, he asserts that, on the basis of the possibility of constructing a series of predictors which strongly interact so that the precision of predictions does not improve as we go through the series, "then Laplace's determinism, and that of others who were influenced by the *prima facie* deterministic character of classical mechanics, is based upon a misinterpretation."[8]

At this stage, Popper's conclusions do not follow from his premises. If we grant that his several arguments about the interaction and self-knowledge problems of predictors are valid, then he has shown that *his* version of determinism is false. But his version of determinism is based on the notion of a prediction machine which is part of the classical physical world, whereas Laplace's demon is not subject to the limitations of Popper's predictors. The most that Popper has done is to cut off one possible way of arguing for Laplace.

It is important to take note of exactly what definition of determinism Popper has adopted in his arguments to refute determinism, because he cites his 1950 article on indeterminism in the 1965 clouds and clocks lecture exactly at the point where he declares that he is an indeterminist.[9] We are, therefore, entitled to conclude that his 1950 indeterminism article is relevant to his 1965 discussion. A borrowed metaphor will aid us in seeing why the 1950 article is not relevant. In the nightmare of determinism, which Popper does take seriously with determinism in the strong sense, we are on a trolley which we cannot get off and whose speed and direction along the single set of tracks are unalterable. The most Popper has shown is that at times, we cannot look forward on the ride, but only look back. So far, Popper hasn't got us off the trolley.

3. Popper and Wiener

It is time to return to "Clouds and Clocks." In an important passage,

---

[7] *Ibid.*, p. 191.
[8] *Ibid.*, p. 193.
[9] "Of Clouds and Clocks," p. 6.

Popper says: "I believe that the only form of the problem of determinism which is worth discussing seriously is exactly that problem . . . which arises from a physical theory which describes the world as a *physically complete* or a *physically closed* system. . . . It is this 'closure' of the system that creates the deterministic nightmare."[10]

Popper has hit upon a point of fundamental significance for the solution of his problems. In discussing this point, it will be helpful to adopt some terminology of David Bohm from his book *Causality and Chance in Modern Physics* (1957).[11] In Bohm's usage, a system described by the quotation above from Popper is *mechanistic*, in that there is a fixed finite set of variables in terms of which a complete description of the system may be written. Among mechanistic systems, there are both deterministic and indeterministic ones. Classical physics, in its Laplacian interpretation, is mechanistic and deterministic. Quantum mechanics according to the way it is put forward by most physicists is mechanistic and indeterministic. However, all deterministic systems are mechanistic.

We will now proceed to the discussion of Popper's proposed solution to the problem of providing a replacement for the closed physical system, a replacement which, he hopes, will permit a solution to the mind-body problem and to the problem of how rational discourse can govern behavior. Very briefly, Popper's scheme is this. To break out of the mechanistic impasse, he must destroy the notion that all systems are completely describable in terms of a closed set of physical variables. First, the false opposition between determinism (complete predictability) and utter randomness or chaos must be refuted. This refutation is accomplished by means of the notion of plastic control. Popper's aim is to describe every organism as "a hierarchical system of *plastic controls*—as a system of clouds controlled by clouds."[12] On the level of ideas, plastic control is exemplified by the regulation of thought by critical discussion and argument, the familiar Popper doctrine of conjecture and refutation, stated here in a more complex form where solutions to problems beget more problems. For a physical example of plastic control Popper chooses the soap bubble, which consists of a gas inside and a fluid film outside which "control each other"—cloud controlling cloud.

Second, other variables besides physical variables are introduced to

[10] *Ibid.*, p. 8.
[11] New York: Harper Torchbooks, 1961.
[12] "Of Clouds and Clocks," p. 25.

counter the mechanistic view of a closed *physical* system. Popper holds an interactionist view of the mind-body relationship. Mental states interact with physical states. The interaction is to be conceived on the pattern of plastic control. The organism, a system of mental states and physical states, is neither wholly a clock nor wholly a cloud, but, as I said, clouds controlling clouds.

Third, a form of the theory of evolution is invoked to explain how mental control of some aspects of the organism came about, and also how organisms alter their behavior patterns. The problem-solving or conjecture-refutation scheme is seen as a speeded-up version of evolution. These three points form the heart of Popper's proposed solution to the problems mentioned above.

Now I propose to compare the general features of Popper's solution to Norbert Wiener's approach in *Cybernetics, or Control and Communication in the Animal and the Machine* (1948).[13] Wiener begins his chapter I with a contrast between meteorology and astronomy as statistical and nonstatistical sciences, just as Popper introduced clouds and clocks. In chapter I, which is called "Newtonian and Bergsonian Time," Wiener is concerned with irreversibility, statistical laws, and information. Thus, we do not expect an exact analogue to Popper's discussion. Yet the ideas of Popper's lecture appear in that chapter, as well as in other writings of Wiener. Wiener says, "To return to the contrast between Newtonian astronomy and meteorology; most sciences lie in an intermediate position, but most are rather nearer to meteorology than to astronomy."[14]

Though Wiener does speak of clouds and clocks (clocks are the seventeenth-century craftsman's image of the heavens), he prefers to speak, instead, of the "transition from a Newtonian, reversible time to a Gibbsian, irreversible time."[15] This transition culminates in Heisenberg's quantum theory. But, rather than offering new hope for biological vitalists, they find that the "chance of the quantum theoretician is not the ethical freedom of the Augustinian, and Psyche is as relentless a mistress as Ananke."[16] Wiener is here referring to the fact that biology has been taken into the sphere of physics, and subsumed under a mechanistic scheme.

[13] 2nd ed. (Cambridge, Mass.: MIT Press, 1965).
[14] *Ibid.*, p. 35.
[15] *Ibid.*, pp. 37–38.
[16] *Ibid.*, p. 38.

To show that my comparison of Popper and Wiener is not just a matter of a few coincidences of phrases, I will demonstrate that there are problems which they both recognize and that their projected solutions have analogous features. First, to connect Popper's use of the theory of evolution with Wiener's, consider another quotation from Wiener: "In tidal evolution as well as in the origin of species, we have a mechanism by means of which a fortuitous variability, that of the random motions of the waves in a tidal sea and of the molecules of the water, is converted by a dynamical process into a pattern of development which reads in one direction."[17] Wiener and Popper both apply the theory of evolution to contexts of physical causality in general.

Second, both are indeterminists. In the second volume of his autobiography, Wiener states, "My early work on probability theory, as exemplified in my studies of the Brownian motion, had convinced me that a significant idea of organization cannot be obtained in a world in which everything is necessary and nothing is contingent."[18] Here, Wiener denies determinism in its strong, ontological form. Third, both break the closure of physical systems by introducing new variables not heretofore included in the physicist's account. In the case of Popper, we have seen that mental states are allowed to interact with physical states. In Wiener's case, the concept of information enters: "Information is information, not matter or energy. No materialism which does not admit this can survive at the present day."[19] Beyond this similarity, however, information for Wiener does not play the same role as ideas do for Popper. Wiener might be called a neo-materialist on the mind-body relation; Popper is an interactionist.

Finally, what about Popper's plastic control? In Wiener's work, plastic control is exemplified by the novel and ingenious statistical methods for which Wiener is famous. One of the great insights of Wiener was the idea that the mathematics of communication and control lies properly in the realm of statistical theory, not in such simple models as the Boolean algebra of switching circuits. Popper, perhaps, has failed to see the basically statistical nature of cybernetics. In "Clouds and Clocks," he rejects the computer models of organisms because "these systems, although incorporating what I have called plastic controls, consist essentially of complex

[17] Ibid., p. 37.
[18] N. Wiener, I Am a Mathematician (Garden City, N.Y.: Doubleday, 1956), p. 322.
[19] Cybernetics, 2nd ed. (Cambridge, Mass.: MIT Press, 1961), p. 132.

relays of master-switches. What I was seeking, however, was a simple physical model of Peircean indeterminism . . ."[20]

The point of comparing Popper and Wiener is to underscore the idea
that Popper's problems can be attacked with mathematical tools as well as
with simple metaphors. Popper's lecture is sketchy, but Wiener's book is
programmatic. Whether there is anything in the program for philosophers
will have to be determined by a careful analysis of the arguments used by
Wiener and his successors among the mathematical statisticians, cyberne-
ticians, and information theorists.

4. Conclusion

In the lecture discussed above, Popper was primarily concerned with the
consequences of Laplace's views for rationality and related problems. My
interest is rather in the problem of what picture of physical science should
replace the Laplacian picture.

An insight of David Bohm will be helpful to us. Bohm sees that philosophers and physicists take general features of prevailing physical theory and
project or extrapolate them as if they were permanent features of science.[21]
For example, Laplace in his time was expressing his faith that the Newtonian program of the eighteenth century would be a success. That is,
scientists would continue to discover new forces, but these forces would
be describable by differential equations allowing exact prediction and ret-
rodiction. In effect, then, Laplace was extrapolating from a certain general feature in Newtonian science.

Wiener reverses Laplace's extrapolation. Instead of emphasizing the
precision of mathematics, he stresses the character of measurements:

No scientific measurement can be expected to be completely accurate, nor
can the results of any computation with inaccurate data be taken as precise. The traditional Newtonian physics takes inaccurate observations,
gives them an accuracy which does not exist, computes the results to which
they should lead, and then eases off the precision of these results on the
basis of the inaccuracy of the original data. The modern attitude in physics
departs from that of Newton in that it works with inaccurate data at the
exact level of precision with which they will be observed and tries to compute the imperfectly accurate results without going through any stage at
which the data are assumed to be perfectly known.[22]

---

[20] "Of Clouds and Clocks," p. 27.
[21] See, for example, *Causality and Chance in Modern Physics* (New York: Harper
Torchbooks, 1957), pp. 69–70.
[22] *I Am a Mathematician*, p. 258.

The suggestion is that, in a general statistical approach to physical phenomena, the mathematics conforms more realistically to the data. But a statistical approach does not have the pictorial appeal or the intuitiveness of mathematical argument of the Laplacian scheme. Thus it seems that, without a clear and compelling account of the statistical approach, we have not yet replaced Laplace's paradigm, but only rejected it.

INDEXES

Name Index

431

Name Index

432

Name Index

Name Index

Name Index

Platt, J. R., 353, 395, 412
Plutarch, 96
Poincaré, H., 3, 15, 147, 227
Poisson, S. D., 407
Polanyi, M., 415
Polefka, J. T., 362
Popper, K. R., 7, 9, 15, 24, 30, 63, 64, 76–81, 90, 95, 99–102, 105, 106, 108, 110, 113, 116, 119, 125, 129, 151, 190, 310, 312, 313, 318, 320, 325, 330–333, 345, 346, 348, 351, 352, 354, 359, 361, 370, 385–388, 403–406, 411–413, 417–426
Postman, L., 393
Pound, E., 91
Prandtl, L., 140, 141
Presswood, S., 412
Ptolemy, C., 119, 121
Putnam, H., 143, 146, 156, 162, 164, 165, 228, 272
Pythagoras, 96

Quine, W. V. O., 9, 10, 160, 161, 226, 229

Radcliff, P., 110
Rado, S., 380
Ramsey, F. P., 146, 149, 181, 182, 189
Rankine, W. J. M., 142
Ratliff, F., 352, 395
Reichenbach, H., 3–5, 15, 16, 147, 149–151, 227, 229, 253, 277, 344, 387, 406
Reines, F., 411
Rescher, N., 386, 401
Riemann, G. F. B., 263, 265, 272
Ritz, W., 10
Robespierre, 23
Róheim, G., 102, 107
Ronchi, V., 104, 117
Rorty, R., 282
Rosen, E., 331
Rosen, Edward, 96, 122
Rosen, N., 128
Rosenfeld, L., 99, 126
Rosenzweig, M. R., 362
Ross, L., 416
Rozeboom, W. W., 201, 204, 220, 256, 411
Rozental, S., 99
Russell, B., 5, 181–185, 187–189, 191, 192, 215, 277, 305, 361
Ruyer, R., 286, 300
Ryle, G., 273, 304

Sakurai, J. J., 117
Salmon, W., 106, 220, 227
Samuel, A., 361

Santayana, G., 409
Sartre, J.-P., 276
Sayre, K., 334
Scarr, S., 393
Scheffé, H., 394
Scheffler, I., 16, 149, 156
Scher, S. C., 380
Schiaparelli, 117
Schilpp, P. A., 95, 100, 164, 319, 320, 340, 412
Schlick, M., 3, 5, 16
Schmitt, C. B., 122, 123
Schneider, E. F., 385, 386
Schopenhauer, A., 276, 303
Schrödinger, E., 352
Schultz, Q. H., 103
Schumacher, D. L., 128
Schwarzschild, K., 117, 118
Scriven, M., 128, 146, 147, 203, 276, 320, 340, 386
Sellars, W., 15, 158, 176–180, 276, 308, 320, 339, 340, 350, 354, 385, 386–388, 404
Sells, S. B., 335
Sen, D. K., 117, 126
Shaffer, J. A., 280
Shankland, R. S., 116
Shapere, D., 156
Shaw, J. C., 361, 362
Sherif, M., 99, 100
Shields, J., 393
Sigerist, H., 124
Simon, G. M., 99
Simon, H. A., 361, 362, 386, 401
Simpson, G. G., 102
Sinisi, V., 218
Sitter, W. de, 10, 11
Skinner, B. F., 311, 312, 331, 404, 406
Slater, E., 393
Smart, J. J. C., 16, 239, 278, 283
Smith, D. E., 272
Soldan, W. G., 122
Staal, J. F., 125, 176, 354
Stachnik, T., 312
Stanley, J. C., 375
Stebbing, L. S., 352
Stegmüller, W., 167, 168
Sternberg, F., 129
Sterne, L., 395
Sternheim, C., 17, 111
Stodolsky, S., 392
Storer, T., 385, 386
Stringfellow, J., 137
Stroud, B., 128
Suchting, W. A., 386, 388

435

Name Index

Summers, M., 125
Suppes, P., 8, 143, 150–152
Swensen, D. F., 105
Synge, J. L., 118, 126, 128

Tarski, A., 116, 143, 205
Thales, 122
Thiessen, D. D., 393
Thompson, W. R., 393
Thorndike, L., 119
Tillyard, E. M. W., 122
Tolman, E. C., 331, 362, 363
Tolstoy, L., 395
Tomonaga, S., 132
Trevor-Roper, H., 101, 104
Trotsky, L., 102, 118

Ullmann, L. P., 312
Ulrich, R., 312

Van Rootselaar, B., 125, 176, 354
Van Vliet, C. M., 220
Vesey, G. A., 274
Vigier, J.-P., 115
Vitruvius, 119
Vives, L., 124
Von Eschenbach, W., 119
Von Gleichen, 124
Von Kármán, T., 141
Von Mises, R., 151
Von Neumann, J., 150

Wallis, J. H., 395

Walters, R. W., 386
Wartofsky, M. W., 106, 128, 129, 156, 158, 176, 412
Washburn, S. L., 103
Watkins, J. W. N., 126
Watling, J., 385, 386
Watson, J. B., 333
Weber, M., 324, 325
Weinberg, J., 385
Weyer, J., 101
Weyl, H., 150
Wheeler, J. A., 84, 126, 127
White, A. D., 101
White, M. G., 161
Whitehead, A. N., 361
Wiener, N., 424–426
Wiener, P. P., 15
Will, F. L., 385
Williams, F. M., 220
Wilson, F., 404
Winston, H., 393
Wittgenstein, L., 215, 286, 291, 358, 359
Wolfe, B. D., 104
Wolff, R. P., 109, 112
Woodger, J. H., 16
Woodworth, R. S., 335
Wright, G. H. von, 318
Wright, O. and W., 137, 140, 243

Xenophanes, 122

Zeno of Elea, 37, 52
Ziff, P., 274, 286
Zilboorg, G., 120, 124

Subject Index

Subject Index

Counter-identicals, 387–389

Counterfactual conditionals, 384–389, 395–396, 399, 401

Counterinduction, 35–36, 43–45, 52, 55, 106

Criticism, 48, 75, 79, 109: of customary concepts, 45; measure of, 45. *See also* Rationalism

Crucial experiment, 247

Cybernetics, 424–426: and mind-body problem, 292–300; model of mind, 279

Data-point representation, structure of, 138–139

Definition, 159, 227, 252: implicit, 3, 5, 149, 151–152, 162; operational, 5–7, 158–159, 250–251; in physics, 230; theoretical, 390

Derivations, length of, 91

Descriptions, 182, 187, 192

Designation, multiple, 217–218

Determinism, 276, 376: Laplacian, 417–427 *passim*; physical, 418; psychological, 310–372 *passim*

Dialectics, 25, 35–36, 115

Dictionary, 146

Direction, absolute, 121–122, 243–245

Disambiguation of utterances, 285–288, 300

Discovery, context of, 4, 70–72, 133–134, 180, 257, **325**

Distance function, see Metrics

Distance relation, conventionality of, 270–271

Doppler effect, transverse, 88

Education: progressive, 27; scientific, 20

Egocentric particulars, 277, 304–305, 307

Electrodynamics: classical, 11, 39, 231; quantum, see Quantum field theory

Electron, 221–223, 230–232: existence of, 224–225

Electron theory of Lorentz, 40

Empiricism, 8: logical, 88, 200–202, 207, 216–217, 415

Entelechies, 279–280, 302

Entropy, 253

Epiphenomenalism, 277

Epistemology, 17–18, 196: anarchistic, 17–130 *passim*

Equipartition principle, 40

Error, theory of, 18–21

Evidence: contaminated, 44; historico-physiologic character of, 44

Evolution, 314, 424–425

Experience, 7, 18, 26, 71–72, 123: change of, 61; immediate, 14; science without, 92–94

Experiments, 65, 88

Explanation: deductive model of, 164–165, 169, 172, 175, 180; probabilistic, 239

Facts, 43, 51–52

Falsifiability, 405, 411–412

Falsification, 42–43: naive, 77

Fermat's Last Theorem, 351, 353–354

Finitude, 33, 35

Flight, 136–137, 140, 243

Fluid mechanics, 131–141 *passim*, 258

Formalization, see Axiomatizations

Frame principles, 223

Freedom of will, 324, 418

Functions, theory of, 139, 259

Fundamental Rule of psychoanalysis, 414

Galileo's law of falling bodies, 166–169

Geodesics, 83

Geometry: Euclidean, 4, 133, 151; physical, 147, 263

Gödel numbering, 421

"Grain" argument, 308

Grammars, phrase structure, 294–300

Hankel's principle, 230

Happiness, principle of maximum, 112

Hegelian cosmology, 32–33

Hilbert's ε-operator, 190

History of science, 20, 95, 147

Holism, 404

Hollerith machine, 346–347

Hydrodynamics, 243, 254, 257

Hypotheses: ad hoc, 22, 39, 63–66, 68–69, 73; self-inconsistent, 22

Hypothetico-deductive systems, 4

Identity thesis, mind-body, 276, 278, 346

Ideologies, 47, 52

Illocutionary force, 287

Impetus theory, 65, 67

Implicit definitions, 3, 5, 149, 151–152, 162

Impressions, Humean, 7

Incommensurability, see Theories, incommensurable

Incorrigibility, of first-person psychological statements, 280–284

Inductive inference, 164–165, 325, 407

Inertia, circular, 62–63, 66

Information, 425

Initial conditions, 83, 237

438

439

Subject Index

Phenomenalism, 183, 185
Phlogiston theory, 232
Physical₂, 276
Physical thing language, 7, 242, 245, 338
Physicalism, and mind-body problem, 273–274, 277–282, 285, 301, 305–309, 329, 345, 355
Physically closed system, 313, 423
Plastic controls, 423, 425
Positivism, see Empiricism, logical
Postulates: independent testability of, 229–230; logical independence of, 10
Pragmatics, 339, 342
Pre-theoretical vocabulary, see Antecedently available vocabulary
Predictability, 419
Prediction machines, 419–422
Prejudices, 48
Probability, 151: of hypothesis, 407; logical, 9; subjective, 9
Problem shifts: degenerating, 77; progressive, 77
Problem-solving, in science, 81
Professionalism, 95–98
Proliferation, principle of, 26–29, 106–107, 112–113
Propaganda, 23–24, 55, 61, 69
Propositional functions: mixed, 185; pure, 185
Protocol sentence, 242
Pseudo-afterimages, 86
Psycholinguistics, 329–330, 333
Psychologism, 323, 339
Purpose, 367–368

Q-models, 170–173, 177–179
Quantum field theory, 39, 41, 132, 134–135, 254, 258
Quantum mechanics, 12, 81, 114, 150, 230–231, 423: relation of classical physics to quantum physics, 247–248

Ramsey sentence, 187–192
Rationalism, 8: critical, 30, 32, 72, 75–76, 79, 82, 100
Rationality, 24–25, 32, 72, 80, 101–102, 312–372 passim
Realism, 30, 83, 85, 185, 192: naive, 49–50, 59, 120
Reasons, and causes, 313–372 passim
Reconstruction, rational, 4, 8, 13
Reduction sentences, 161
Referential opacity, 283
Referring, 182, 184, 187, 191

Relativity, 11, 81–82, 84, 88–89, 149–150, 229–230, 233, 235–236, 250–251, 254: dynamical and kinematic, 64; general theory of, 37, 40–41, 117–118, 126–128, 238; principle of, 62; special theory of, 37, 128
Renormalization, 40, 132
Revolutions, 103–104
Rules, methodological, 17–130 passim

Scanning mechanisms, and mind-body problem, 292–300
Schema (1), of Hempel, 143, 253
Schema (2), of Hempel, 146, 158, 252
Schwarzschild solution, 41
Scientific practice, 18
Second-order properties, 188
Self: existence of, 273, 291; nature of, 273–276, 304–305
Self-embedded sentences, 298–299
Self-selection of organisms, 374, 376
Semantic content, excess, 65
Semantic relations, 209–217
Semantic rules, 13, 226, 350, 360
Sense-data language, 198, 242, 245
Simplicity, inductive, 264, 266
Simultaneity, 250–251
Sine square law, 136
Snell's law, 11, 255
Space: continuity of, 263–265, 269, 271; discrete, 263–264, 268, 271–272; intrinsic metrical amorphousness of, 268; nondenumerable dense, 265–268, 270; relative and absolute, 64; topological structure of, 263–265, 268–270
Spurious influence, 379, 394
Statistical laws, 391
Stimulus, and response, 334–337, 355
Structural properties, 340–345: and intrinsic properties, 188–189, 192
Sufficient reason, principle of, 417
Syntax, 338–340
Systematic unmatching, 376–377

Telescope, invention of, 96
Theoretical language, 7
Theoretical principles, 224
Theoretical scenario, 142, 145, 149, 153–154, 253
Theoretical terms, 85, 145–146, 149, 153–154, 158–159, 162–163, 175–177, 181: defining of, 190; meaning of, 7, 134, 162–163, 164, 170, 172, 174, 181, 187, 192, 196–198, 202–209, 213, 217, 219, 226